ASTROPHYSICS AND SPACE SCIENCE LIBRARY

A SERIES OF BOOKS ON THE RECENT DEVELOPMENTS OF SPACE SCIENCE AND OF GENERAL GEOPHYSICS AND ASTROPHYSICS PUBLISHED IN CONNECTION WITH THE JOURNAL SPACE SCIENCE REVIEWS

Editorial Board

J. E. BLAMONT, *Laboratoire d'Aeronomie, Verrières, France*

R. L. F. BOYD, *University College, London, England*

L. GOLDBERG, *Kitt Peak National Observatory, Tucson, Ariz., U.S.A.*

C. DE JAGER, *University of Utrecht, The Netherlands*

Z. KOPAL, *University of Manchester, England*

G. H. LUDWIG, *NOAA, Environmental Research Laboratories, Boulder, CO, U.S.A.*

R. LÜST, *President Max-Planck-Gesellschaft zur Förderung der Wissenschaften, München, F.R.G.*

B. M. McCORMAC, *Lockheed Palo Alto Research Laboratory, Palo Alto, Calif., U.S.A.*

H. E. NEWELL, *Alexandria, Va., U.S.A.*

L. I. SEDOV, *Academy of Sciences of the U.S.S.R., Moscow, U.S.S.R.*

Z. ŠVESTKA, *University of Utrecht, The Netherlands*

VOLUME 106

PROCEEDINGS

DYNAMICAL TRAPPING AND EVOLUTION IN THE SOLAR SYSTEM

DYNAMICAL TRAPPING AND EVOLUTION IN THE SOLAR SYSTEM

PROCEEDINGS OF THE 74TH COLLOQUIUM OF THE INTERNATIONAL ASTRONOMICAL UNION HELD IN GERAKINI, CHALKIDIKI, GREECE, 30 AUGUST – 2 SEPTEMBER, 1982

Edited by

VASSILIS V. MARKELLOS

University of Patras, Greece

and

YOSHIHIDE KOZAI

Tokyo Astronomical Observatory, Japan

D. REIDEL PUBLISHING COMPANY

A MEMBER OF THE KLUWER ACADEMIC PUBLISHERS GROUP

DORDRECHT / BOSTON / LANCASTER

Library of Congress Cataloging in Publication Data

CIP

International Astronomical Union. Colloquium (74th : 1982 : Gerakini, Greece)
Dynamical trapping and evolution in the solar system.

(Astrophysics and space science library ; v. 106)
Includes indexes.
1. Solar system–Congresses. 2. Orbits–Congresses.
I. Markellos, Vassilis V. II. Kozai, Yoshihide, 1928– .
III. Title. IV. Series.
QB500.5.157 1982 521'.6 83-16100
ISBN 90-277-1650-1

Published by D. Reidel Publishing Company,
P.O. Box 17, 3300 AA Dordrecht, Holland.

Sold and distributed in the U.S.A. and Canada
by Kluwer Academic Publishers,
190 Old Derby Street, Hingham, MA 02043, U.S.A.

In all other countries, sold and distributed
by Kluwer Academic Publishers Group,
P.O. Box 322, 3300 AH Dordrecht, Holland.

Printed in The Netherlands

TABLE OF CONTENTS

Introduction ix

List of Speakers and Participants xi

PART I - SATELLITES AND PLANETS

J. KOVALEVSKY / High Order Resonances in the Evolution of the Lunar Orbit 3

A.T. SINCLAIR / A Re-Consideration of the Evolution Hypothesis of the Origin of the Resonances Among Saturn's Satellites 19

G. DULINSKI and A.J. MACIEJEWSKI / Orientation of a Satellite Located at the Libration Point in the Restricted Three-Body Problem 27

M. DUBOIS-MOONS / Theory of the Libration of the Moon (Abstract) 37

R. CID, S. FERRER and A. ELIPE / Regularization of the Equations of Motion in a Central Force-Field. Application to the Zonal Earth Satellite 39

E.M. STANDISH / The JPL "Long Ephemeris", DE102/LE51 47

J.H. LIESKE / A Collection of Galilean Satellite Eclipses 1652-1982 51

ZHANG JIA-XIANG / The Study of Planetary Secular Perturbations 61

TONG FU and CHEN ZHEN / Perturbations due to the Asteroid Belt 73

PART II - COMETS AND METEOR STREAMS

I.P. WILLIAMS / Physical Processes Affecting the Motion of Small Bodies in the Solar System and their Application to the Evolution of Meteor Streams 83

K. FOX / The Orbital Evolution of the Perseid and Quadrantid Meteor Streams 89

T. NAKAMURA / Steady State Number of the Extinct Comets in High-Inclination Orbits 97

M. KRESAKOVA, A. CARUSI and G.B. VALSECCHI / Ejection of Particles from Comet Lexell: The Gravitational Influence of Jupiter 105

D. BENEST, R. BIEN and H. RICKMAN / Capture of the Comet P/Boethin by Jupiter

PART III - ASTEROIDS

Y. KOZAI / Families of Asteroids 117

V. SZEBEHELY, R. VICENTE and J. LUNDBERG / Regions of Stability of Asteroids 123

R. O. VICENTE / The Stability of Some Asteroids 137

J.D. HADJIDEMETRIOU and S. ICHTIAROGLOU / On the Stability of Resonant Asteroid Orbits 141

R. BIEN and J. SCHUBART / Long Periods in the Three-Dimensional Motion of Trojan Asteroids 153

H. SCHOLL and C. FROESCHLE / Resonant Asteroidal Motion in the Kirkwood Gaps: A Three-Dimensional Study (Abstract) 163

B. ERDI / Orbital Evolution of Trojan Asteroids 165

V. ZAPPALA, P. FARINELLA and P. PAOLICCHI / Collisional Origin of Asteroid Families: Effects of the Target's Gravity 177

A. LEMAITRE / Analysis of a Simple Mechanism to Deplete the Kirkwood Gaps 189

M. YUASA / On the Ages of Asteroid Families 203

PART IV - PERIODIC ORBITS

I.A. ROBIN and V.V. MARKELLOS / The Mechanism of Branching of Three-Dimensional Periodic Orbits from the Plane 213

A. MILANI / Stability and Bifurcations of Symmetric Periodic Orbits in the Restricted 3-Body Problem 225

C. G. ZAGOURAS and V.V. MARKELLOS / Resonant Three-Dimensional Periodic Solutions About the Triangular Equilibrium Points in the Restricted Problem 235

A. TSOUROPLIS and C.G. ZAGOURAS / Asymmetric Periodic Orbits in the Three-Body Problem and their Stability 249

C. EDELMAN / Construction of Periodic Orbits, Problems of Stability and Period Determination, in the Elliptical Non-Planar Restricted Problem 257

J. CASASAYAS and J. LLIBRE / Symmetric Periodic Orbits in the Anisotropic Kepler Problem 263

N. CARANICOLAS / Characteristics of Periodic Orbits in Elliptical Galaxies 271

PART V - TRAPPED MOTION IN THE THREE-BODY PROBLEM

A.E. ROY / Asymptotic Approach to Mirror Conditions as a Trapping Mechanism in N-Body Hierarchical Dynamical Systems 277

R. MEIRE / New Results for the Linear Stability of the Triangular Points in the Elliptic Restricted Problem 289

A. MILANI and A.M. NOBILI / On Topological Stability in the General Three and Four-Body Problem 301

M. DELVA / Boundaries for the Equipotential Curves in the Elliptic Restricted Three-Body Problem 317

G. GOMEZ and J. LLIBRE / Capture Escape Boundary in the Collinear Restricted Three-Body Problem 325

E. PERDIOS / Doubly Asymptotic Orbits at the Unstable Equilibrium in the Elliptic Restricted Problem 339

PART VI - MISCELLANEOUS DYNAMICS

B.C. XANTHOPOULOS and G. BOZIS / The Planar Inverse Problem for Autonomous Systems 253

T.B. OMAROV and M.J. MINGLIBAEV / Analytical Theory of a Trapping in a Two-Body Problem of Variable Mass 369

A. CARUSI, E. PEROZZI and G.B. VALSECCHI / Low Velocity Encounters of Minor Bodies with the Outer Planets 377

R. GONCZI, CH. FROESCHLE and C. FROESCHLE / Trapping Time of Resonant Orbits in Presence of Poynting - Robertson Drag 397

H. VARVOGLIS / Degenerate Dynamical Systems and the Disappearance of (K.A.M.-Type) Integrals of Motion 411

INDEX OF NAMES 417

INDEX OF SUBJECTS 421

INTRODUCTION

The papers comprising this volume were presented at Colloquium No 74 of the International Astronomical Union, on "Dynamical Trapping and Evolution in the Solar System", which was held in Gerakini, Chalkidiki, Greece, from August 30 through September 2, 1982, a few days after the IAU General Assembly of Patras, Greece. The Scientific Organizing Committee consisted of C.L. Goudas, J.D. Hadjidemetriou, Y. Kozai (Chairman), L. Kresák, V.V. Markellos, P.J. Message, A.E. Roy and V. Szebehely. To the Local Organizing Committee consisting of G. Bozis, J.D. Hadjidemetriou (Chairman), V.V. Markellos, C. Zagouras, and M. Michalodimitrakis, was due the success of the local arrangements.

There were 62 participants from 17 countries and 42 papers were presented (in the same order as given in this volume) on a variety of topics of Solar System Dynamics at seven sessions chaired by Y. Kozai, J. Kovalevsky, A.E. Roy, V.V. Markellos, V. Szebehely, J.D. Hadjidemetriou, and V.A. Brumberg; the seven sessions being on Satellites and Planets, Comets and Meteor Streams, Asteroids (two sessions), Periodic Orbits, Trapped Motion in the Three-Body Problem, and Miscellaneous Dynamics. Since the Colloquium overlapped, unfortunately, with the IAU Colloquium No 75 on "Planetary Rings" held at Toulouse, France, no papers were presented at Chalkidiki on the dynamics of planetary rings and several people who wanted to attend the Chalkidiki Colloquium could not come.

In spite, or because, of the relatively small number of participants the Colloquium at Chalkidiki was successful and enjoyable. Many stimulating discussions on the papers presented and on the participants' current researches were carried out between sessions, the formal sessions of each day thus being informally extended into the evening. The Colloquium was marked, in particular, by the presence of many brilliant young dynamical astronomers who have already contributed significantly to our understanding of the aspects of dynamical trapping in the Solar System's dynamical evolution and related topics.

V.V. Markellos and Y. Kozai
Editors of the Proceedings

LIST OF SPEAKERS AND PARTICIPANTS

Antonakopoulos, G.	Dept. of Astronomy, University of Patras, Patras, Greece
Barbanis, V.	Dept. of Astronomy, University of Thessaloniki, Thessaloniki, Greece
Benest, D.	Observatoire de Nice, B.P. 252, 06007 Nice Cedex, France
Bozis, G.	Dept. of Theoretical Mechanics, University of Thessaloniki, Thessaloniki, Greece
Brumberg, V.A.	Institute of Theoretical Astronomy, 191187 Leningrad, U.S.S.R.
Caranicolas, N.	Dept. of Astronomy, University of Thessaloniki, Thessaloniki, Greece
Casasayas, J.	Departamento de Teoria de Functiones, Facultad de Matematicas, Universidad de Barcelona, Barcelona 7, Spain
Christides, T.	Dept. of Theoretical Mechanics, University of Thessaloniki, Thessaloniki, Greece
Delva, M.	Institut für Astronomie, Karl-Franzens-Universität Graz, A-8010 Graz, Universitätsplatz 5, Austria
Dubois-Moons, M.	Rue de Ecoles 60, 5170 ARRBRE, Belgium
Dulinski, G.	Institute of Astronomy, Nicolaus Copernicus University, Chopina 12/18, P1-87-100 Torun, Poland

Edelman, C.	Bureau des Longitudes, 77, avenue Denfert-Rochereau, 75014 Paris, France
Erdi, B.	Dept. of Astronomy, LORÁND EÖTVÖS UNIVERSITY, Budapest, Hungary
Farinella, P.	Istituto di Matematica "LEONIDA TONELLI", Università di Pisa, Piazza dei Cavalieri 2, I-56100 Pisa, Italy
Ferrer, S.	Departamento de Astronomia, Facultad de Ciencias, Universidad ZARAGOZA, Spain
Fox, K.	Dept. of Applied Mathematics, Queen Mary College, University of London, Mile End Road, London, E1 4NS, U.K.
Froeschlé, C.	Observatoire de Nice, B.P. 252, 06007 Nice Cedex, France
Gonczi, R.	Observatoire de Nice, B.P. 252, 06007 Nice Cedex, France
Goudas, C.	Dept. of Mechanics, University of Patras, Patras, Greece
Grigorelis, F.	Dept. of Theoretical Mechanics, University of Thessaloniki, Thessaloniki, Greece
Hadjidemetriou, J.	Dept. of Theoretical Mechanics, University of Thessaloniki, Thessaloniki, Greece
Ichtiaroglou, S.	Dept. of Theoretical Mechanics, University of Thessaloniki, Thessaloniki, Greece

Katopodis, K.	Dept. of Theoretical Mechanics, University of Thessaloniki, Thessaloniki, Greece
Kovalevsky, J.	Cerga, Avenue Copernic, 06130 Grasse, France
Kozai, Y.	Tokyo Astronomical Observatory, Mitaka, Tokyo, 181, Japan
Kuzmanoski, M.	Institute of Astronomy, Faculty of Sciences, P.O. Box 550, 11000 Belgrade, Yugoslavia
Lamy, P.	Laboratoire de Astronomie Spatiale, Les Trois Lucs, 13012 Marseille, France
Lemaitre, A.	Departement de Mathématique, Facultés Universitaires de Namur, Rempart de la Vierge, 8 5000 Namur, Belgium
Lichtenegger, H.	Institut für Astronomie, Karl-Franzens-Universität Graz, A-8010 Graz, Universitätsplatz 5, Austria
Lieske, J.	JET Propulsion Laboratory, California Inst. of Technology, 4800 Oak Grove Drive, Pasadena, California 91109, U.S.A.
Llibre, J.	Departamento de Matematicas, Facultad de Ciències, Universidad Autónoma de Barcelona, Bellaterra, Barcelona, Spain
Markellos, V.V.	Dept. of Engineering Science, University of Patras, Patras, Greece
Meire, R.	Astronomical Observatory, Ghent State University, Krijgslaan 281, 9000 Ghent, Belgium

Milani, A.	Istituto di Matematica "LEONIDA TONELLI", Università di Pisa, Piazza dei Cavalieri 2, I-56100 Pisa, Italy
Michalodimitrakis, M.	Dept. of Theoretical Mechanics, University of Thessaloniki, Thessaloniki, Greece
Nakamura, T.	Dodaira Station, Tokyo Astronomical Observatory, Tokigawa, Hiki-Gun, Saitama 355-05, Japan
Nobili, A.	Istituto di Matematica "LEONIDA TONELLI", Università di Pisa, Piazza dei Cavalieri 2, I-56100 Pisa, Italy
Omarov, T.B.	The Astrophysical Institute, Academy of Science of Kazakh, Alma-Ata, 480068 U.S.S.R.
Papadakis, C.	Dept. of Engineering Science, University of Patras, Patras, Greece
Pauwels, T.	Groenpark 17, B-9720 de Pinte, Belgium
Perdios, E.	Dept. of Engineering Science, University of Patras, Patras, Greece
Pinotsis, A.	Dept. of Astronomy, University of Athens, Athens, Greece
Polymilis, C.	Dept. of Astronomy, University of Athens, Athens, Greece
Rickman, H.	Observatoriet Astronomiska, Box 515, S-75120 Uppsala, Sweden

Roy, A.	Dept. of Astronomy, University of Glasgow, Glasgow G 12 8QQ, U.K.
Schmidt, D.	Dept. of Mathematical Sciences University of Cincinnati, Cincinnati, Ohio 45221, U.S.A.
Scholl, H.	Astronomisches Rechen-Institut, Mönchhofstrasse 12-14, D-6900 Heidelberg 1, F.R. Germany
Schubart, J.	Astronomisches Rechen-Institut, Mönchhofstrasse 12-14, D-6900 Heidelberg 1, F.R. Germany
Sinclair, A.T.	Royal Greenwich Observatory Herstmonceux Castle, Hailsham, Sussex, U.K.
Standish, J.R.	JET Propulsion Laboratory, California Inst. of Technology, 4800 Oak Grove Drive, Pasadena, California 91109, U.S.A.
Szebehely, V.	Dept. of Aerospace Engineering and Engineering Mechanics, University of Texas at Austin, Austin, Texas 78712, U.S.A.
Terzidis, C.	Dept. of Astronomy, University of Thessaloniki, Thessaloniki, Greece
Tong Fu	Purple Mountain Observatory Academia Sinica, Nanking, China
Tsouroplis, A.	Dept. of Mechanics, University of Patras, Patras, Greece
Valsecchi, G.B.	IAS Reparto Planetologia, Viale Università 11, 00185 Roma, Italy

Varvoglis, H.	Dept. of Astronomy, University of Thessaloniki, Thessaloniki, Greece
Vicente, R.	R. Mestre Aviz, 30, R/c 1495 Lisboa, Portugal
Williams, I.P.	Dept. of Applied Mathematics, Queen Mary College, University of London, Mile End Road, London, E1 4NS, U.K.
Yuasa, M.	Dept. of Mathematics and Physics, Kinki University, Higashi-Osaka, Osaka, 577, Japan
Zachilas, L.	Dept. of Astronomy, University of Athens, Athens, Greece
Zagouras, C.	Dept. of Mechanics, University of Patras, Patras, Greece
Zhang Jia-xiang	Purple Mountain Observatory, Academia Sinica, Nanking, China

PART I

SATELLITES AND PLANETS

HIGH ORDER RESONANCES IN THE EVOLUTION OF THE LUNAR ORBIT

J. KOVALEVSKY
CERGA, Grasse, France

ABSTRACT

This paper deals with the long term evolution of the motion of the Moon or any other natural satellite under the combined influence of gravitational forces (lunar theory) and the tidal effects. We study the equations that are left when all the periodic non-resonant terms are eliminated. They describe the evolution of the mean elements of the Moon. Only the equations involving the variation of the semi-major axis are considered here. Simplified equations, preserving the Hamiltonian form of the lunar theory are first considered and solved. It is shown that librations exist only for those terms which have a coefficient in the lunar theory larger than a quantity A which is function of the magnitude of the tidal effects. The solution of the general case can be derived from a Hamiltonian solution by a method of variation of constants. The crossing of a libration region causes a retardation in the increase of the semi-major axis. These results are confirmed by numerical integration and orders of magnitude of this retardation are given.

I. INTRODUCTION

The secular acceleration of the Moon has been studied for many years. It is well known that the lunar orbit undergoes a secular acceleration due to the Earth's tidal deformation. Presently, this acceleration is estimated to $dn/dt = -25''/\text{century}^2$ (see, for instance, Calame and Mulholland, 1978, Ferrari et al., 1980 or Cazenave and Daillet, 1981). As it is shown by Lambeck (1978), this value seems to have been surprisingly constant throughout the last 500 million years as inferred from the paleontological evidence from fossile corals or bivalves (e.g. Johnson and Nudds, 1974).

In the general theory of tides, it is customary to introduce a lag due to the viscosity of the Earth between the direction of the Moon and the main axis of the tidal bulge (Melchior, 1973). This angle corresponds to a delay Δt of about 10 minutes between the excitation and the deforma-

V. V. Markellos and Y. Kozai (eds.), Dynamical Trapping and Evolution in the Solar System, 3–17.

tion. It is principally a function of the second order Love number k_2.

Observational evidence quoted above allows us to assume that this time-lag was more or less constant in the past. In this case, it is possible to integrate the equation giving the variations of the semi-major axis of the lunar orbit (MacDonald, 1966). It is found that, when the solution is extended in the past, one has $a = 0$ at about -1.8×10^{-9} year. This number is 2 1/2 times too small in comparison with the actual age of the Earth-Moon system. In order to interpret this difference, it can be alleged that in the earlier past, Love numbers were different so that the time lag was smaller, allowing a much slower tidal evolution.

The goal of the work which is reported here is to attempt to find other - purely dynamical - effects that might affect the speed of the secular increase of the lunar semi-major axis.

II. EQUATIONS

The equations that describe the evolution of the Earth-Moon system are obtained from the combination of the usual equations of the lunar theory and the equations that describe the tidal evolution of the Moon. In the present work, the latter were taken from a series of papers published by Mignard on the evolution of the Earth-Moon system (Mignard, 1979 and 1980) that seem to be the most complete ever published. We introduce the following parameter :

$$k = - \frac{3Gm^2 \; k_2 \; R^5 \quad \Delta t}{\mu} \tag{1}$$

where G is the geocentric constant of gravitation, k_2 is the Love number describing the second order term of the Earth tidal potential, R is the radius of the Earth and m is the mass of the Moon, the Earth's mass M being taken as unity.

$$\mu = mM \; / \; (m + M)$$

The numerical value of k is $-2.3 \quad 10^{-14}$ if the unit of mass is the Earth's mass, the unit of length, the semi-major axis of the lunar orbit and the unit of time, $1/2\pi$ the period of the lunar orbit.

We introduce also σ/n, ratio of the rotational speed of the Earth to the mean motion of the Moon n.

Mignard (1980) gives the general expressions of the components of the tidal acceleration due to a Love number k_2 of any order ℓ in function of the osculating elements of the lunar orbit. Taking $\ell = 2$ and substituting the expressions in the Gaussian equations of motion (see for instance, Kovalevsky, 1967), one gets the basic equations giving the variation of the osculating elements in function of time. Let us reproduce the equations for the metric elements.

$$\frac{da}{dt} = \frac{2k}{a^7} \frac{1}{1-e^2} \left(\frac{a}{r}\right)^8 (1 + 2e^2 + 2e \cos v - e^2 \cos 2v)$$
$$- \frac{2\sigma}{n} \frac{k}{a^7} \sqrt{1-e^2} \left(\frac{a}{r}\right)^8 (\cos \varepsilon \cos i + \sin \varepsilon \sin i \cos \Omega) \tag{2}$$

$$\frac{de}{dt} = \frac{k}{a^8} \left(\frac{a}{r}\right)^8 (3e + 2 \cos v - e \cos 2v)$$
$$- \frac{\sigma}{n} \frac{k}{a^8} \frac{1}{\sqrt{1-e^2}} \left(\frac{a}{r}\right)^6 \left(\frac{3e}{2} + 2\cos v + \frac{e}{2}\cos 2v\right)(\cos\varepsilon \cos i + \sin\varepsilon \sin i \cos \Omega) \tag{3}$$

$$\frac{di}{dt} = \frac{1}{4} \frac{\sigma}{n} \frac{k}{a^8} \frac{1}{\sqrt{1-e^2}} \left(\frac{a}{r}\right)^6 \left\{ \begin{array}{l} 2 \sin i \, (1 + 2 \cos 2(\omega + v)) \\ + 2 \sin \varepsilon \cos i \cos \Omega \\ + \sin\varepsilon \, (1+\cos i) \cos(2\omega + 2v + \Omega) \\ - \sin\varepsilon \, (1 - \cos i) \cos (2\omega + 2v - \Omega) \end{array} \right. \tag{4}$$

The classical osculating elements are noted a, e, i, Ω, ω, ℓ. Furthermore, r is the radius vector, v is the true anomaly, ε is the obliquity of the ecliptic.

When developed in trigonometric functions of ℓ, the right-hand members will be even series of ℓ, Ω and ω. Similar equations can be derived for the angular elements Ω, ω and ℓ. The right-hand members are odd in v (and, hence, in ℓ) and therefore do not produce secular effects. We do not have to consider them in the present study. Equations with the same properties may be derived if Delaunay variables are chosen instead of the elements.

The terms originated by the solar perturbation derive from a disturbing function of the form :

$$R = n'^2 a^2 \sum_{ijkh} A\left(e, e', \sin\frac{i}{2}, \frac{a}{a'}, \mu\right) \cos(i\Omega + j\omega + k\ell + h\ell') \tag{5}$$

where the primed quantities refer to the apparent ellipse described by the Sun around the Earth and are considered as being constant in the problem. The contribution to the right-hand members is a series of odd terms in the case of equations (2) to (4), and even terms in the equations relative to angular elements.

III. Elimination of terms

The complete equations, written in a pseudo-hamiltonian form using conjugate variables ξ_j, η_j ($j = 1$ to 3) have the following form :

$$\begin{cases} \dfrac{d\xi_j}{dt} = \dfrac{\partial F}{\partial \eta_j} + X_j \\[2ex] \dfrac{d\eta_j}{dt} = \dfrac{\partial F}{\partial \xi_j} + Y_j \end{cases} \qquad (6)$$

where F is the Hamiltonian of the main lunar problem and X_j, Y_j are the tidal terms. Periodic terms may be eliminated by Delaunay method or any of the derived methods (von Zeipel, Lee series, etc...). As shown by Brouwer and Hori (1961), the elimination can also be applied to the terms that have not the hamiltonian form. Following their theory, after the elimination of all the periodic terms of the lunar theory, equations (6) become :

$$\frac{d\xi'_j}{dt} = X_j(\xi',\eta') + \sum_i \left(\frac{\partial X_j}{\partial \xi'_i} \delta\xi_i + \frac{\partial X_j}{\partial \eta'_i} \delta\eta_i \right)$$

$$\frac{d\eta'_j}{dt} = -\frac{\partial F'(\xi')}{\partial \xi'} + Y_j(\xi',\eta') + \sum_i \left(\frac{\partial Y_j}{\partial \xi'_i} \delta\xi_i + \frac{\partial Y_j}{\partial \eta'_i} \delta\eta_i \right)$$

where the primed quantities indicate the new Hamiltonian or variables after the elimination has been performed. The periodic terms of X_j and Y_j can be similarly eliminated. Because of the smallness of k, a first order solution is sufficient, so that it may just be added to the solution. After this, taking into account the fact that Y_j are odd functions of the arguments, the equations become :

$$\begin{cases} \dfrac{d\xi'_j}{dt} = \overline{X_j(\xi')} \\[2ex] \dfrac{d\eta'_j}{dt} = \dfrac{\partial F'(\xi')}{\partial \xi'_j} \end{cases} \qquad (7)$$

Their solution gives the secular-variations of all the six variables.

IV. EQUATIONS FOR THE RESONANT CASE

This procedure fails if, among the terms of the lunar theory, there exist one or several terms whose period is so large that they cannot be eliminated and that resonance theory has to be applied. Let us consider one of such arguments :

$$\theta = i\Omega + j\omega + k\ell + h\ell'$$

The corresponding mean motion is

$$n_\theta = i n_\Omega + j n_\omega + k n_\ell + h n'$$

where the indexed n are functions of the mean elements a", e" and i". If one takes the present values of these elements, there is no known combination of integers i, j, k, h that leads to a sufficiently small n_θ. But since in the past a", e" and i" were slowly varying because of the tidal effects, many such combinations may have existed. Let us consider such a case and let θ be the corresponding critical argument. The elimination procedure being applied to all the other - non critical - periodic terms, we end up with equations of the form :

$$\begin{cases} \dfrac{d\xi'_j}{dt} = \overline{X_j}(\xi') + x_j \sin\theta \quad , \\[2ex] \dfrac{d\eta'_j}{dt} = -\dfrac{\partial F'(\xi')}{\partial \xi'_j} + y_j \cos\theta \end{cases} \tag{8}$$

It is possible, by a canonical tranformation, to have θ as one of the angular variables so that the other two become ignorable, and we are left with three equations in ξ' and one in η' = θ . As a final transformation, we shall come back to the mean elliptic elements and the equations will have the following form :

$$\begin{cases} \dfrac{da}{dt} = \alpha_1 + \beta_1 \sin\theta \\[2ex] \dfrac{de}{dt} = \alpha_2 + \beta_2 \sin\theta \\[2ex] \dfrac{di}{dt} = \alpha_3 + \beta_3 \sin\theta \\[2ex] \dfrac{d\theta}{dt} = n_\theta + y_\theta \cos\theta \end{cases} \tag{9}$$

where the coefficients are all functions of a, e and i. Note that, at this point, we have dropped the primes and that now on, unprimed quantities are used for the mean elements, solutions of (9).

V. REDUCED EQUATIONS IN THE LUNAR CASE

In this paper, we shall neglect the equations in e and i. In other terms, we shall assume that the variations in e and i are sufficiently small so as not to introduce in the first and last equations sizeable effects. This is clearly a simplifying assumption made in order to study the behaviour of the solution. Later, the equations (9) should be discussed as a whole. Using (2), one obtains :

$$\begin{cases} \dfrac{da}{dt} = \dfrac{2k}{a^7} - \dfrac{2\sigma}{\sqrt{GM}} \dfrac{k}{a^{11/2}} \cos\varepsilon \cos i + Q \sin\theta \\ \\ \dfrac{d\theta}{dt} = n_\theta\ (a) + Q' \cos\theta \end{cases} \tag{10}$$

where Q and Q' are the coefficients of the resonant terms coming from the lunar theory.

In a last change of variables, let us put

$$a = a_o + x \tag{11}$$

where a_o is the value of such that

$$n_\theta\ (a_o) = 0$$

If now, we put

$$A = \frac{2k}{a_o^7} - \frac{2\sigma}{\sqrt{GM}} \frac{k}{a^{11/2}} \cos\varepsilon \cos i$$

$$B = \left(\frac{\partial A}{\partial a}\right)_{a_o} = \frac{-14k}{a_o^8} + \frac{11\sigma\ k \cos\varepsilon \cos i}{\sqrt{GM}\ a_o^{13/2}}$$

$$n_\theta = n_\theta\ (a_o) + \left(\frac{\partial n_\theta}{\partial a}\right)_o x = 2\ Gx$$

the equations (10) become

$$\begin{cases} \dfrac{dx}{dt} = A - Bx + Q \sin\theta \\ \\ \dfrac{d\theta}{dt} = 2\ Gx + Q' \cos\theta \end{cases}$$

In order to have a feeling of the actual order of magnitude of the coefficient, the present values of A and B are :

$$A = 1.10\ 10^{-12}$$

$$B = 5.95\ 10^{-12}$$

Q and Q' can be any term taken from the lunar theory , nothing can be said about their size. However, the present study has dealt with values of Q smaller than 10^{-9} or 10^{-10}. Finally, G is normally a finite number. This means that Q' is negligible with respect to 2 Gx. So, we shall effectively neglect it and reduce the equations to :

$$\begin{cases} \dfrac{dx}{dt} = A - Bx + Q \sin \theta \\ \\ \dfrac{d\theta}{dt} = 2\ Gx \end{cases} \tag{12}$$

Equations (12) describe the behaviour of the semi-major axis a and of the critical argument θ when a is close to the resonant situation. It is important to note that (12) is not a hamiltonian system.

VI. STUDY OF SOME PARTICULAR CASES

Let us first describe the solution of some particular cases of the system (12). This will help to understand the behaviour of the general system.

6.1. $Q = 0$

Equations (12) describe simply the secular effect of the tides on the semi-major axis. The equations separate and one gets the general solution

$$t - t_o = \frac{1}{B} \text{Log} \left|\frac{A - Bx}{A}\right|$$

or

$$x = \frac{A}{B} \left[1 - \exp\left(-B(t - t_o)\right)\right] \tag{13}$$

This is equivalent to the general solution as given by MacDonald (1966). For small values of x, this solution may be approximated by a linear function of time

$$x = A\ (t - t_o) \tag{14}$$

6.2. $A = B = 0$

The system (12) describes a resonant situation in the lunar theory. It has a Hamiltonian

$$H = -\ Gx^2 - Q \cos \theta$$

The variations of x in function of θ are described in the $x - \theta$ plane by the integral $H = C$. The equilibrium points correspond to the maxima and minima of $H=C$ in the $x^2 - \theta$ plane. For various values of C, one gets classical libration or circulation orbits and the limiting cases of a stable equilibrium or asymptotic orbits (see fig. 1).

6.3. $B = 0,\ A \neq 0$

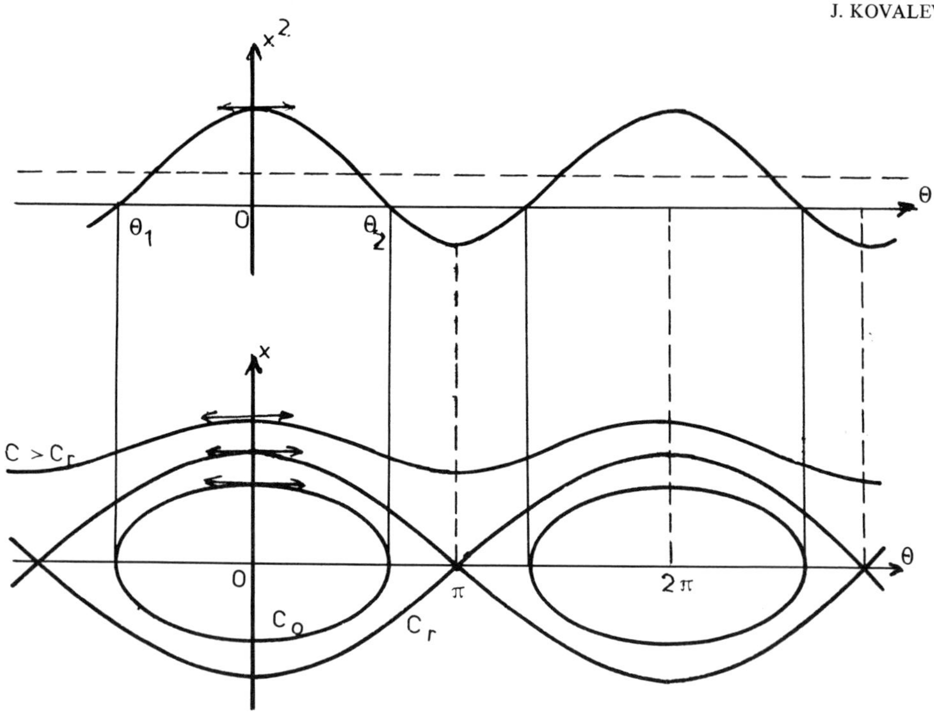

Figure 1. Solution of the A = B = 0 case in the x −θ plane.

The system (12) now describes the motion under a resonant situation with a tidal effect that does not depend upon the semi-major axis. It has the form

$$\begin{cases} \dfrac{dx}{dt} = A + Q \sin \theta \\ \\ \dfrac{d\theta}{dt} = 2\ Gx \end{cases} \tag{15}$$

and has the Hamiltonian $H = A\theta - Q \cos \theta - Gx^2$, so that the integral $H = C$ exists and may be used to discuss the motion. In the $x^2 - \theta$ plane, curves $H = C$ are sinusoidal curves constructed with respect to an inclined axis.

$$x^2 = \frac{C}{G} + \frac{A}{G}\theta - \frac{Q}{G}\cos\theta \tag{16}$$

As in the preceding case, equilibrium points correspond to maxima or minima of (15) in the $x' - \theta$ plane. They are given by

$$\frac{\partial H}{\partial x} = 0 \quad \text{or } x = 0 \quad ; \quad \frac{\partial H}{\partial \theta} = 0 \quad \text{or } \sin\theta = -\frac{A}{Q}$$

Their existence depends upon the value of the ratio A/Q.

a) $|A| < |Q|$. The situation is described in figure 2. The integral curves in the x - θ plane have several components and, depending upon the initial value of θ, one may be trapped in a libration orbit or be on a circulation orbit.

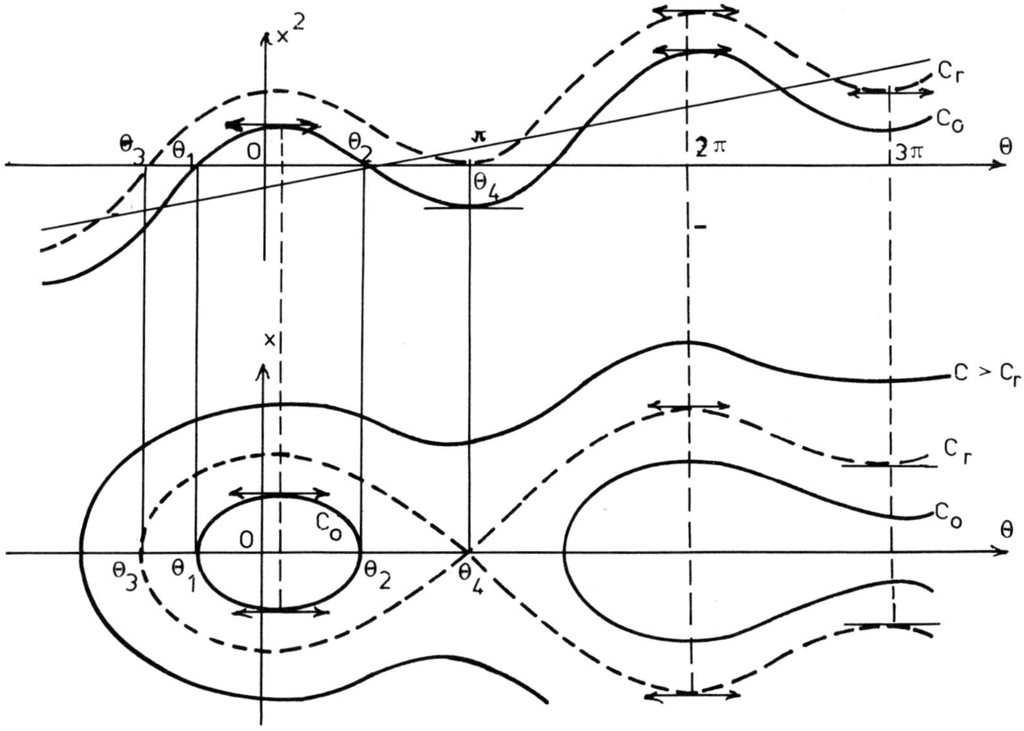

Figure 2. Solution for the B = 0 case in the x - θ plane.

The entire picture has a periodicity of 2π in θ corresponding to 2πA in C. However, physically, there is continuity when C varies and one should consider only the curves that originate from a given stable equilibrium point. When C increases, starting from such a point, one has successively libration orbits, a limiting asymptotic orbit and then, symetric circulation orbits.

The libration period is given by :

$$P = \frac{2}{\sqrt{G}} \int_{\theta_1}^{\theta_2} \frac{d\theta}{\sqrt{C + A\,\theta - Q\cos\theta}}$$

where θ_1 and θ_2 are the values of θ surrounding the stable equilibrium in θ_o and for which $x^2 = 0$ (see figure 2).

b) $|A| > |Q|$. The $x^2 = f(\theta)$ curve has no horizontal tangent and it crosses the x^2 axis in a single point P (figure 3). The corresponding orbit is a circulation orbit with continuously increasing x. So, in practice, if the terms $Q \sin \theta$ disturb the purely tidal evolution of the system, they do not change the general evolutionary behaviour.

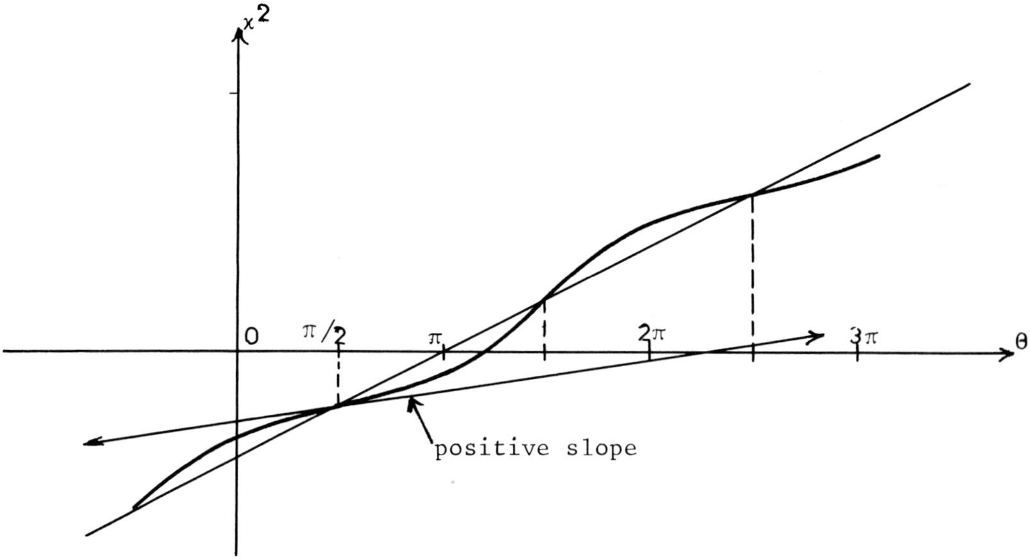

Figure 3. Constant C curves in the $x^2 - \theta$ plane when $A > |Q|$

c) $A = |Q|$. One has the same situation as above, except that there exist unstable equilibrium points for $\theta = \pi/2 + k\pi$

VII. EXTENSION TO THE GENERAL CASE

Let us now consider the complete equations (12), assuming $B \neq 0$. There is no more integral of the type that was used for Hamiltonian systems studied in the last section.

Qualitatively, one may consider that equations (12) are represented by (15) where A is allowed to be slightly modified by Bx. In practice, A and B are of the same order of magnitude whereas x in the libration region is smaller than 10^{-4}. It is therefore possible to consider (12) as a perturbed case of (15) where A is replaced by $A_o - Bx$, Bx remaining very small as compared with A_o. The equations that are obtained are analogous to those studied by Burns (1979) for the rotation of Mercury. It is also possible to apply the adiabatic invariant theory (Henrard, 1982).

Let us consider a solution of the reduced system (15), solution that can explicitely be obtained, since the system is integrable.

$$\begin{cases} x_o = x_o\ (t,\ C,\ p) \\ \theta_o = \theta_o\ (t,\ C,\ p) \end{cases} \tag{17}$$

where C and p are the two independent integration constants. Let us construct the general solution of (12) in the form (17) where increments of C and p are now considered as functions of time.

$$\begin{cases} C = C_o + \Delta C(t) \\ p = p_o + \Delta p(t) \end{cases} \tag{18}$$

The solution of (12) has the form :

$$\begin{cases} x = x_o(t, C_o, p_o) + \left(\dfrac{\partial x_o}{\partial C}\right)_o \Delta C + \left(\dfrac{\partial x_o}{\partial p}\right)_o \Delta p \\ \theta = \theta_o(t, C_o, p_o) + \left(\dfrac{\partial \theta_o}{\partial C}\right)_o \Delta C + \left(\dfrac{\partial \theta_o}{\partial p}\right)_o \Delta p \end{cases} \tag{19}$$

In order to obtain the differential equations in C and p, let us write the derivatives of (19) :

$$\begin{cases} \dfrac{dx}{dt} = \dfrac{dx_o}{dt} + \dfrac{\partial}{\partial C}\left(\dfrac{dx_o}{dt}\right) \Delta C + \dfrac{\partial}{\partial p}\left(\dfrac{dx_o}{dt}\right) \Delta p + \dfrac{\partial x_o}{\partial C}\dfrac{d\Delta C}{dt} + \dfrac{\partial x_o}{\partial p}\dfrac{d\Delta p}{dt} \\ \dfrac{d\theta}{dt} = \dfrac{d\theta_o}{dt} + \dfrac{\partial}{\partial C}\left(\dfrac{d\theta_o}{dt}\right) \Delta C + \dfrac{\partial}{\partial p}\left(\dfrac{d\theta_o}{dt}\right) \Delta p + \dfrac{\partial \theta_o}{\partial C}\dfrac{d\Delta C}{dt} + \dfrac{\partial \theta_o}{\partial p}\dfrac{d\Delta p}{dt} \end{cases}$$

and substitute whenever possible the right-hand members of (15) for dx_o/dt and $d\theta_o/dt$. Then we equate them to the right-hand members of (12) where x and θ are replaced by (19) and where second order terms in ΔC and Δp are neglected. This permits to linearize the system in ΔC and Δp that describes locally the general solution in the form (17). After some algebra, calling

$$\begin{cases} y = \dfrac{\partial x_o}{\partial C} \Delta C + \dfrac{\partial x_o}{\partial p} \Delta p\ , \\ z = \dfrac{\partial \theta_o}{\partial C} \Delta C + \dfrac{\partial \theta_o}{\partial p} \Delta p \end{cases} \tag{20}$$

one finally obtains the following system :

$$\frac{dy}{dt} - By - Qz \cos \theta_o = Bx_o \quad ; \quad \frac{dz}{dt} - 2Gy = 0 \tag{21}$$

The system (21) is equivalent to (12), the solution being written under the form (17). It describes how the actual perturbed orbit can be considered as slowly varying orbits of the kind studied in VI-3.

If we consider the solutions trapped in the libration region, they will only very slightly differ from the orbits given in figure 2. In particular, in the case B = 0, the mean value in time of x is zero, because of the symmetry dx/dt and of the orbit with respect to the axis x = 0. At present, one has, neglecting higher order terms :

$$\frac{dx}{dt} = \frac{dx_o}{dt} - Bx_o$$

Figure 4 shows simultaneously the orbit $x_o = f(\theta)$, the curve dx_o/dt and dx/dt. It results that the mean value $\bar{x}$ of x over a period of θ is positive, since the time spent in the upper half of the orbit is larger than in the lower. However, $\bar{x}$ is of the order of $B|x_o|$ and, therefore, is very small. Consequently, while the body is trapped in libration, the mean value of the semi-major axis is practically constant.

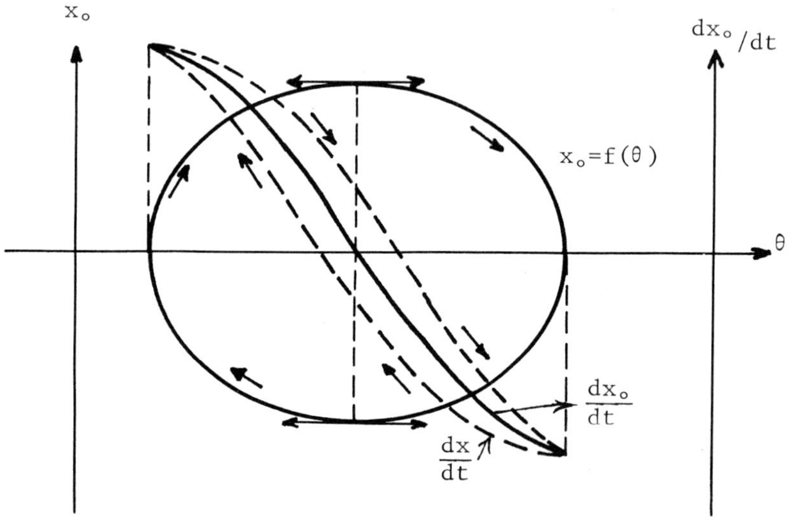

Figure 4. Variations of x_o, dx_o/dt and dx/dt for a libration orbit.

VIII. EVOLUTION OF THE ORBIT

Using the preceding results, it is possible to present a qualitative description of the evolution with time of the semi-major axis of the orbit (i.e. x).

8.1. Before entering the libration region.

Outside the libration region, it is still possible to eliminate the term $Q \sin \theta$ from the equations (6), so that finally only the secular terms are left in the reduced equations. The evolution of the orbits is described by the case VI-1. There is a continuous increase of x given by (13) or (14).

8.2. While crossing the libration region.

Two cases are to be considered.

a) $|Q| \leqq A$. There exist no libration regions (see VI-3-b) so that the evolution of x continues while the critical argument n_θ crosses the value zero. It is however to be expected that this evolution is somewhat perturbed by this term, especially if $|Q|$ is not too much smaller than A.

b) $|Q| > A$. The orbit is trapped in the libration region and evolves in it as described in section VII. In this first approach to the problem, we have not evaluated the time during which it is trapped nor did we consider the capture probabilities. If it is captured, during a certain time ΔT, the semi-major axis does not change.

8.3. After leaving the libration region.

The situation is the same as in the first case. The $Q \sin \theta$ term can again be eliminated and formulae (13) or (14) are again valid.

Finally, if $|Q| \leqq |A|$, the critical term does not affect the general increase of the semi- major axis. If $|Q| > |A|$, this increase is stopped during the time ΔT defined above.

IX. NUMERICAL SIMULATIONS

A number of numerical integrations was performed over the system (12). It led to confirm some of the results given in the present paper.

9.1. Effects of $|Q| / A$.

Figure 5 shows the results of numerical integrations of the equations (12) with $A = 1.1\ 10^{-12}$ and $B = 5.95\ 10^{-12}$, when Q varies between 10^{-13} and 5.10^{-11}. The dotted line represents the variation of x with $Q = 0$ (case VI-1). The separation between the trapped and untrapped orbits appears very clearly for $Q_o = 1.113\ 10^{-12}$. The difference with the theoretical value A comes from the disturbing effect of B (figure 5).

All these orbits - and many others that were calculated - started from $x = 0$, so that they correspond initially to a position in the libration region. All the orbits with $Q > Q_o$ that were initially trapped, escaped after a certain time. No example of orbit that could not escape was found.

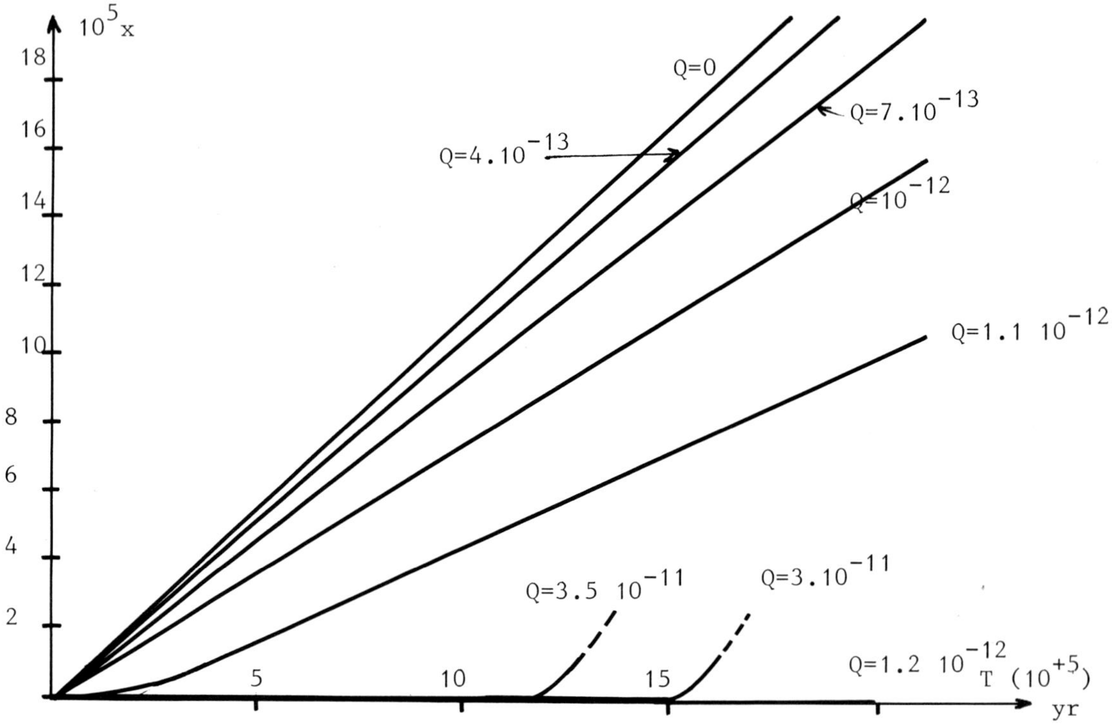

Figure 5. Evolution of x for various values of Q. Curves are stopped when the effect of Q becomes short-periodic and affects the integration scheme.

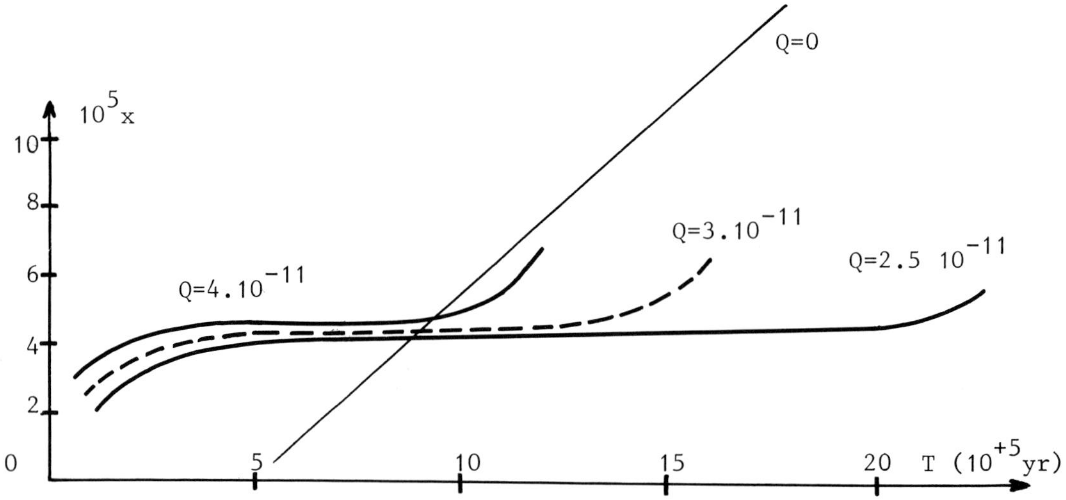

Figure 6. Some evolutionary tracks crossing the libration region. Curves are stopped when the effect of Q becomes short-periodic and affects the integration scheme.

9.2. Case when $Q < Q_o$

It can also be seen that the mean slopes of evolutionary tracks decrease when Q increases for $Q < Q_o$. This effect also plays a role in the lengthening of the evolution time. Several runs have been made also to study haw some orbits arrive in the libration region, get trapped and then, escape. Figure 6 gives some examples of such evolutionary tracks. They illustrate the general features that were described in section VII. It would be of interest to explain analytically more detailed features that appear in these curves.

REFERENCES

Brouwer, D. and Hori, G., 1961, Astron. J., 66, p.193.

Burns, T.J., 1979, Celestial Mechanics, 19, p. 297.

Calame, O. and Mulholland, J.D., 1978, Science, 199, p. 977.

Cazenave, A. and Daillet, S., 1981, J. Geophys. R., 86, p. 1659.

Ferrari, A.J., Sinclair, W.S., Sjogren, W.L., Williams, J.G. and Yoder, C.F., 1981, J. Geophys. R., 85, p. 3939.

Henrard, J., 1982, Celestial Mechanics, 27, p. 3.

Johnson, G.A.L. and Nudds, J.R., 1974, in "Growth, Rythms and the History of the Earth's Rotation", G.D. Rosenberg and S.K. Runcorn ed., John Wiley and sons, London, p. 27.

Kovalevsky, J., 1967, "Introduction to Celestial Mechanics", D. Reidel Publ. Co, p. 116.

Lambeck, K., 1978, in "Tidal Friction and the Earth's Rotation", P. Brosche and J. Sündermann ed., Springer-Verlag, Berlin, p. 145.

MacDonald, G.J.F., 1966, in "The Earth-Moon System", B.G. Marsden and A.G.W. Cameron ed., Plenum Press, New-York, p. 165.

Melchior, P., 1973, "Physique et dynamique planétaires", Vander, Louvain, vol. 4, p. 4.

Mignard, F., 1979, The Moon and the Planets, 20, p. 301.

Mignard, F., 1980, The Moon and the Planets, 23, p. 185.

A RE-CONSIDERATION OF THE EVOLUTION HYPOTHESIS OF THE ORIGIN OF THE RESONANCES AMONG SATURN'S SATELLITES

A.T. Sinclair
Royal Greenwich Observatory
Hailsham, Sussex, England.

Abstract

The hypothesis that the origin of the resonances among Saturn's satellites is due to orbital evolution is critically reviewed. It is concluded that the hypothesis provides a plausible explanation of the origin of the Mimas-Tethys resonance, but it is unsatisfactory for Enceladus-Dione, since their resonance is having little effect on the relative evolution rate of these satellites at present.

1. INTRODUCTION

It was shown by Roy & Ovenden (1954) that there are more occurrences of pairs of satellites with mean motions close to the ratio of two small integers than would be expected by chance. These are referred to as commensurabilities of mean motions or, more simply, as resonances. Goldreich (1965) suggested that these resonances were formed as the result of orbital evolution caused by dissipation of the tidal energy generated in the planet by the satellite. He suggested that the satellite orbits would evolve until a resonance was encountered, and showed that once a resonance had been entered by some means it would not be disrupted by subsequent tidal evolution. Allan (1969) worked out further details of the evolution of a resonance, and Sinclair (1972) and Greenberg (1973) described mechanisms whereby a resonance would be entered as the satellite orbits evolved towards it. At this stage it appeared that the tidal evolution hypothesis would satisfactorily explain most of the features of the three resonances in the Saturn system, namely those of Mimas-Tethys, Enceladus-Dione and Titan-Hyperion. Peale (1976) has written a comprehensive review of this subject.

In this paper we review the tidal hypothesis as applied to Mimas-Tethys and Enceladus-Dione. For Mimas-Tethys it appears to be a satifactory explanation, and is consistent with their anomalously high inclinations. For Enceladus-Dione it is not a satisfactory explanation of the present ratio of the mean motions.

V. V. Markellos and Y. Kozai (eds.), Dynamical Trapping and Evolution in the Solar System, 19–25.

For Titan-Hyperion the tidal hypothesis is dynamically possible (Greenberg 1973), but there is some doubt (Goldreich 1965) as to whether tidal dissipation is large enough to cause signficant evolution of Titan's orbit. An alternative explanation (Sinclair 1972) is that Hyperion is one of many small bodies that were formed near Titan, and only Hyperion has remained, as it happened to be formed in a resonance that protected it from close approaches to Titan.

2. RESONANCES AT A 2:1 COMMENSURABILITY

Suppose that a pair of satellites is close to a 2:1 resonance, so that the mean motion of the inner satellite is approximately twice that of the outer satellite. We take the equatorial plane of the planet as the reference plane, and denote the orbital elements of the inner satellite by $a, e, i, \lambda, \tilde{\omega}, \Omega$, where a is the semi-major axis, e is the eccentricity, i is the inclination to the equatorial plane, λ is the mean longitude, $\tilde{\omega}$ is the longitude of the pericentre, and Ω is the longitude of the ascending node on the equatorial plane. The orbital elements of the outer satellite are denoted by similar primed symbols. We denote by n, n' the mean motions, and by m, m' the ratios of the masses of the satellites to the mass of the planet.

At a 2:1 commensurability the following critical combinations of angular elements occur in the disturbing functions of the satellites, and can cause resonance effects:-

$$\theta = 2\lambda' - \lambda - \tilde{\omega}$$

$$\theta' = 2\lambda' - \lambda - \tilde{\omega}'$$

$$\phi = 2\lambda' - \lambda - \Omega$$

$$\phi' = 2\lambda' - \lambda - \Omega'$$

The oblateness of the planet causes $\tilde{\omega}, \tilde{\omega}', \Omega, \Omega'$ to vary at significantly different rates, and so the arguments $\theta, \theta', \phi, \phi'$ become critical (ie have a very small rate of change) for a range of values of $2n' - n$. Combinations of these arguments also occur in the disturbing functions, but the only combination that can cause significant resonance effects is $\phi + \phi'$. The resonances of the arguments ϕ, ϕ' and $\phi + \phi'$ involve the inclinations and nodes of the orbits, and are referred to as inclination-type. The resonances of θ and θ' involve the eccentricities and apses, and are referred to as eccentricity-type.

It has been shown by Yoder (1973: see Peale, 1976 p 236) that as a pair of satellites evolves towards a resonance it is only possible for them to be trapped in the resonance if the direction of evolution is such that a/a' is increasing. If so the resonance arguments would normally be encountered in the order $\phi, \phi + \phi', \phi', \theta', \theta$, assuming that the oblateness of Saturn is the only significant contribution to the rates of change of the nodes and apses (Sinclair 1972). However it is possible for $\phi + \phi'$ to be encountered earlier in the sequence if $i \ll i'$, for θ' to be encountered earlier if e' is small, and for θ to be encountered earlier if e is small. This is because in these circumstances the resonances will have a significant effect on the rates of change of the nodes and apses. However resonances have very little effect on the evolution rate when encountered far from their normal position in this way, and can essentially be ignored until the normal position is approached.

Before a resonance is encountered the argument concerned will have a negative rate of change (since the direction of evolution is such that $2n' - n$ is increasing from negative values). This rate of change becomes zero at the exact resonance, and positive if the resonance is passed through. If the satellites become trapped in the resonance then the argument will have on average a zero rate of change, and will oscillate, or librate, about a certain value. There are two mechanisms by which the satellites can enter a libration. One of these is a capture process, with usually a fairly low capture probability, and leads to a large amplitude libration. It occurs very close to the normal position of the resonance, and there is an abrupt change in the evolution rates of the mean motions as the libration commences, so that $d/dt\ (n - 2n')$ drops virtually to zero. The second mechanism operates if one of the eccentricities or inclinations is very small, when entry into the libration is automatic, and results in a small amplitude libration. Entry into libration can occur some distance from the normal position of the resonance, and initially the libration has little effect on the evolution rates of n and n'. Instead it forces the apse or node of one of the orbits to vary at such a rate as to maintain the resonance condition. As the exact resonance is approached the libration has an increasing effect on the evolution rates of n and n'. This second mechanism can operate at the $\phi + \phi'$, θ' and θ resonances only.

3. MIMAS-TETHYS RESONANCE

Mimas and Tethys are in a libration at the $\phi + \phi'$ resonance, in which the argument $\phi + \phi'$ oscillates about 0° with an amplitude of 97°. This is the only inclination-type resonance known in the solar system, whereas there are several eccentricity-type resonances.

The hypothesis that this resonance is due to tidal evolution (Allan 1969, Sinclair 1972) supposes that initially Mimas' orbit lay inside the position of resonance with Tethys. The tidal forces would cause Mimas to evolve outwards relatively faster than Tethys, so that a/a' increased. The ϕ resonance was encountered first, where the probability of capture into resonance was 7%. Capture did not occur, and the evolution continued until the $\phi + \phi'$ resonance was encountered. The probability of capture here was 4%, and capture did occur. A libration was formed with amplitude very close to the limiting value of 180°. At this stage the inclinations were $i = 0^{\circ}.42$ and $i' = 1^{\circ}.05$, and these would have been the approximate original inclinations as the evolution up to this point would not have affected them significantly. After the formation of the libration the rate of change of n − 2n' would drop to about 10^{-5} of its value before entering the libration. The libration amplitude would gradually decrease, and the inclinations increase, so that after a time of about 2 x 10^8 years the present situation would be reached, with amplitude 97° and i = 1°.52, i' = 1°.09. (The quantity $(a'/a)^{\frac{1}{2}}\, m'i'^2/m - i^2$ remains constant during the evolution (Sinclair 1974)).

This appears to be a plausible explanation of the origin of the resonance, but there are a few unsatisfactory features. The probability involved is very small, but this is perhaps acceptable since this is the only resonance for which we have to invoke a probability mechanism. The recent formation of the resonance 2 x 10^8 years ago is worrying. This is a lower estimate of the age, derived by assuming that Mimas was formed just above synchronous height, but it must lie close to this limit if we suppose that appreciable evolution has occurred. However Smith et al (1982) deduce from the crater density observed on Iapetus from Voyager 2 that the inner satellites must have suffered far greater impact rates, and may have possibly been disrupted and re-accreted several times. If so, then a recent formation of the resonance would be reasonable.

The present inclinations of Mimas and Tethys (1°.52 and 1°.09) are rather large compared with those of the other inner satellites (Rhea − 0°.34, Titan − 0°.31, Enceladus and Dione ~ 0°.03). The tidal hypothesis suggests that the initial inclination of Mimas was a more typical value of 0°.42, and only Tethys had a somewhat anomalously large initial value of 1°.05. The probability of capture into the $\phi + \phi'$ resonance is larger for small values of i/i', and so the anomalously high value of i' resulted in a reasonable probability of capture into the resonance.

Any satisfactory theory of the origin of this resonance must relate the present anomalously high inclinations of Mimas and Tethys to the existence of this unique inclination resonance, and we see that the tidal hypothesis meets this condition.

4. ENCELADUS-DIONE RESONANCE

Enceladus and Dione are in a libration at the θ resonance, in which the argument θ oscillates about 0° with amplitude 1°.4 (Kozai, 1957, p. 95. An erroneous value of 20' is often quoted for the amplitude). The hypothesis that the resonance is due to tidal evolution (Sinclair 1972, 1974, Greenberg 1973) supposes that initially the orbit of Enceladus was inside the resonance with Dione. The tidal forces would cause Enceladus to evolve outwards relatively faster than Dione, so that a/a' increased. The ϕ, $\phi + \phi'$, ϕ' and θ' resonances would be encountered first, in this order. For suitable initial values of i, i' and e' there would be a low probability of capture into each of these resonances. Having avoided capture into any of these resonances the system would approach the θ resonance with a value of e (the eccentricity of Enceladus) smaller than a critical value 0.019 so that entry into the libration was certain. A very small initial value of e of about 0.0001 would explain the present libration amplitude.

Peale (1976, p 243) argues that this small value of e would mean that the θ resonance would be encountered before the other resonances, and this would explain the failure to enter the other resonances. This is not so. If θ were in a libration a long way from the exact resonance then the only quantities affected by the resonance would be e and $\tilde{\omega}$. These are not involved in the other four resonances, and so the θ resonance would not influence the process of passing through or being captured in the other resonances. In fact it is possible for more than one resonance to exist simultaneously between a pair of satellites, as occurs for Io and Europa (Sinclair 1975).

There is a serious problem with the above description of the origin of the θ resonance. Suppose that in the absence of the libration the rate of evolution due to tidal forces would be $d/dt\,(n - 2n') = -K$. Then it can be shown from the equations given by Sinclair (1972, p 178) that, if the system is in a small amplitude libration of the argument θ, then as the system evolves into resonance the variations of $n - 2n'$ and e are given by

$$de/dt = Kq\,e^2/(pe^3 + q^2) \qquad (1)$$

$$d/dt\,(n - 2n') = -\,Kq^2/(pe^3 + q^2) \qquad (2)$$

where $$p = (3n^2\alpha m' + 12n'^2\,m)\,A$$

$$q = n\alpha m'A$$

where $A = \frac{1}{2}\,(4 + \alpha D)b_{\frac{1}{2}}^{(2)}\,(\alpha)$, $b_{\frac{1}{2}}^{(2)}$ is a Laplace coefficient, and D denotes differentation wrt α. For Enceladus-Dione, $\alpha = 0.63$, $A = 1.19$, $n = 2n'$ (on RHS of equations), $m = 1.27 \times 10^{-7}$, $m' = 1.82 \times 10^{-6}$, $e = 0.0044$. Hence

$$d/dt\,(n - 2n') = -\,0.83\,K.$$

So at present the resonance is having little effect on the rate of evolution of the system. An idea of the time scale involved can be obtained by assuming a value $Q = 6.7 \times 10^4$ for the tidal dissipation function of Saturn, which was used by Allan (1969) and Sinclair (1972). (See these papers for a definition of Q). This gives $K = 1.55 \times 10^{-14}$ rad/(time unit)^2, where the time unit is such that $n' = 1$ rad/time unit (about 0.436 days).

Equations (1) and (2) have been integrated (equation 2 numerically), taking $n - 2n' = 0$ and $e = 0.0044$ at $t = 0$. The results are plotted in Figure 1. The approximate positions of the other resonances are marked relative to the present position of the system, although any

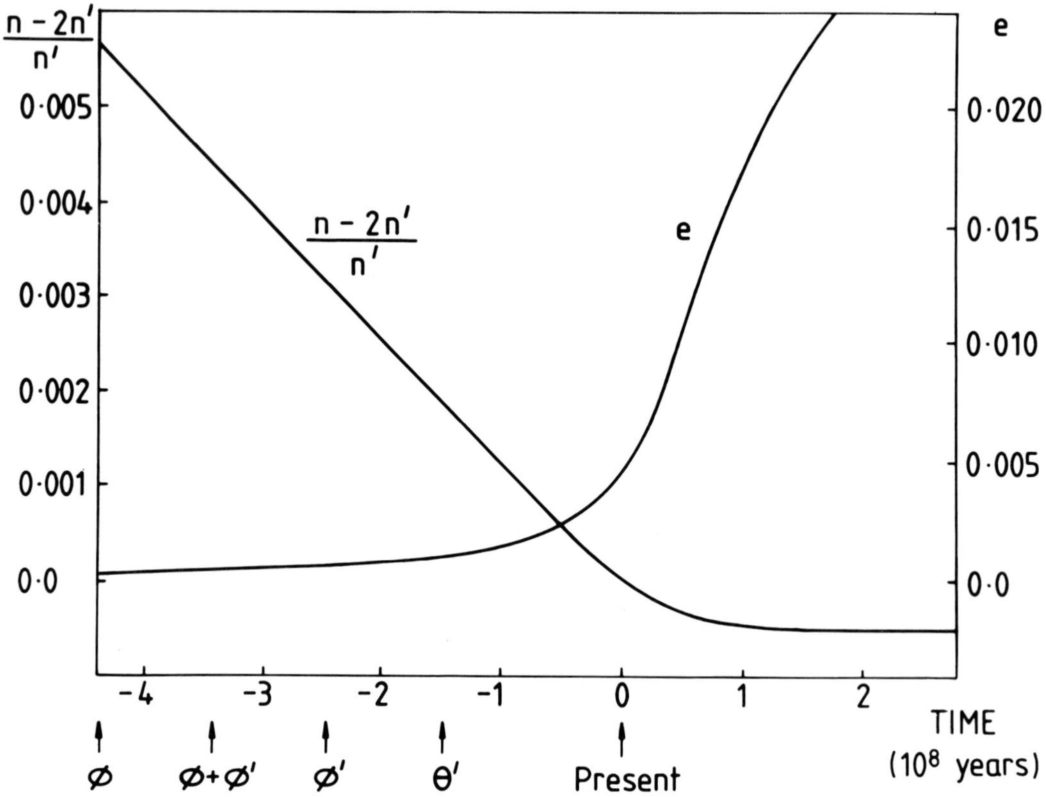

FIGURE 1. Variation of n - 2n' and e caused by θ resonance of Enceladus and Dione.

effects of these resonances are ignored in this calculation. It can be seen that according to the tidal hypothesis the satellites first encountered the general 2:1 resonance about 4.4×10^8 years ago. The θ resonance has had a significant effect on the eccentricity of Enceladus, but has had very little effect on the evolution rate of the system. However in about another 10^8 years time the resonance will have slowed the evolution rate virtually to a standstill.

This is clearly a most unlikely state in which to find the system. The idea of the tidal hypothesis is that we happen to see many objects in resonances because these are trapped states in which the objects spend a large proportion of their lifetimes. This can not be used to explain why we see the Enceladus-Dione system in a state through which it is passing at a virtually unhindered rate.

Hence we must conclude either that the tidal hypothesis does not provide an explanation of the origin of the Enceladus-Dione resonance, or that our model of tidal evolution is inadequate. The latter is probably the case. Yoder (1979) has considered the effect on orbital evolution of tidal dissipation within the satellite, generated by the forced eccentricity of an orbit caused by a resonance. He considers this effect in addition to the evolution caused by dissipation within the planet. He has applied this mechanism particularly to Io in order to explain the origin of the Laplace resonance among Io, Europa and Ganymede. He also considers the possible effect on Enceladus, which also has a significant forced eccentricity (0.0044) and a very small proper eccentricity (0.0001); such a state is indicative of tidal dissipation within the satellite. His calculations suggest that Enceladus would reach a forced eccentricity of 0.022 in a fully evolved state, and he concludes that it is unlikely that the pair have reached a steady-state configuration.

The Voyager 2 images of Saturn (Smith et al. 1982) show that Enceladus has had a complex geological history, with tidal heating caused by the resonance being the most likely cause. Hence it is probable that orbital evolution driven by dissipation in the planet has also acted, in order to maintain the satellites in a resonant situation. The complete description of the system is probably far more complex than the simple models so far proposed.

5. CONCLUSIONS

The hypothesis that orbital resonances between satellites are caused by evolution driven by tidal dissipation within the planet satisfactorily explains many of the features of the Mimas-Tethys resonance. The hypothesis fails to explain the present state of the Enceladus-Dione system, where the resonance is at present having very little effect on the evolution rate. It is probable that tidal dissipation with Enceladus is also involved.

6. REFERENCES

Allan, R.R. 1969. Astr. J., 74, pp 497-506.
Goldreich, P., 1965. MNRAS., 130, pp 159-181.
Greenberg, R., 1973. Astr. J., 78, pp 338-346.
Kozai, Y., 1957. Ann. Tokyo Obs., Ser. 2, 5, pp 73-106.
Peale, S.J. 1976. Annual Rev. Astron. & Astrophys., 14, pp 215-246.
Roy, A.E. & Ovenden, M.W., 1954. MNRAS., 114, pp 232-241.
Sinclair, A.T., 1972. MNRAS., 160, pp 169-187.
Sinclair, A.T., 1974. MNRAS., 166, pp 165-179.
Sinclair, A.T., 1975. MNRAS., 171, pp 59-72.
Smith, B.A., et al (29 authors). Science, 215 pp. 499-537.
Yoder, C.F., 1973. PhD Thesis. Univ. California, Santa Barbara, 303 pp.
Yoder, C.F., 1979. Nature, 279, pp 767-770.

ORIENTATION OF A SATELLITE LOCATED AT THE LIBRATION POINT IN THE RESTRICTED THREE-BODY PROBLEM

Grzegorz Duliński, Andrej J. Maciejewski
Institute of Astronomy
Nicolaus Copernicus University
Chopina 12/18, 87-100 Toruń, Poland

1. INTRODUCTION

In this paper the initial results of an investigation of the motion of a rigid body located at the libration point in the planar, restricted three-body problem are given. This problem was analyzed in part by Kane and Marsh (1971), Markeev (1967a,b). However the present investigation is formulated in terms of hamiltonian mechanics. The final results will by used to study nonlinear effects connected with the gravitational influence of the "second" central body.

2. EQUATIONS OF MOTION

Let m_1, m_2, m_3 designate rigid bodies of mass m_1, m_2 and m_3 respectively. We assume that m_3 is so small in comparison with m_1 and m_2 that it has no effects on the motions of m_1 and m_2. Furthermore let m_1 and m_2 have spherically symmetric mass distributions so that the gravitational effects of m_1 and m_2 on other bodies are equivalent to those of point masses. The motion of the mass centers of these two massive bodies are the same as in the elliptic 2-body problem. Then we introduce the following righthanded, orthonormal coordinate systems: inertial, orbital and principal axes (see fig. 1).

For describing an orientation of a third coordinate system relative to the second one we use Euler´s angles: φ, Θ, ψ in 1-2-1 sequence (Markley 1978). Next we assume that the third rigid body-satellite- is axially symmetric and is located at the triangular L_4 libration point.

The Lagrange function has the form

$$L = \frac{1}{2}\,\omega^T I\omega - V \tag{1}$$

where

$$\omega^T = (\omega_1, \omega_2, \omega_3), \quad \omega_i = \omega_i(\varphi, \Theta, \psi, \dot{\varphi}, \dot{\Theta}, \dot{\psi}), \quad \dot{} = \frac{d}{dt}$$

V. V. Markellos and Y. Kozai (eds.), Dynamical Trapping and Evolution in the Solar System, 27–35.

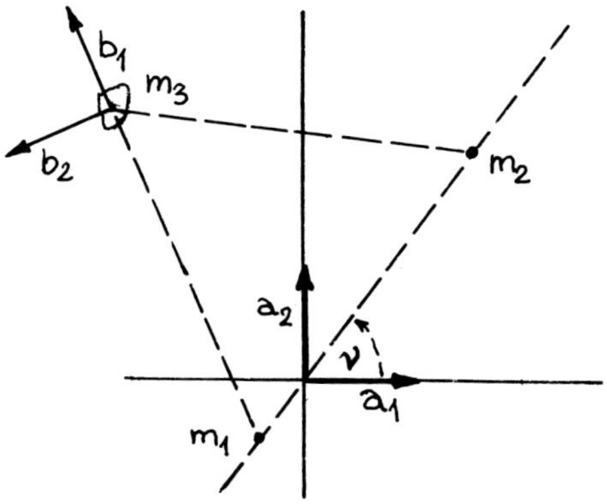

Fig. 1. Geometry of the 3-body problem.
(a_1, a_2, a_3) - inertial coordinate system
(b_1, b_2, b_3) - orbital coordinate system

are the components of the absolute angular velocity in the third coordinate system.

$$I = \begin{bmatrix} I_1 & 0 & 0 \\ 0 & I_2 & 0 \\ 0 & 0 & I_2 \end{bmatrix},$$

is an inertia tensor $(I_1, I_2$ - are the moments of inertia),

$$\omega = N\dot{q} + b,$$

$$N = \begin{bmatrix} c\Theta & 0 & 1 \\ s\psi\, s\Theta & c\psi & 0 \\ c\psi\, s\Theta & -s\psi & 0 \end{bmatrix}, \quad b = \begin{bmatrix} -s\Theta\, c\varphi \\ c\psi\, s\varphi + s\psi\, c\Theta\, c\varphi \\ -s\psi\, s\varphi + c\psi\, c\Theta\, c\varphi \end{bmatrix},$$

$s\Theta = \sin\Theta$, $c\Theta = \cos\Theta$, etc.

$$\dot{\nu} = \frac{C}{r^2}, \quad C = G(m_1 + m_2)p, \quad r = \frac{p}{1 + e\cos\nu},$$

G - gravitational constant, p - orbit parameter,
e - eccentricity, r - radius vector, ν - true anomaly
T - transposition of the matrix,
$\dot{q} = (\dot{\varphi}, \dot{\Theta}, \dot{\psi})$,

$$V = \frac{3}{2}\omega_0^2 (1-e^2)^{-3} (1 + e\cos\nu)^3 (I_1 - I_2) \times \left[(1-\mu)c^2\Theta + \frac{1}{4}\mu(c\Theta - \sqrt{3}\, s\Theta\, s\varphi)^2 \right], \tag{2}$$

$$\omega_0 = \frac{2\pi}{P}, \quad P \text{ - orbital period of the massive bodies,}$$

$$\mu = \frac{m_2}{m_1 + m_2}, \qquad 0 \leqq \mu \leqq \frac{1}{2}.$$

Introducing now:

$$P_i = \frac{\partial L}{\partial \dot{q}_i}, \quad i = 1,2,3, \quad \dot{q}_i = \dot{\varphi}, \dot{\Theta}, \dot{\psi},$$

where L is given by expression (1), we get the hamiltonian

$$H = \frac{p_\varphi^2}{2 I_2 s^2\Theta} + \frac{p_\Theta^2}{2I_2} + \frac{1}{2}\left(\frac{ctg^2\Theta}{I_2} + \frac{1}{I_1}\right) p_\psi^2 - \frac{ctg\,\Theta}{I_2 s\Theta} p_\psi p_\varphi$$
$$-\dot{\nu}\, ctg\Theta\, c\varphi\, p_\varphi - \dot{\nu}\, s\varphi p_\Theta + \dot{\nu}\frac{c\varphi}{s\Theta} p_\psi + V. \tag{3}$$

It is seen that ψ is a cyclic coordinate, therefore:

$$P_\psi = I_1\omega_1 = \text{constant}.$$

After substitution into equations (3) of the true anomaly $-\nu$ as independent variable instead of time $-t$, and introduction of dimensionless momenta defined by:

$$p'_\varphi = \frac{p_\varphi}{\sigma}, \quad p'_\Theta = \frac{p_\Theta}{\sigma}, \quad \text{where: } \sigma = I_2(1-e^2)^{-3/2}\omega_0,$$

we derive the following form for Hamilton's function:

$$H = \frac{p'^2_\varphi}{2(1 + e\cos\nu)^2 s^2\Theta} + \frac{p'^2_\Theta}{2(1 + e\cos\nu)^2} + \frac{\gamma\, ctg\Theta\, p'_\varphi}{(1 + e\cos\nu)^2 s\Theta}$$
$$- ctg\Theta\, c\varphi\, p'_\varphi - s\varphi p'_\Theta + \gamma\frac{c\varphi}{s\Theta} + \frac{\gamma^2\, ctg^2\Theta}{(1 + e\cos\nu)^2} \tag{4}$$
$$+ \frac{3}{2}(\alpha-1)(1 + e\cos\nu)\left[(1-\mu)c^2\Theta + \frac{1}{4}\mu(c\Theta - \sqrt{3}\, s\Theta\, s\varphi)^2\right]$$

where:

$$\alpha = \frac{I_1}{I_2}, \quad \beta = \frac{\omega_1}{\omega_0}, \quad \gamma = \alpha\beta(1-e^2)^{3/2}.$$

For $\mu = 0$ we have obtained the same expression as that derived by Markeev (1967b) for the one central body case.

3. PARTICULAR SOLUTION AND THE STABILITY ANALYSIS

Hamilton's equations of motion are as follows:

$$\dot{\Theta} = \frac{\partial H}{\partial p_\Theta}, \quad \dot{\varphi} = \frac{\partial H}{\partial p_\varphi}, \quad \dot{p}_\Theta = -\frac{\partial H}{\partial \Theta}, \quad \dot{p}_\varphi = -\frac{\partial H}{\partial \varphi}, \tag{5}$$

where:

˙ - denotes differentation with respect to ν,
H is given by equation (4).

The above equations have only one particular, stationary solution such that:

$$\Theta = \frac{\pi}{2}, \quad \varphi = \pi, \quad p_\Theta = 0, \quad p_\varphi = 0. \tag{6}$$

This is the only known particular solution of equations (5) even though we get e = 0. To discuss the properties of the solutions near this particular solution, it is necessary to investigate its stability. This can be done in the e = 0 case by analytic methods.

For small e, but larger than zero, we can do it and calculate instability regions studying the parametric resonance in this system (Arnold 1978, Markeev 1978). We introduce new variables q_1, q_2, p_1, p_2 defined as follows:

$$q_1 = \Theta - \frac{\pi}{2}, \quad q_2 = \varphi - \pi, \quad p_1 = p_\Theta, \quad p_2 = p_\varphi. \tag{7}$$

Then the hamiltonian takes the form

$$H(q, p) = H_2(q, p) + H_3(q, p) + H_4(q, p) + \dots \tag{8}$$

For e = 0 (circular problem) we may write

$$H = \frac{q_1^2}{2}\left[\alpha^2\beta^2 - \alpha\beta + \frac{3}{4}(\alpha-1)(4-3\mu)\right] + \frac{q_2^2}{2}\left[\frac{9}{4}(\alpha-1)\mu + \alpha\beta\right]$$
$$+ \frac{p_1^2}{2} + \frac{p_2^2}{2} + q_1 q_2\left[\frac{-3\sqrt{3}}{4}(\alpha-1)\mu\right] + p_1 q_2 + q_1 p_2(\alpha\beta - 1). \tag{9}$$

For this hamiltonian the equations of motion are linear and may be written:

$$\frac{dX}{dt} = I\,H\,X \tag{10}$$

where:

$$X^T = (q_1, q_2, p_1, p_2),$$

$$IH = \begin{bmatrix} 0 & 1 & 1 & 0 \\ a & 0 & 0 & 1 \\ b & c & 0 & -a \\ c & d & -1 & 0 \end{bmatrix},$$

$$a = \alpha\beta - 1, \quad b = \alpha\beta(1-\alpha\beta) - \frac{3}{4}(\alpha-1)(4-3\mu),$$

$$c = \frac{3\sqrt{3}}{4}(\alpha-1)\mu, \quad d = -\frac{9}{4}(\alpha-1)\mu - \alpha\beta,$$

$$I = \begin{bmatrix} 0 & E_n \\ -E_n & 0 \end{bmatrix}, \quad \text{where: } E_n \text{ - nxn unit matrix} \quad (n = 2).$$

The characteristic equation of matrix I H has the form

$$\lambda^4 + a'\lambda^2 + b' = 0, \tag{11}$$

where

$$a' = \alpha^2\beta^2 - 2\alpha\beta + 3\alpha - 1,$$

$$b' = (\alpha\beta - 1)(3\alpha + \alpha\beta - 4) + \frac{27}{4}\mu(\mu-1)(\alpha-1)^2 .$$

Stability conditions are as follows:

$$a' > 0, \quad b' > 0, \text{ and } \quad a'^2 - 4b' > 0 . \tag{12}$$

On the parametric plane (γ,α), $(\gamma = \alpha\beta)$, the above conditions designate the stability regions of the solution (6) in linear approximation of the equations of motion. After comparison of these results with those obtained for the one central body case we see that the discussed regions are slightly different. To study the stability of the solution (6) of the complete equations of motion it is necessary to transform the hamiltonian H into normal form. We can do this in the following way.

Let λ_i be a root of the characteristic equation (11). Then we have :

$$\lambda_1 = i\omega_1\varepsilon_1, \quad \lambda_2 = i\omega_2\varepsilon_2, \quad \lambda_3 = -i\omega_1\varepsilon_1, \quad \lambda_4 = -i\omega_2\varepsilon_2,$$

where:

$$\omega_1 > \omega_2 > 0, \quad \varepsilon_i = \pm 1, \text{ a sign that will be fixed later.}$$

However, λ_i are the eigenvalues corresponding to the eigenvectors b_i, of the matrix I H, given by:

$$b_i = \frac{1}{\alpha_i} \begin{bmatrix} \lambda_i(1-a) + c \\ \lambda_i^2 - a^2 - b \\ -a\lambda_i^2 + c\lambda_i + a^2 + b \\ \lambda_i^3 - \lambda_i(a+b) - ac \end{bmatrix},$$

where a, b, c are the same as in equation (10), and α_i is a normalized factor defined below. Now we may write b_i as the sum of two vectors:

$$b_j = r_j + i\, s_j, \quad j = 1,2,3,4, \quad i^2 = -1, \tag{13}$$

Linear transformation of the form

$$X = A\,Y$$

where:

$$X^T = (q_1, q_2, p_1, p_2), \qquad Y^T = (y_1, y_2, y_3, y_4)$$

$$A = 2\left[-s_1, -s_2, r_1, r_2\right],$$

transforms the hamiltonian H into

$$H_2 = \sum_{i=1}^{2} \varepsilon_i\omega_i\,(y_i^2 + y_{i+2}^2) \tag{15}$$

Since the matrix A is sympletic, so finally we may write the following relations:

$$(r_k, I\,s_k) = \frac{1}{4}\alpha_k, \quad k = 1,2, \tag{16}$$

where (,) denotes the scalar product.

Then we have a relation for the normalized factor α_i as follows:

$$\frac{1}{4}\alpha_i^2 = \varepsilon_i\omega_i\, f_i(\alpha, \beta, \mu), \tag{17}$$

where:

$$f_i(\alpha, \beta, \mu) = \omega_i^2 \left[(\alpha\beta - 2)^2 - \frac{3}{2}(\alpha-1)(2-3\mu)\right]$$
$$+ (\alpha\beta - 1)(\alpha\beta - 2)^2 + 3(\alpha-1)\left[\alpha^2\beta^2 - 3\alpha\beta + 3\alpha\right]$$
$$+ \frac{9}{4}(\alpha-1)\mu\left[(\alpha\beta-1)^2 + 9\alpha - 8\right] + \frac{27}{2}(\alpha-1)\mu^2$$

One can find that $f_i(\alpha,\beta,\mu) > 0$ in the stability region I (see fig. 2). Therefore on this region $\varepsilon_i = +1$, H_2 is a positive definite function and sufficient stability conditions are satisfied on this region. On region II (see fig. 2) we have $f_2(\alpha,\beta,\mu) < 0$ and $f_1(\alpha,\beta,\mu) > 0$, and we choose $\varepsilon_1 = +1$, $\varepsilon_2 = -1$. Therefore on this region the hamiltonian H_2 has not a fixed sign.

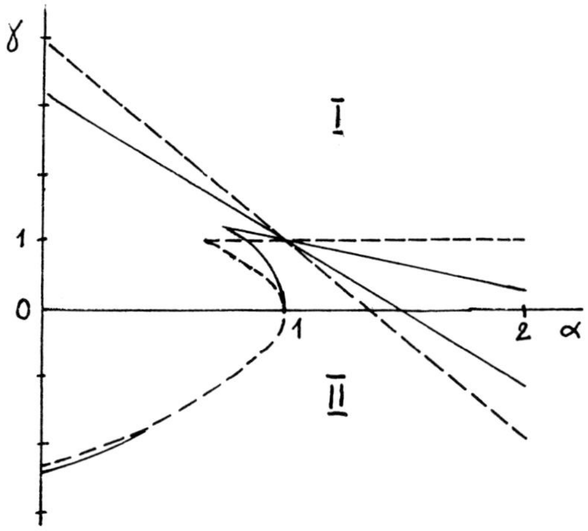

Fig. 2. Stability regions for $\mu = 0$ (---) and for $\mu = \frac{1}{2}$ (——).

To make an exact study of the stability problem in this case we may take into account the H_4 term ($H_3=0$) in equation (8). This term may be written as:

$$H_4(q,p) = \sum_{n_1+n_2+n_3+n_4=4} h_{n_1n_2n_3n_4}\, q_1^{n_1} q_2^{n_2} p_1^{n_3} p_2^{n_4}, \quad (18)$$

where:

$$n_i \geqq 0, \quad n_i \in N, \quad (i = 1,2,3,4),$$

$$h_{4000} = 8\alpha^2\beta^2 - 5\alpha\beta - 3(\alpha-1),$$

$$h_{0400} = -\alpha\beta - 9(\alpha-1)\mu,$$

$$h_{2020} = 1,$$

$$h_{3001} = 5\alpha\beta - 2,$$

$$h_{0310} = -1,$$

$$h_{2200} = \alpha\beta - 9(\alpha-1)\mu - \frac{3\sqrt{3}}{4}(\alpha-1)\mu,$$

$$h_{1201} = 1,$$

and the other terms $h_{n_1 n_2 n_3 n_4} = 0$.

Assuming $i_1\omega_1 + i_2\omega_2 \neq 0$ for $0 < |i_1| + |i_2| \leqq 4$,

which in our case corresponds to

$$\omega_1 \neq \omega_2, \quad \omega_1 \neq 2\omega_2, \quad \omega_1 \neq 3\omega_2,$$

and making the transformation (14) as well as the appropriate Birkhoff transformation we get the following for the hamiltonian H:

$$H = \omega_1(q_1^2 + p_1^2) - \omega_2(q_2^2 + p_2^2) + a_{11}(p_1^2 + q_1^2)^2 + a_{12}(p_1^2 + q_1^2)(q_2^2 + p_2^2) + a_{22}(p_2^2 + q_2^2)^2 + \dots \qquad (19)$$

where:

$$a_{ij} = a_{ij}(\alpha,\beta,\mu),$$

$+ \dots$ - stands for the higher terms in expression (19).

Using Arnold-Moser theorem (Markeev 1978) we notice that on region II (see fig. 2) the solution (6) will be stable everywhere except those points for which the following relations are satisfied:

$$\omega_1 = \omega_2, \quad \omega_1 = 2\omega_2 \text{ and } \omega_1 = 3\omega_2, \text{ and} \qquad (20a)$$

$$f(\alpha,\beta,\mu) = a_{11}\omega_1^2 + a_{12}\omega_1\omega_2 + a_{22}\omega_2^2 = 0 . \tag{20b}$$

The further investigation including nonlinear effects of higher orders will be presented in another paper(Duliński and Maciejewski,1983).

ACKNOWLEDGEMENT

This research was partly supported by a grant of the Polish Academy of Sciences.

REFERENCES

Arnold, V.I.: 1978, *Mathematical Methods in Classical Mechanics*, Springer Verlag, New York

Diliński, G. and Maciejewski, A.: 1983, in preparation

Kane, T.R. and Marsh, L.E.: 1971, *Celes. Mech.* 4, pp. 78-91

Markeev, A.P.: 1967a, *Kosm. Issled.* 5, pp. 365-376

Markeev, A.P.: 1967b, *Kosm. Issled.* 5, pp. 530-540

Markeev, A.P.: 1978, *Libration Points in Celestial Mechanics and Astrodynamics*, Nauka, Moscow (in Russian)

Markley, F.L.: 1978, in J.R. Wertz (ed.), D. Reidel Publ. Company, ASSL Vol. 73

THEORY OF THE LIBRATION OF THE MOON

M. DUBOIS-MOONS
Dept. of Mathematics, F.N.D.P., Namur, Belgium

ABSTRACT

The paper presents a new theory of the libration of the Moon, completely analytical with respect to the harmonic coefficients of the lunar gravity field. This field is represented through its fourth degree harmonics for the torque due to the Earth (the second degree for the torque due to the Sun). The Moon is assumed to be rigid and its orbital motion is described by the ELP 2000 solution (Chapront and Chapront-Touzé 1981) for the main problem of lunar theory with planetary perturbations and influence of the non-sphericity of the Earth. Comparisons with other theories (Migus 1980 and Eckhardt 1981) are also presented.

V. V. Markellos and Y. Kozai (eds.), Dynamical Trapping and Evolution in the Solar System, 37.

REGULARIZATION OF THE EQUATIONS OF MOTION IN A CENTRAL FORCE-FIELD. APPLICATION TO THE ZONAL EARTH SATELLITE.

R. Cid, S. Ferrer, and A. Elipe
Departamento de Astronomía. Universidad de Zaragoza. Spain

Abstract. Within the framework of linear and regular celestial mechanics, we revise a recent method of Belen'kii (1981). We generalize some of his results, giving a new regularizing function.

We make an application to the zonal earth satellite, considering the hamiltonian function through the harmonic J_4. After the angular variable u has been removed, we introduce a new time and we reduce the problem to a linear equation.

1. INTRODUCTION

In this paper, the method of regularization given by Belen'kii (1981) is revised. We propose a function g(r) that generalizes the one studied by him. Then an application to the zonal earth satellite, considering harmonics through J_4, is made.

We use the canonical set of variables (P_r, P_u, P_h, r, u, h) of Hill (1913) and, in order to apply that regularization, the angular variable u is eliminated (Caballero, 1975) using von Zeipel's method. As a consequence the new hamiltonian is

$$\bar{H}(\bar{P}_r, \bar{P}_u, \bar{P}_h, \bar{r}, -, -) = \frac{1}{2}(\bar{P}_r^2 + \frac{\bar{P}_u^2}{\bar{r}^2}) - (\frac{a_1}{\bar{r}} + \frac{a_3}{\bar{r}^3} + \frac{a_4}{\bar{r}^4} + \frac{a_5}{\bar{r}^5})$$

where $\bar{P}_u$, a_i are constant.

We make a transformation of time $d\tau = g_5^{-1}(\bar{r})\ dt$ that reduces the problem to a linear equation.

Other analytical theories have been proposed based on canonical elements associated with a suitable time regularization (Kustaanheimo-Stiefel, 1965; Scheifele-Graf, 1974; Deprit, 1981). In particular, regularizations linearizing the equations, that have also applications in other dynamics problems, have been considered by Stiefel-Scheifele (1971), Belen´kii (1981), Szebehely (1976).

V. V. Markellos and Y. Kozai (eds.), Dynamical Trapping and Evolution in the Solar System, 39–46.

2. BELEN'KII REGULARIZATION

In certain problems of Celestial Mechanics, the hamiltonian of the relative motion of a particle in a central force-field has the form

$$H = \frac{1}{2}\left(P_r^2 + \frac{1}{r^2} P_\phi^2\right) + V_0(r) \tag{1}$$

where $P_r = \dot{r}$ denotes the radial velocity, $P_\phi = r^2\dot{\phi} = c$ is the angular momentum and

$$V_0 = - \sum_1^n \frac{a_i}{r^i} \tag{2}$$

is the potential function.

The energy integral $H = h$, may be written as

$$\frac{1}{2}\left(\frac{dr}{dt}\right)^2 = h - \left\{ V_0(r) + \frac{c^2}{2r^2}\right\} = h - V(r) \tag{3}$$

and Belen'kii introduces a new independent variable τ, by means of the relation

$$d\tau = g^{-1}(r)\, dt \tag{4}$$

with $g(r) > 0$ and $g(r) \in C^{(1)}$. Then, (3) can be written in the form

$$\frac{1}{2}\left(\frac{dr}{d\tau}\right)^2 = g^2(r)\{ h - V(r)\} \tag{5}$$

Differenciating (5) with respect to τ, and after dividing by the nonzero factor $dr/d\tau$, Belen'kii equals the result to a linear expression, obtaining

$$\frac{d^2r}{d\tau^2} = \frac{d}{dr}\left(g^2(r)\{h - V(r)\}\right) = 2c_1 r + c_2 \tag{6}$$

where we have written $2c_1$ for subsequent simplifications.

A full study of the linear equation (6) for the three cases $c_2/c_1 \gtreqless 0$, has been given by Belen'kii (1981.a; Section 2).

Integrating (6), the regularizing function must satisfy the relation

$$g^2(r)\{ h - V(r)\} = c_1 r^2 + c_2 r + c_3 \tag{7}$$

Belen'kii has applied (7) to the potentials

$$V_0 = V_2 = - \frac{a_1}{r} - \frac{a_2}{r^2} \quad ; \quad V_0 = V_3 = - \frac{a_1}{r} - \frac{a_2}{r^2} - \frac{a_3}{r^3}$$

where the corresponding regularizing functions are

$$g_2(r) = r \quad ; \qquad g_3(r) = r^{3/2}\,(1 + \beta r)^{-1/2}$$

respectively. The parameter β depends on a_1, a_2, a_3 and h.

Likewise, Ferrer and Elipe (1982), studying these potentials have considered the following regularizing function

$$g_3(r) = r^{3/2}\,(\beta + r)^{-1/2}$$

which allows the treatment of these cases in a more uniform manner.

3. A NEW REGULARIZING FUNCTION

In a more general problem, with the potential

$$-V_0 = -V_n = \frac{a_1}{r} + \frac{a_3}{r^3} + \ldots + \frac{a_n}{r^n}$$

we have

$$V = V_0 + \frac{c^2}{2r^2} = -\sum_1^n \frac{a_i}{r^i} \tag{8}$$

where $a_2 = -c^2/2$. In this case we propose the following regularizing function

$$g_n(r) = r^{n/2}\{r^{n-2} + \alpha_1 r^{n-3} + \alpha_2 r^{n-4} + \ldots + \alpha_{n-2}\}^{-1/2} \tag{9}$$

where $\alpha_1, \alpha_2, \ldots \alpha_{n-2}$, are parameters which depend on a_i, h, and must be suitably chosen to have $g_n(r) > 0$.

Inserting (8), (9) in (7), we arrive at the equation

$$hr^n + \sum_1^n a_i\, r^{n-i} = \{r^{n-2} + \sum_1^{n-2} \alpha_k r^{n-k-2}\}\,\{c_1 r^2 + c_2 r + c_3\}$$

Equating the coefficients of the same powers of r in both sides, we have the system

$$\begin{array}{l} h = c_1 \;;\quad a_1 = c_1\alpha_1 + c_2 \;;\quad a_2 = c_1\alpha_2 + c_2\alpha_1 + c_3 \\ \text{--- --- --- --- --- ---} \\ a_i = c_1\alpha_i + c_2\alpha_{i-1} + c_3\alpha_{i-2} \qquad (i = 3, \ldots, n-2) \\ \text{--- --- --- --- --- ---} \\ a_{n-1} = c_2\alpha_{n-2} + c_3\alpha_{n-3} \;;\quad a_n = c_3\alpha_{n-2} \end{array} \tag{10}$$

Solving (10) with respect to the coefficients $c_1, c_2, c_3, \alpha_1, \alpha_2, \ldots \alpha_{n-2}$, we get the expression of the coefficients in terms of $a_1, a_2, \ldots a_n$, h.

The study of this last system is difficult and it seems that the more practical way of solving it is by a numerical method.

In particular, we have studied the system (10) for $n = 5$, taking

$$g_5(r) = r^{5/2} (r^3 + \alpha r^2 + \beta r + \gamma)^{-1/2} \tag{11}$$

In this case, the equations of that system are given by

$$\begin{aligned} &h = c_1 && (12_1) \qquad && a_3 = c_1\gamma + c_2\beta + c_3\alpha && (12_4) \\ &a_1 = c_1\alpha + c_2 && (12_2) \qquad && a_4 = c_2\gamma + c_3\beta && (12_5) \\ &a_2 = c_1\beta + c_2\alpha + c_3 && (12_3) \qquad && a_5 = c_3\gamma && (12_6) \end{aligned} \tag{12}$$

From (12_1), (12_2), (12_3) we obtain

$$\begin{aligned} &c_1 = h \quad ; \quad c_2 = a_1 - h\alpha = c_2(\alpha) \\ &c_3 = a_2 - a_1\alpha + h\alpha^2 - h\beta = c_3(\alpha, \beta) \end{aligned} \tag{13}$$

From (12_6), if $a_5 \neq 0$, we get: $c_3 \neq 0$, $\gamma \neq 0$. Then $\gamma = a_5/c_3(\alpha, \beta)$.

Finally, substituting the above expressions in (12_4), (12_5), we have the system

$$\begin{aligned} &C\beta^2 + B\beta + A = 0 \\ &D'\beta^3 + C'\beta^2 + B'\beta + A' = 0 \end{aligned} \tag{14}$$

where

$$C = h^2\alpha^5 - 2ha_1\alpha^4 + (a_1^2 + 2ha_2)\alpha^3 - (2a_1a_2 + a_3h)\alpha^2 + (a_2^2 + a_1a_3)\alpha + ha_5 - a_2a_3.$$

$$B = - h(2h + 1)\alpha^3 + 4ha_1\alpha^2 - (3ha_2 + a_1^2)\alpha + a_1a_1 + ha_3$$

$$A = h(2h - a_1)$$

$$D' = h^2$$

$$C' = - 2h(h\alpha^2 - a_1\alpha + a_2)$$

$$B' = h^2\alpha^4 - 2ha_1\alpha^3 + (a_1^2 + 2ha_2)\alpha^2 - 2a_1a_2\alpha + a_2^2 + ha_4$$

$$A' = - a_4h\alpha^2 + (a_1a_4 - ha_5)\alpha + a_1a_5 - a_2a_4$$

Eliminating β in (14) we get an equation of the form $P(\alpha) = 0$ where $P(\alpha)$ is a polynomial in α of eighteenth degree. Thus it seems convenient to solve (14) by numerical methods.

4. SOME PARTICULAR CASES FOR THE NEW REGULARIZING FUNCTION

i) $a_5 = 0$

In this case, from (12_6) it follows that we can take $\gamma = 0$. Then, the last three equations of (12) reduce to

$$\begin{aligned} a_3 &= c_2\beta + c_3\alpha \\ a_4 &= c_3\beta \end{aligned} \qquad (15)$$

Then, substituting c_3, given by (13), in (15_1), we get

$$\beta = \frac{a_3 - a_2\alpha + a_1\alpha^2 - h\alpha^3}{a_1 - 2h\alpha} \qquad (16)$$

and substituting β in (15_2), we have a sixth degree equation

$$\sum_{0}^{6} A_n\alpha^n = 0$$

where

$$\begin{aligned} A_6 &= 2h^3 \\ A_5 &= -5a_1h^2 \\ A_4 &= 4(a_1^2 + ha_2)h \\ A_3 &= -\{a_1^3 + 6a_1a_2h + (2a_3 + 1)h^2\} \\ A_2 &= 2a_1^2a_2 + (3a_1a_3 - 2a_2^2 - a_1)h - 4a_4h^2 \\ A_1 &= -\{(a_1a_3 - a_2^2)a_1 + (4a_1a_4 + 2a_2a_3 + a_2)h\} \\ A_0 &= a_1a_2a_3 - a_1^2a_4 - a_3h \end{aligned}$$

Then, the regularizing function is

$$g_4(r) = r^2(r^2 + \alpha r + \beta)^{-1/2}$$

and we must take the values α, β in such a way that $r^2 + \alpha r + \beta > 0$ or else, we must find the range of r for which the regularization is well defined.

ii) $a_4 = a_5 = 0$

Again, it is sufficient to take $\gamma = \beta = 0$ in (12). Then, the parameter α verifies the cubic equation

$$h\alpha^3 - a_1\alpha 2 + a_2\alpha - a_3 = 0 \tag{17}$$

The regularizing function is now

$$g_3(r) = r^{3/2}(r + \alpha)^{-1/2} \tag{18}$$

A study and application of (17) and (18) has been made by Ferrer-Elipe (1982).

iii) $a_3 = a_4 = a_5 = 0$

In this case it is sufficient to take $\gamma = \beta = \alpha = 0$. Then, the system (12) reduces to

$$c_1 = h \quad ; \quad c_2 = a_1 \quad ; \quad c_3 = a_2$$

where

$$g_2(r) = r$$

is the regularizing function of Sundman.

5. AN APPLICATION TO THE ZONAL EARTH SATELLITE

I.- It is well known that the kinetic energy T and the potential V of an artificial zonal satellite of the Earth, in the canonical set of variables (P_r, P_u, P_h, r, u, h) of Hill (1913), are given by the equations

$$T = \frac{1}{2}\left(P_r^2 + \frac{P_u^2}{r^2}\right)$$

$$V = -\frac{\mu}{r}\left\{1 - \sum_{n>2} J_n \left(\frac{1}{r}\right)^n P_n(\sin\phi)\right\}$$

The corresponding hamiltonian with the harmonics J_2, J_3, J_4 is given by the expression

$$H = H(P_r, P_u, P_h, r, u, -) = H_0 + H_1 + H_2$$

where

$$H_0 = \frac{1}{2}\left(P_r^2 + \frac{P_u^2}{r^2}\right) - \frac{\mu}{r}$$

$$H_1 = -\frac{\mu}{r^3} J_2 (B_{20} + B_{22} \cos 2u)$$

$$H_2 = \frac{\mu}{8r^4} J_3 \sqrt{1 - \theta^2}\{3(1 - 5\theta^2)\sin u - 5(1 - \theta^2)\sin 3u\} +$$

$$\frac{3\mu}{8r^5} J_4 \left\{\left(\frac{3}{8} - \frac{15}{4}\theta^2 + \frac{35}{8}\theta^4\right) + \left(-\frac{5}{6} + \frac{20}{3}\theta^2 - \frac{35}{6}\theta^4\right)\cos 2u +\right.$$

$$\frac{35}{24} (1 - \theta^2)^2 \cos 4u \}$$

and where we use the notation

$$B_{20} = -\frac{1}{4} (1 - 3\theta^2) \quad ; \quad B_{22} = \frac{3}{4} (1 - \theta^2) \quad ; \quad \theta = \frac{P_h}{P_u} = \cos I$$

The equatorial radius of the Earth, has been taken as unity.

The elimination of the variable u has been done by Caballero (1975) using the method of von Zeipel. The new hamiltonian

$$\bar{H} = \bar{H} (\bar{P}_r, \bar{P}_u, \bar{P}_h, \bar{r}, -, -)$$

takes the form

$$\bar{H} = \frac{1}{2} (\bar{P}_r^2 + \frac{\bar{P}_u^2}{\bar{r}^2}) - (\frac{a_1}{\bar{r}} + \frac{a_3}{\bar{r}^3} + \frac{a_4}{\bar{r}^4} + \frac{a_5}{\bar{r}^5})$$

or

$$\bar{H} = \frac{1}{2} (\frac{d\bar{r}}{dt})^2 - \sum_1^5 \frac{a_i}{\bar{r}^i} \tag{19}$$

where

$$a_1 = \mu \; ; \quad a_2 = \frac{\bar{P}_u^2}{2} \; ; \quad a_3 = J_2 \mu B_{20} + \frac{J_2^2 \mu^3 B_{22}}{16 \bar{P}_u^2} (3 - 7\theta^2)$$

$$a_4 = \frac{J_2^2 \mu^2 B_{22}}{48 \bar{P}_u^2} (-21 + 69\theta^2) \quad ; \quad a_5 = -\frac{9 J_4 \mu}{64} (1 - 10\theta^2 + \frac{35}{3} \theta^4)$$

Since $\bar{u}$ and $\bar{h}$ are cyclic, $\bar{P}_u$, $\bar{P}_h$ are constant. Hence the coefficients a_i are constant too. Then we can apply to (19) the study made in section 3. (Cid et al., 1982).

II.- As we have said, the solution of $P(\alpha) = 0$ as well as the effective calculation of the values of α, β, γ which determines a regularizing function $g_5(r)$ with $g_5(r) > 0$, seems to need numerical methods.

Now we give two numerical examples, obtained through the system (12), which show the feasibility of the method proposed. The existence of $g_5(r)$ for a wide set of values for the orbital parameters a, e, I, remains to be analyzed.

Data:

$\mu = 0.00553$ $\qquad J_2 = 1.082631 \; 10^{-3}$ $\qquad J_4 = -1.65 \; 10^{-6}$

example 1: $a = 2$ $e = 0.1$ $I = 80°$
example 2: $a = 4$ $e = 0.1$ $I = 80°$

Results:

	Case 1	Case 2
γ	-0.295732	-0.1425387
β	$-0.400184\ 10^{1}$	$-0.8003057\ 10^{1}$
α	$0.245787\ 10^{3}$	$0.4726564\ 10^{2}$
c_1	$-0.2765\ 10^{-2}$	$-0.1382\ 10^{-2}$
c_2	$-0.2553\ 10^{-5}$	$-0.2113\ 10^{-5}$
c_3	$-0.3076\ 10^{-8}$	$-0.6383\ 10^{-8}$

We have also checked that the variation of the eccentricity e in the range $0.01 \leqslant e \leqslant 0.3$ has small influence on the values of the last table. It is easy to see that in the two cases considered we have $g_5(r) > 0$, because $r \geqslant 1$.

REFERENCES.

Belen'kii, I.M.: 1981a, Celes. Mech. 23, 9-31

Belen'kii, I.M.: 1981b, PMM U.R.S.S. 45, 24-29

Caballero, J.A.: 1975, Tesis Doctoral. Universidad Zaragoza

Cid, R., Ferrer, S., Elipe, A.: 1982, IX Jornadas Hisp-Lusas de Mat. Salamanca (to appear)

Deprit, A.: 1981, Celes. Mech., 23, 299-305

Ferrer, S., Elipe, A.: 1982, IX Jornadas Hisp-Lusas de Mat. Salamanca (to appear)

Hill, G.W.: 1913, Astron. J., 27, 171

Kustaanheimo, P., Stiefel, E.: 1965, J.R.A.M., 218, 204-219

Scheifele, G., Stiefel, E.: 1971, Linear and Regular Celestial Mechanics Springer-Verlag, Berlin.

Szebehely, V.: 1976, Celes. Mech. 14, 499-508.

THE JPL "LONG EPHEMERIS", DE102/LE51

E. Myles Standish, Jr.
Jet Propulsion Laboratory
California Institute of Technology
Pasadena, California 91109 U.S.A.

ABSTRACT A number of applications exist in astronomical research for planetary and lunar ephemerides covering an extended length of time. This paper discusses such a set of ephemerides, DE102/LE51, produced at JPL, covering the time 1411 B.C. to 3001 A.D. The ephemerides are dynamically self-consistent, in that the equations of motion were integrated simultaneously. They also represent the most accurately known positions covering such a time span. They have already been used by a number of different users in a variety of different applications.

I. INTRODUCTION

In 1977, it was decided to create a set of planetary and lunar ephemerides covering an extended length of time by integrating the equations of motion using the best set of initial conditions then available. The result of this long integration is described briefly in this paper. A full documentation of all aspects of the ephemeris (DE102/LE51) is now in press (see Newhall et al., 1982). In the present paper, a few features of DE102/LE51 are discussed and mention is made of possible applications with already existing examples being given.

II. THE CREATION OF DE102

The actual integration of DE102/LE51 was performed during many weekends (low computer rates) in the years 1977 and 1978. At the end, the individual files were merged and transformed into the final output format, producing a readable ephemeris covering the time span 1411 B.C. to 3001 A.D. This represents about 1.6×10^6 days and required approximately 5.6×10^6 integration steps. The program used a variable-order, variable-step-size integrator; the average step-size was about 0.29 days. Nearly 350 hours of Univac 1108 computer time was required.

V. V. Markellos and Y. Kozai (eds.), Dynamical Trapping and Evolution in the Solar System, 47–50.

A. Initial Conditions

The starting epoch of the integration was June 28, 1969 (JD 2440400.5), the ephemeris being integrated both forward and backward from this date. The initial conditions were the best available at that time and represented a least squares adjustment to a variety of observational data types. These included: 1) Lunar-laser ranging; 2) Mariner 9 and Viking Orbiter spacecraft ranging; 3) radar-ranging to Mercury, Venus, and Mars; and 4) Meridian circle optical data. These are described in detail in the paper cited above.

B. Equations of Motion

The equations of motion used in the integration included: 1) the n-body forces of the sun, moon, and the nine major planets; 2) the lunar librations; 3) isotropic, PPN-relativistic formulation; and 4) the perturbations from five asteroids. Though a number of the inherent constants have subsequently been modified, it is of importance to mention that the form of the equations of motion in DE102/LE51 has not been changed in any of the more recent ephemerides produced at JPL.

III. VARIOUS USES OF A LONG EPHEMERIS

There are a number of applications for which a long integrated ephemeris is a useful tool. Some of these are mentioned here.

A. Validation and Fit of an Analytical Theory

One may compare an analytical theory to an integrated ephemeris in order to locate incorrect or missing terms in the theory. Subsequently, one may fit the theory to the ephemeris instead of fitting to actual data. These methods have the advantages that: 1) equal mesh points are available, useful for spectral and Fourier analyses; 2) there is no observational error, in that the equations of motion are known exactly; and 3) the tedious process of data reduction is avoided. Such examples are the works of Stumpff (1981) and of Bretagnon (1980, 1981, and 1982). Also, Stephenson and Houlden (1981) have compared Tuckerman's Tables with DE102/LE51 in order to assess their accuracy.

B. Reduction of "Ancient" Observations

A computerized reduction of an observation is a relatively simple process with an integrated ephemeris. An ancient observation often contains valuable astronomical information when reduced with an accurate ephemeris. Such examples are the eclipse observations discussed by Muller (1975) and by Fiala (referred to by Rothschild, 1982), and the observation of Neptune by Galileo as discovered by Kowal and Drake (1980).

C. Quantitative Parameter Determinations

One sometimes wishes the value of an associated ephemeris parameter, averaged over a long period of time. Standish (1982) has used DE102/LE51 in determinations of the mean obliquity and dynamical equinox.

D. Integration of One-Body Orbits

The integration of the orbit of an asteroid or comet, for example, may be quickly accomplished using the pre-computed positions of the major planets as input data. Scholl (1982) has reported a reduction of nearly an order of magnitude in computer time for such a program as opposed to integrating the full n-body system.

IV. IMPROVEMENT OF DE102/LE51

Since the creation of DE102, there have been some significant improvements made to the initial conditions of subsequent ephemerides of JPL. As mentioned before, however, the equations of motion remain unchanged. As a result, it is possible to modify the coordinates of DE102 itself in order to reflect these improvements. Using the SET III formulation of Brouwer and Clemence (1961), one may improve DE102 to more accurately agreee with JPL's latest ephemeris, DE118. The SET III corrections have been determined by fitting the two ephemerides over the two centuries covered by DE118, 1850-2050. Extrapolation of the correction process seems to be valid. A more complete description is given in the cited reference (Newhall et al., 1982).

V. EXPORT TAPES

Various JPL ephemerides have been sent to over 100 users throughout the world. We now have the capability of producing computer tapes in directly machine-readable format for a variety of different computers including UNIVAC, IBM, CYBER, VAX, PDP, VAX, MODCOMP, and Honeywell. The complete package contains a user's guide and software for reading and interpolating the ephemerides, as well as the ephemerides themselves.

Anyone wishing a copy should contact the author.

BIBLIOGRAPHY

Bretagnon, P., (1980) Astron. and Astrophys., 84, 329.

Bretagnon, P., (1981) Astron. and Astrophys., 101, 342.

Bretagnon, P., (1982) Astron. and Astrophys., 108, 69.

Brouwer, D. and Clemence, G. M. (1961) "Methods of Celestial Mechanics", Academic Press, New York.

Kowal, C. T. and Drake, S. (1980), "Nature", 287, 311.

Muller, P. M. (1975) "An Analysis of the Ancient Astronomical Observations with the Implications for Geophysics and Cosmology", Thesis, University of Newcastle upon Tyne, School of Physics, Fiddes Litho Press.

Newhall, X.X, Standish, E. M., and Williams, J. G. (1982), Astron. and Astrophys. (in press).

Rothschild, R. F. (1982) Sky and Telescope, 63, No. 6, 558.

Scholl, H. (1982) private communication.

Standish, E. M. (1982) Astron. and Astrophys., in press.

Stephenson, F. R. and Houlden, M. A. (1981), J. Hist. Astron., 12, part 2, 133.

Stumpff, P. (1981), Astron. and Astrophys., 101, 52.

A COLLECTION OF GALILEAN SATELLITE ECLIPSES 1652-1982

Jay H. Lieske
Jet Propulsion Laboratory
Pasadena, California 91109 U.S.A.

I. INTRODUCTION

It is known that early (i.e. 17th-18th century) visual observations of Jupiter's Galilean satellites are approximately as accurate as modern visually observed eclipses (Lieske 1982). In the early days the clocks were the primary source of error while in present days problems related to the Earth's and Jupiter's atmospheres have become the primary limiting factor. The early observations were generally made in local apparent time (i.e. related to the real hour angle of the sun). After the development by Huygens in 1656 of the pendulum clock, these early observations, when reduced to a modern UT system, have been shown to be quite accurate.

The early data are of great interest, both historically and scientifically. From the historical standpoint, the data are of interest: for example, in exploring the work of Ole Roemer and his determination of the finite speed of light; or, in the observations of Neptune by Galileo (Kowal and Drake 1980). Also, many of the early geographical charts were developed by employing satellite eclipses to determine longitudes. The problem of Galilean satellite eclipses was one of the great areas of research in 17th century astronomy.

Scientifically, the early data are still of great interest because of the long time-base of observations with essentially uniform accuracy. These data could be used to explore questions related to ΔT (Ephemeris Time minus Universal Time) in the early days. Current investigations of ΔT (e.g. Morrison, 1979 Morrison, et al 1981) generally employ lunar occultations and must deal with the secular deceleration of the moon. Galilean satellite eclipses do not depend on the lunar tidal terms and should provide a useful independent check for 17th century values of ΔT, although there may be tidal effects from the Jupiter system which must be considered.

The evolution of the Laplace commensurability and of the periods of the satellites is also of great current interest. The studies by

V. V. Markellos and Y. Kozai (eds.), Dynamical Trapping and Evolution in the Solar System, 51–59.

Peale et al (1979, 1980), Greenberg (1980), Yoder (1979), and Yoder and Peale (1981) all present scenarios which would benefit greatly from analyses which can only effectively be done by employing the old data over a long time-span.

So the value of old eclipse observations is readily apparent. The question is: do they still exist? The purpose of this paper is to introduce two treasures (in the true sense of the word) which still exist and which do contain the old data which for many years were thought to have disappeared.

II. THE DELAMBRE COLLECTION

It is generally believed that Delambre (1749-1822) collected over 6000 eclipse observations of the Galilean satellites prior to approximately 1808. Sampson (1910), who initiated an exhaustive study of manuscripts at the Bureau des Longitudes found part of the "computer" records (the personal computers of those days were humans, rather than Apple II's), for Satellites II and IV. These records formed part of the "Delambre" collection which Sampson described as:

> ". . . forming part of Delambre's great collection, which, so far as I know, astronomers have hither to supposed to be lost. Probably no such collection could now be made nor would anyone attempt it with so much thoroughness. Much of it seems to have been gathered from correspondents. [p. 200]
>
> . . . But the original achievement of Delambre, who collected and reduced some 6000 observations . . . fully deserves Laplace's [1749-1827] praise as 'un des principaux titres de ce savant illustre à la reconnaissance des astronomes.' [p. 199]
>
> Delambre regarded his collection as exhaustive; in his Tables of 1817 (Introduction, p. i) he says, 'j'ai donc receulli soigneusement tout ce que les éphémerides, les collections académiques, les ouvrages des astronomes et ma correspondence m'ont fait connaître; and further (pp. liv, lv) he mentions that this collection of eclipses consisted of 3439 eclipses of I, 1100 of II, 590 of III, and 334 of IV. If those could be recovered, it was clear that they would form a more valuable collection than any that could now be made. Tisserand, who was in a position to know, seems to imply that they were lost: 'le travail dans lequel Delambre avait discuté plus de 6000 observations, et dont Laplace parle à plusiers reprises, a malheureusement été pardu' (Méc. Cél., t. iv, p. 84). [p. 215]."

These partial records for Satellites II and IV, which were found by Sampson, have been shown (Lieske 1982) to be quite accurate especially for Satellite II (for which the records contained the observer's location). Hence, I became interested in pursuing the possibility of locating further old documents and looking for the lost "Delambre" collection.

III. THE PINGRÉ TREASURE

While staying in 1980 at the Astronomisches Rechen-Institut in Heidelberg as an Alexander von Humboldt Senior American Scientist awardee, I was able to further investigate the possible sources and existence of old data. From L. Morrison of the Royal Greenwich Observatory, I received a clue to examine a Pingré book. This volume, A.-G. Pingré: Annales Célestes du dix-septième siècle by Bigourdan (1901), [hereafter referred to as Pingré] proved to be extremely interesting. Alexandre-Gui Pingré (1711-1796) was a French cleric and astronomer who for 30 years collected and compiled a history of seventeenth century astronomy. According to Gillispie (1970 X, p. 615):

> "In 1791 Le Monnier and Lalande persuaded the Academy to vote a large sum for its publication but the printer was slow and Pingré's death in 1796, coupled with devaluation the preceding year, led the printer to abandon the project and to sell the printed sheets as waste paper. Worse still, the manuscript was lost. Almost a century later, however, a Parisian bibliophile found in a country town what turned out to be LeMonnier's set of sheets and the remainder of the manuscript was discovered in the archives of the Paris observatory. In 1898, at the instigation of C. G. Bigourdan, the Academy again decided to publish; and the volume appeared in 1901."

This work, which had been thought lost for 100 years contains comments on 17th century astronomy, as well as observations. It makes for extremely interesting reading. Since the Galilean satellites were a prime focus of 17th century astronomy, eclipse observations of them occur with regularity. Some interesting examples are as follows. We find the entry for September 1689:

"Sept. 21 11^h 39' . . . émersion du 1er. Malaca. les P. Comille et de Bèze. Anc. Mém t VII, p. 756.

. . .
. . .

28 13 37 . . . émersion du 1er:: Malaca, les mêmes que le 21.ibid. Les observateurs étoiant en prison; il ne leur étoit guères possible de s'assurer de l'heure avec quelque précision." [p. 469].

This series of observations (there are others) were made at Malaca, Malaysia in 1689 by two people who were in prison and, hence, couldn't accurately set up a clock! However, when reduced from apparent time to universal time and compared with the E-2 ephemeris (Lieske 1980), the observations are quite good. The moral here might be that everyone can make a valuable contribution to science and humanity!

Another example from the same year (1689) is found:

"Nov. 1 6^h 1'20" . . . suivant la correction du P. Gouye. Anc. Mém. t. VII, p. 783. Le P. Noël remarque que ses observations ont été faites dans la partie orientale de Hoai-Ngan, grande ville. . . La lunette du P. Noël portoit un objectif de 13 1/2 pieds, et un oculaire de 2 1/2 pouces: elle étoit, dit-il, excellente" [p. 470].

This particular example is interesting for the following reason. As has been mentioned, one of the prime objectives of 17th century astronomy was the determination of longitudes. A difficulty that a modern analyst frequently finds is that the names of cities (and countries) have changed over the centuries. In addition, the Pingré book generally contains French phonetic spellings of place-names. One needs to know the longtiude of the locality in order to make the reduction from apparent to universal time. When I first consulted my Chinese colleagues regarding "Hoai-Ngan" they immediately informed me that that was not a Chinese name (because of the "ng"). Since I had 100 of these observations and they all were excellent, giving the same residuals for an assumed location in Vietnam (where I was told the combination "ng" occurs in spelling), I applied the modern ephemerides to the 17th century task of determining observatory longitude. From the residuals I calculated a longitude correction which led me to a longitude of $119\overset{\circ}{.}11$ East where I found Hwaian (Kiangsu, China) and which my Chinese colleagues pronounced in the same manner as my French colleagues pronounced "Hoai-Ngan" or "Hoyaingan" as it also is spelled in some of the Delisle manuscripts. This gives an idea of some of the difficulties and also the rewards of reducing the old data.

The Pingré work is certainly the main source for Delambre's 17th century collection. It is quite remarkable that Sampson, who published his research on the old data in 1910, was not aware of this treasure which was published in 1901! I highly recommend the Pingré book to anyone interested in historical astronomy and scientific attitudes during the 17th century.

IV. THE DELISLE TREASURE

The second treasure which I would like to mention is the manuscript collection of J. N. Delisle (1688-1768), which is probably the basic source for both Pingré and for the "Delambre" collection of Galilean satellite eclipses. Delisle (also spelled De l'Isle) was a student astronomer of Maraldi who collected reports of eclipses and occultations for the Mémoires. According to Gillispie (1970, IV, p. 22-25),

> "Delisle's growing reputation brought him, in 1721, an offer from Peter the Great to found an observatory and an associated school of Astronomy in Russia . . . . Planned for four years, Delisle's stay in Russia lasted twenty-two years. . . . One of Delisle's long-standing activities had been the amassing of vast amounts of geographical and astronomical material through an extraordinarily extensive correspondence, through inheritance, and through laborious copying."

After his return to Paris

> ". . . he obtained a new observatory at the Hôtel de Cluny. It was there, in 1759, that his pupil and assistant, Charles Messier, observed the return of Halley's comet."

Pingré became associated with the project headed by Delisle to observe the transit of Venus in 1761. It is, therefore, quite probable that the original source of the Pingré collection of eclipses and the "Delambre" collection is Delisle.

While visiting the Paris observatory, I mentioned my interest in old manuscripts to Mme A-M. de Narbonne, Conservateur of the Observatory of Paris Library. She showed me several volumes of the Delisle manuscripts and made it possible for me to study them. In the Inventaire Général by Bigourdan (1897) they are classified under the heading A 5 1 through A 5 8 and described on pages F11-F12 of the Inventaire Général. Again it is surprising that Sampson was unaware of this publication.

The inscription in Delisle's handwriting on one page of his collection is reproduced in Figure 1. It reads: "Eclipses of the satellites of Jupiter observed in Paris after my return from Russia in 1747 where the apparent times have been, for the most part, calculated by Mr. Messier for my journal."

The "Mr. Messier" referred to here is the same Charles Messier mentioned earlier. A second moral to this paper might be that today's assistant will become tomorrow's leader since Messier became famous in his own right.

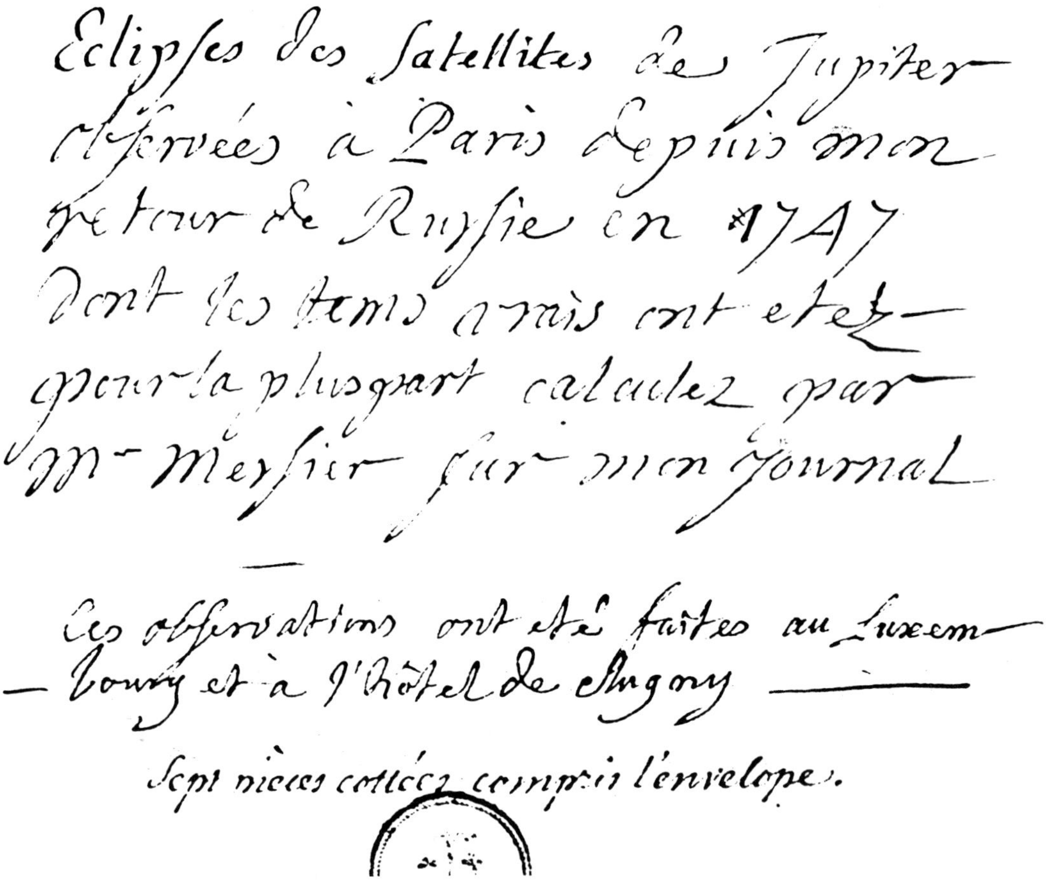
Eclipses des Satellites de Jupiter
observées à Paris depuis mon
retour de Russie en 1747
dont les tems ainsi ont eté
pour la plusgpart calculez par
Mr Messier sur mon Journal

Ces observations ont eté faites au Luxem-
-bourg et à l'hôtel de Cluny

Sept pieces cotées compris l'enveloppe.

Figure 1. Inscription in Delisle Manuscript

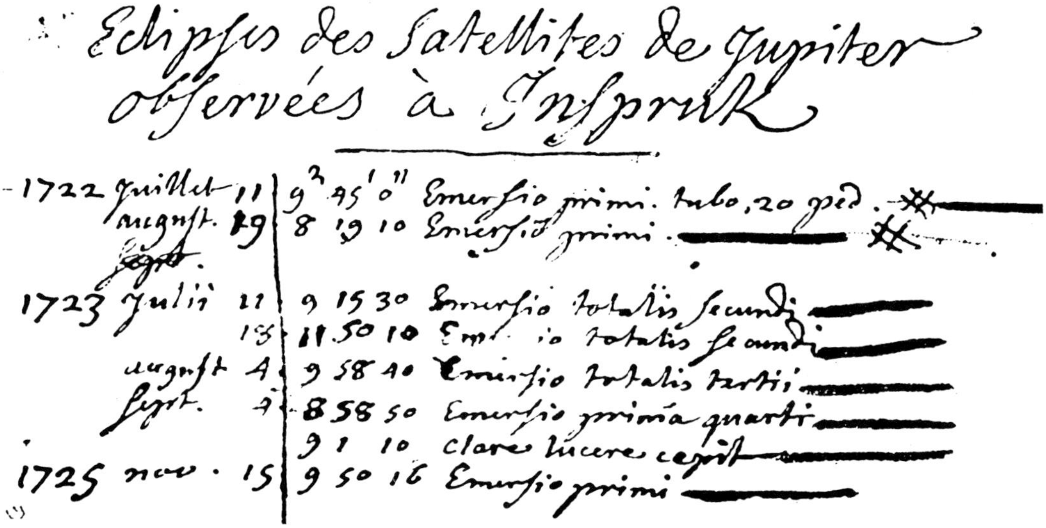
Eclipses des Satellites de Jupiter
observées à Inspruk

1722	juillet 11	9h 45' 0"	Emersio primi. tubo, 20 ped.
	august. 19	8 19 10	Emersio primi
	sept.		
1723	Julii 11	9 15 30	Emersio totalis secundi
	18	11 50 10	Em[illegible]io totalis secundi
	august 4	9 58 40	Emersio totalis tertii
	sept. 4	8 58 50	Emersio prima quarti
		9 1 10	clare lucere cepit
1725	nov. 15	9 50 16	Emersio primi

Figure 2. Observations from "Inspruk"

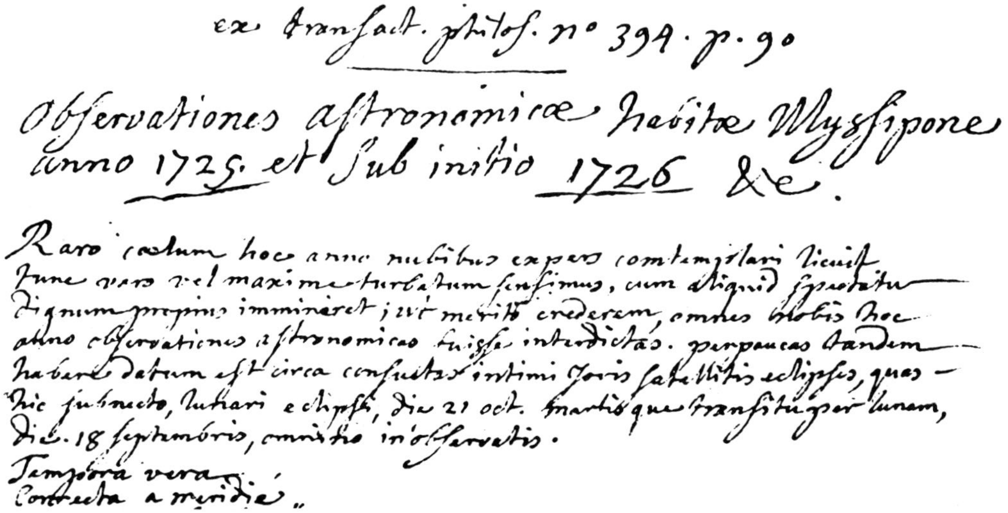
ex transact. philos. no 394. p. 90

Observationes astronomicae habitae Ulyssipone
anno 1725. et sub initio 1726 &c.

Raro caelum hoc anno nubibus expers contemplari licuit
Tunc vero vel maxime turbatum sensimus, cum aliquid spectatu
dignum propius immineret; ut merito crederem, omnes nobis hoc
anno observationes astronomicas fuisse interdictas. perpaucas tandem
habere datum est circa consuetas intimi Jovis satellitis eclipses, quas
hic subnecto, luminari eclipsi, die 21 oct. martisque transitu per lunam,
die 18 septembris, omnino inobservatis.
Tempora vera
Correcta a meridie.

Figure 3. Observations from "Ulysʃipone"

The handwritten manuscripts are generally in French or Latin and contain eclipse observations which Delisle made or recorded from his correspondence. There are numerous deciphering problems which occur, similar to that cited regarding Hoai-Ngan. In Figure 2 is shown a relatively easy one where Innsbruck, Austria is spelled Inspruk. In Figure 3 the more difficult Ulysſipone is Lisbon, Portugal upon close study.

V. SUMMARY

I have examined the Delisle manuscript and the Pingré book and have recorded and reduced the observations listed there. We now have more than 6800 eclipse observations before 1800 and over 16,000 prior to 1982. Hence, we may be certain that the lost "Delambre" collection has been found and that it probably should be called the Delisle-Pingré collection. It has been an exciting treasure hunt, and I hope to publish the reduced observations on microfiche so that these valuable data will not be lost to future generations.

The collected eclipse observations will become the basis for exploring the evolution of the Laplace commensurability as well as for determining ephemerides, masses, libration parameters, etc. for the upcoming Galileo mission.

This paper represents the results of one phase of research carried out at the Jet Propulsion Laboratory, California Institute of Technology, under Contract No. NAS7-100, sponsored by the National Aeronautics and Space Administration. I am indebted to L. Morrison of the Royal Greenwich Observatory for introducing me to the Pingré book and to Mme. A-M de Narbonne of the Paris Observatory for granting me access to the Delisle manuscripts.

REFERENCES

Bigourdan, G., 1897, "Inventaire général et sommaire des manuscrits de la bibliothéque de l'observatoire Paris", Ann. de l'Obs. Paris 21, F1-F60.

Bigourdan, G., 1901, A.-G. Pingré: Annales Célestes du dix-septième siècle, Gauthier-Villars, Paris [Pingré book].

Gillispie, C. C. (ed), 1970, Dictionary of Scientific Biography, C. Scribner, NY.

Greenberg, R., 1980, "Orbital Evolution of the Galilean Satellites" in The Satellites of Jupiter, D. Morrison, ed., IAU Colloquium No. 57, Academic Press.

Kowal, C. T., Drake, S., 1980, Nature, 287, 311.

Lieske, J. H., 1980, Astron. Astrophys. 82, 340.

Lieske, J. H., 1982, Cel. Mech. 26, 257.

Morrison, L. V., 1979, Monthly Notices Roy. Astron. Soc. 187, 41.

Morrison, L. V., Lukac, M. R., Stephenson, F. R., 1981, R. Greenwich Obs. Bull., (in press).

Peale, S. J., Cassen, P., Reynolds, R. T., 1979, Science 203, 892.

Peale, S. J., Cassen, P., Reynolds, R. T., 1980, Icarus 44, 234.

Sampson, R. A., 1910, Mem. Roy. Astron. Soc. 59.

Yoder, C. F., 1979, Nature 279, 767.

Yoder, C. F., Peale, S. J., 1981, Icarus 47, 1.

THE STUDY OF PLANETARY SECULAR PERTURBATIONS*

Zhang Jia-xiang
Purple Mountain Observatory, Academia Sinica, Nanjing, China

Summary. The methods of the first-order planetary secular perturbations are discussed in this paper. On the basis of Gauss' method, some concrete applicable formulae are derived. And then, orbital evolutions of the nine major planets in solar system during about 2,100,000 years are investigated numerically. The method applied here is in principle suitable for the orbits of arbitrary eccentricities and inclinations.

FOREWORD

In order to make an approach to the secular perturbations which some celestial bodies with rather special orbits (for instances, the Mercury, the Pluto, asteroids and comets) have undergone, the way in the classical method to solve the non-periodic part of disturbing function is not suitable, because it depends on developing the eccentricities and inclinations regarded as small quantities. We tried to introduce here the Gauss' method. Making use of its complete analytical solution for part of the problem and also the combination of the analytical and numerical method, we study the orbital evolution of this kind of special celestial bodies during a comparatively long period.

I. ON THE SECULAR PERTURBATIONS

There is no secular perturbation on semi-major axis, and the secular inequality of mean anomaly is of no consequence. Therefore, only the variations of the remaining four elements need to be considered. Let

$$\begin{aligned} h &= e\sin(\omega+\Omega), \\ k &= e\cos(\omega+\Omega), \\ p &= \sin i \sin\Omega \\ q &= \sin i \cos\Omega \end{aligned} \qquad (1)$$

*This paper has been published in Acta Astronomica Sinica, Vol. 23 (1982), No.1, 56, in Chinese.

V. V. Markellos and Y. Kozai (eds.), Dynamical Trapping and Evolution in the Solar System, 61–71.

For the sake of simplicity, we limit the problem to the disturbed planet and the disturbing planet. Their masses be designated by m and m_1 respectively. Developed up to square terms of the eccentricities and inclinations, the related non-periodic part of the disturbing function is[1]

$$N = Gm_1D[h^2 + k^2 + h_1^2 + k_1^2 - p^2 - q^2 - p_1^2 - q_1^2 + 2pp_1 + 2qq_1] \\ - 2Gm_1E(hh_1 + kk_1). \quad (2)$$

Accordingly, the secular inequalities of the related elements can be expressed in the following simple form:

$$[\dot{h}] = \frac{2Gm_1}{na^2}(kD - k_1E); \quad [\dot{k}] = -\frac{2Gm_1}{na^2}(hD - h_1E);$$
$$[\dot{p}] = \frac{2Gm_1D}{na^2}(q_1 - q); \quad [\dot{q}] = \frac{2Gm_1D}{na^2}(p - p_1); \quad (3)$$

In which G is the gravitational constant, n the mean motion of disturbed planet; D and E are symmetrical functions of the two semi-major axes.

If we exchange m, m_1 and their corresponding elements, the $[\dot{h}_1]$, $[\dot{k}_1]$, $[\dot{p}_1]$ and $[\dot{q}_1]$ may be obtained in the same form as (3).

With the simple form of (3), replacing h, k, p, q by [h], [k], [p] and [q], we can solve simultaneously the perturbation equations of all the major planets analytically, though it is very long and complicated. That is why for a long time it has been used as the classical method for studying secular perturbations of the major planets.

The main disadvantage of this method is that only terms of second order in the eccentricities and inclinations are considered. It seems not accurate enough for rather special orbits.

Gauss proposed another way to calculate the secular perturbations of elements avoiding the complication of dependence on the development of small quantities[1,2].

We denote by σ any one of the elements. Its perturbation equation is known as

$$\frac{d\sigma}{dt} = A_0 + \sum_i \sum_j B \cos(iM + jM_1 + Q),$$

Where A_0 corresponds to the secular perturbation on the left of (3), the second part on the right being periodic terms, M and M_1 are mean anomalies of the two planets and Q is a function of other angular elements.

The basic concept of Gauss' method is to calculate directly the $[\dot{\sigma}]$ according to the following formula:

$$[\dot{\sigma}] = A_0 = \frac{1}{4\pi^2}\int_0^{2\pi}\int_0^{2\pi} \frac{d\sigma}{dt}\, dM\, dM_1. \quad (4)$$

Regarding the mass of disturbing planet m_1 as a velocity-density distribution along its orbit, i.e., every mass $d\mu$ on the element arc of the orbit is proportional to the time dt taken by the planet to pass through the very arc, using coordinate transformation and introducing elliptical integral, Gauss obtained the analytical solution in (4) for dM_1, i.e. the first integral. As for the second integral, it can be integrated by using numerical method of harmonic analysis.

In order to compare the classical method with Gauss' method, taking a planet of Jupiter type (m_1=0.001, a_1=5.20, e_1=0.05, i_1=1°.2) as a disturbing body, we have made many kinds of computation for a disturbed body (a=2.75) with various eccentricities and inclinations according to the two methods respectively (the concrete formulae of Gauss' method are mentioned below). Now, we list briefly the results about $[\dot{h}]$ in Table 1, in each row and column, the value above is obtained from classical method (3) while the result in parenthesis is from Gauss' method. The latter is correct.

It is thus evident that both results agree well in the case of small eccentricities and inclinations; the larger they are, the divergence becomes more prominent until beyond recognition.

Table 1 $10^6 \times [\dot{h}]$ (for each day)

e \ i	0°.000	5°.730	11°.459	17°.189	22°.918	28°.648	34°.377
0.0	-0.024 (-0.024)	-0.024 (-0.022)	-0.024 (-0.017)	-0.024 (-0.011)	-0.024 (-0.005)	-0.024 (0.000)	-0.024 (+0.004)
0.1	-0.056 (-0.058)	-0.056 (-0.051)	-0.056 (-0.032)	-0.056 (-0.006)	-0.056 (+0.021)	-0.056 (+0.044)	-0.056 (+0.062)
0.2	-0.088 (-0.094)	-0.088 (-0.082)	-0.088 (-0.045)	-0.088 (+0.002)	-0.088 (+0.049)	-0.088 (+0.089)	-0.088 (+0.121)
0.3	-0.121 (-0.134)	-0.121 (-0.114)	-0.121 (-0.056)	-0.121 (+0.015)	-0.121 (+0.082)	-0.121 (+0.136)	-0.121 (+0.178)
0.4	-0.153 (-0.180)	-0.153 (-0.149)	-0.153 (-0.062)	-0.153 (+0.036)	-0.153 (+0.120)	-0.153 (+0.185)	-0.153 (+0.233)
0.5	-0.185 (-0.237)	-0.185 (-0.187)	-0.185 (-0.056)	-0.185 (+0.070)	-0.185 (+0.165)	-0.185 (+0.233)	-0.185 (+0.282)
0.6	-0.218 (-0.314)	-0.218 (-0.223)	-0.218 (-0.029)	-0.218 (+0.118)	-0.218 (+0.213)	-0.218 (+0.277)	-0.218 (+0.321)

II. SOME COMPUTATION FORMULAE

We adopt an appropriate system of units so that the gravitational constant is equal to 1.

It was proved long ago that indirect perturbation doesn't contain secular terms. Let S, T, and W denote respectively the direct perturbations along the disturbed planet's radius direction, transverse direction and normal to its orbital plane. Gauss' method can be summed up in calculating the following integrals:

$$S_0 = \frac{m_1}{2\pi}\int_0^{2\pi} S\,dM_1, \qquad T_0 = \frac{m_1}{2\pi}\int_0^{2\pi} T\,dM_1, \qquad W_0 = \frac{m_1}{2\pi}\int_0^{2\pi} W\,dM_1. \tag{5}$$

According to the principle of Gauss method, we made many trial computations and arrived at the conclusion that the following method is correct and practicable.

1. General condition

Let $\boldsymbol{P}, \boldsymbol{Q}, \boldsymbol{R}$ and $\boldsymbol{P}_1, \boldsymbol{Q}_1, \boldsymbol{R}_1$ denote the unit vectors along the directions of the orbital principal axes of the disturbed planet and the disturbing planet respectively. We refer to the heliocentric system corresponding to the directions thus mentioned as orbital coordinate system; denote by $\boldsymbol{r}$ the vector of radius direction of the disturbed planet; let a_1 and b_1 be the semi-major axis and semi-minor axis of the disturbing planet respectively and e_1 its eccentricity. Let

$$\begin{aligned}
&\alpha = \boldsymbol{r}\cdot\boldsymbol{P}_1, \quad \beta = \boldsymbol{r}\cdot\boldsymbol{Q}_1, \quad \gamma = \boldsymbol{r}\cdot\boldsymbol{R}_1,\\
&\psi_0 = a_1^2 b_1^2 \gamma^2,\\
&\psi_1 = a_1^2 b_1^2 - a_1^2\beta^2 - b_1^2(\alpha + a_1 e_1)^2 - (a_1^2 + b_1^2)\gamma^2,\\
&\psi_2 = (\alpha + a_1 e_1)^2 + \beta^2 + \gamma^2 - a_1^2 - b_1^2.
\end{aligned}$$

With the basic parameter λ, the following equation should be satisfied:

$$\lambda^3 + \psi_2\lambda^2 + \psi_1\lambda + \psi_0 = 0. \tag{6}$$

After solving the cube equation (6) by usual method, we approach it once again in order to be more accurate:

$$\lambda_i = \lambda_{i,0} - \frac{\lambda_{i,0}^3 + \psi_2\lambda_{i,0}^2 + \psi_1\lambda_{i,0} + \psi_0}{3\lambda_{i,0}^2 + 2\psi_2\lambda_{i,0} + \psi_1}, \quad i = 1,2,3. \tag{7}$$

Of the three values of λ two are positive, and one negative. According to an order from large to small, we write them as λ_1, λ_2, λ_3.

To increase the accuracy, the smallest in absolute value should be calculated once again according to $\lambda_2 = -\psi_0/\lambda_1\lambda_3$ or $\lambda_3 = -\psi_0/\lambda_1\lambda_2$.

In the orbital coordinate system of disturbing planet, the unit vectors along directions of the principal axes of the transformed coordinate system that can be integrated are

$$\boldsymbol{d}_i = [-C\lambda_i(b_1^2-\lambda_i)(\alpha+a_1e_1),\ -C\lambda_i(a_1^2-\lambda_i)\beta,\ C(a_1^2-\lambda_i)(b_1^2-\lambda_i)\gamma], \tag{8}$$

here

$$C = \{[\lambda_i(b_i^2-\lambda_i)(\alpha+a_1e_1)]^2+[\lambda_i(a_i^2-\lambda_i)\beta]^2 +[(a_i^2-\lambda_i)(b_i^2-\lambda_i)\gamma]^2\}^{-1/2}, \quad i=1,2,3.$$

In order to be accurate and reliable, the $\boldsymbol{d}$ corresponding to the smallest λ in absolute value should also be computed once again according to $\boldsymbol{d}_2 = \boldsymbol{d}_3\times\boldsymbol{d}_1$ or $\boldsymbol{d}_3 = \boldsymbol{d}_1\times\boldsymbol{d}_2$

Let $k^2 = (\lambda_1-\lambda_2)/(\lambda_1-\lambda_3)$, the corresponding effective integrals are

$$\left.\begin{aligned} G_x &= 4[E(k)-F(k)]/[(\lambda_1-\lambda_2)\sqrt{(\lambda_1-\lambda_3)}], \\ G_y &= 4\sqrt{(\lambda_1-\lambda_3)}\cdot[(1-k^2)F(k)-E(k)]/[(\lambda_1-\lambda_2)(\lambda_2-\lambda_3)], \\ G_z &= 4\cdot E(k)/[(\lambda_2-\lambda_3)\sqrt{(\lambda_1-\lambda_3)}] = -(G_x+G_y), \end{aligned}\right\} \tag{9}$$

where

$$F(k) = \int_0^{\pi/2}\frac{d\theta}{\sqrt{1-k^2\sin\theta}} = \frac{\pi}{2}\Big[1+\Big(\frac{1}{2}\Big)^2k^2+\Big(\frac{1\cdot3}{2\cdot4}\Big)^2k^4 +\Big(\frac{1\cdot3\cdot5}{2\cdot4\cdot6}\Big)^2k^6+\cdots\cdots\Big],$$

$$E(k) = \int_0^{\pi/2}\sqrt{1-k^2\sin\theta}\,d\theta = \frac{\pi}{2}\Big[1-\Big(\frac{1}{2}\Big)^2k^2 -\Big(\frac{1\cdot3}{2\cdot4}\Big)^2\cdot\frac{1}{3}k^4-\Big(\frac{1\cdot3\cdot5}{2\cdot4\cdot6}\Big)^2\cdot\frac{1}{5}k^6-\cdots\cdots\Big]$$

are the first and second standard elliptical integrals respectively. Let

$$\left.\begin{aligned} \boldsymbol{S}^* &= [\alpha/\gamma,\ \beta/\gamma,\ \gamma/\gamma], \\ \boldsymbol{T}^* &= \boldsymbol{W}^*\times\boldsymbol{S}^*, \\ \boldsymbol{W}^* &= [\boldsymbol{R}\cdot\boldsymbol{P}_1,\ \boldsymbol{R}\cdot\boldsymbol{Q}_1,\ \boldsymbol{R}\cdot\boldsymbol{R}_1]. \end{aligned}\right\} \tag{10}$$

Finally we have

$$[S_0,T_0,W_0] = -\frac{m_1}{2\pi}[G_x\boldsymbol{r}\cdot\boldsymbol{d}_1\ G_y\boldsymbol{r}\cdot\boldsymbol{d}_2\ G_z\boldsymbol{r}\cdot\boldsymbol{d}_3]\begin{bmatrix}\boldsymbol{S}^*\cdot\boldsymbol{d}_1 & \boldsymbol{T}^*\cdot\boldsymbol{d}_1 & \boldsymbol{W}^*\cdot\boldsymbol{d}_1\\ \boldsymbol{S}^*\cdot\boldsymbol{d}_2 & \boldsymbol{T}^*\cdot\boldsymbol{d}_2 & \boldsymbol{W}^*\cdot\boldsymbol{d}_2\\ \boldsymbol{S}^*\cdot\boldsymbol{d}_3 & \boldsymbol{T}^*\cdot\boldsymbol{d}_3 & \boldsymbol{W}^*\cdot\boldsymbol{d}_3\end{bmatrix} \tag{11}$$

2. Special condition

When $a_1/r \ll 1$ or $r/a_1 \ll 1$ (e.g. the mutual condition beteen Mercury and Pluto), the solution of Gauss' method given above will be determined with great difficulty and the results thus obtained might be distorted seriously. Actually, the two bodies' orbits this time are a great distance apart. Compared with the attractive influence of their adjacent planets, the relative contribution of perturbations among them are very small. In this condition the two heliocentric radius vectors are very different in length, it is easy for using developing method to deal with.

Being equivalent to the first integral (5) in Gauss' method, the three components along the orbital coordinate system of disturbing planet can be written as

$$F_\alpha = \frac{m_1}{2\pi}\frac{\partial}{\partial\alpha}\int_0^{2\pi}\frac{dM_1}{\Delta}, \quad F_\beta = \frac{m_1}{2\pi}\frac{\partial}{\partial\beta}\int_0^{2\pi}\frac{dM_1}{\Delta}, \quad F_\gamma = \frac{m_1}{2\pi}\frac{\partial}{\partial\gamma}\int_0^{2\pi}\frac{dM_1}{\Delta}, \tag{12}$$

here α, β and γ, as defined before, are the heliocentric coordinates of the disturbed planet in the orbital coordinate system of disturbing planet; Δ is the distance between them.

Expanding $1/\Delta$ up to the cube of r_1/r or r/r_1, we integrate the expansion, partially differentiate it, put it in order and finally obtain the result as follows:
When $a_1/r \ll 1$ (the disturbing attraction of an inner planet on a far outer planet),

$$\left.\begin{aligned}
F_\alpha/m_1 = & -(\alpha + \tfrac{3}{2}a_1e_1)/r^3 + [\tfrac{15}{4}a_1^3e_1(1+\tfrac{3}{4}e_1^2) \\
& + 3a_1^2(1+\tfrac{11}{4}e_1^2)\alpha + \tfrac{9}{2}a_1e_1\alpha^2]/r^5 \\
& - \{\tfrac{75}{16}a_1^3e_1[7(1+e_1^2)\alpha^2 + (1-e_1^2)\beta^2] \\
& + \tfrac{15}{4}[(a_1^2\alpha^2 + b_1^2\beta^2) + 4a_1^2e_1^2\alpha^2]\alpha\}/r^7 \\
& + \tfrac{175}{16}a_1e_1\alpha^2[3(a_1^2\alpha^2 + b_1^2\beta^2) + 4a_1^2e_1^2\alpha^2]/r^9, \\
F_\beta/m_1 = & -\beta/r^3 + \beta[\tfrac{3}{2}(a_1^2+b_1^2) + \tfrac{9}{4}a_1^2e_1^2 + \tfrac{9}{2}a_1e_1\alpha]/r^5 \\
& - \beta\{\tfrac{75}{16}a_1^3e_1(6+e_1^2)\alpha + \tfrac{15}{4}[(a_1^2\alpha^2+b_1^2\beta^2) \\
& + 4a_1^2e_1^2\alpha^2]\}/r^7 + \tfrac{175}{16}a_1e_1\alpha\beta[3(a_1^2\alpha^2+b_1^2\beta^2) \\
& + 4a_1^2e_1^2\alpha^2]/r^9, \\
F_\gamma/m_1 = & -\gamma/r^3 + 3\gamma(\tfrac{1}{2}a_1^2 + \tfrac{3}{4}a_1^2e_1^2 + \tfrac{3}{2}a_1e_1\alpha)/r^5 \\
& - \gamma\{\tfrac{75}{4}a_1^3e_1(1+\tfrac{3}{4}e_1^2)\alpha + \tfrac{15}{4}[(a_1^2\alpha^2+b_1^2\beta^2) \\
& + 4a_1^2e_1^2\alpha^2]\}/r^7 + \tfrac{175}{16}a_1e_1\alpha\gamma[3(a_1^2\alpha^2+b_1^2\beta^2) \\
& + 4a_1^2e_1^2\alpha^2]/r^9;
\end{aligned}\right\} \tag{13}$$

When $r/a_1 \ll 1$ (the disturbing attraction of a far outer planet on an inner planet),

$$\begin{aligned}
F_\alpha/m_1 &= \alpha/2b_1^3 + 3e_1(3\alpha^2+\beta^2-4\gamma^2)/[8a_1b_1^3(1-e_1^2)], \\
F_\beta/m_1 &= \beta/2b_1^3 + 3e_1\alpha\beta/[4a_1b_1^3(1-e_1^2)], \\
F_\gamma/m_1 &= -\gamma/b_1^3 - 3e_1\alpha\gamma/[a_1b_1^3(1-e_1^2)].
\end{aligned} \tag{14}$$

It is easy from (13) or (14) to obtain

$$[S_0,\ T_0,\ W_0] = [\mathbf{F}\cdot\mathbf{S}^*, \mathbf{F}\cdot\mathbf{T}^*, \mathbf{F}\cdot\mathbf{W}^*], \tag{15}$$

in which $\boldsymbol{S}^*, \boldsymbol{T}^*, \boldsymbol{W}^*$ are defined as before, and

$$\boldsymbol{F} = [F_\alpha, F_\beta, F_\gamma].$$

After S_0, T_0 and W_0 have been obtained, there is only an integral for dM left in (4):

$$[\dot{\sigma}] = \frac{1}{2\pi}\int_0^{2\pi}\left(\frac{d\sigma}{dt}\right)_0 dM, \tag{16}$$

here $(d\sigma/dt)_0$ represents $d\sigma/dt$ in which S_0, T_0 and W_0 are used in place of the usual S, T and W.

Let P and f denote the semi-parameter and the true anomaly of the disturbed planet respectively. Write

$$\begin{aligned} S_1 &= S_0 \sin f, & T_1 &= T_0(p + r)\sin f, & W_1 &= W_0 r \sin f, \\ S_2 &= S_0 \cos f, & T_2 &= T_0(p + r)\cos f, & W_2 &= W_0 r \cos f, \\ & & T_3 &= T_0 r. \end{aligned}$$

The perturbation equation of the elements can be changed as

$$\left.\begin{aligned} \left(\frac{dh}{dt}\right)_0 &= S_1\sqrt{p}\ \sin(\omega+\Omega) - S_2\sqrt{p}\ \cos(\omega+\Omega) \\ &\quad + [T_1\cos(\omega+\Omega) + T_2\sin(\omega+\Omega) + T_3 e\sin(\omega \\ &\quad +\Omega)]/\sqrt{p} + [W_1 e\cos(\omega+\Omega)\cos\omega\tan\frac{i}{2} \\ &\quad + W_2 e\cos(\omega+\Omega)\sin\omega\tan\frac{i}{2}]/\sqrt{p}, \\ \left(\frac{dk}{dt}\right)_0 &= S_1\sqrt{p}\cos(\omega+\Omega) + S_2\sqrt{p}\sin(\omega+\Omega) \\ &\quad -[T_1\sin(\omega+\Omega) - T_2\cos(\omega+\Omega) - T_3 e\cos(\omega \\ &\quad +\Omega)]/\sqrt{p} - [W_1 e\sin(\omega+\Omega)\cos\omega\tan\frac{i}{2} \\ &\quad + W_2 e\sin(\omega+\Omega)\sin\omega\tan\frac{i}{2}]/\sqrt{p}, \\ \left(\frac{dp}{dt}\right)_0 &= \left\{W_1\left[\cos(\omega+\Omega) + 2\sin^2\frac{i}{2}\sin\Omega\sin\omega\right]\right. \\ &\quad \left. + W_2\left[\sin(\omega+\Omega) - 2\sin^2\frac{i}{2}\sin\Omega\cos\omega\right]\right\}/\sqrt{p}, \\ \left(\frac{dq}{dt}\right)_0 &= \left\{W_1\left[-\sin(\omega+\Omega) + 2\sin^2\frac{i}{2}\cos\Omega\sin\omega\right]\right. \\ &\quad \left. + W_2\left[\cos(\omega+\Omega) - 2\sin^2\frac{i}{2}\cos\Omega\cos\omega\right]\right\}/\sqrt{p}. \end{aligned}\right\} \tag{17}$$

In the condition of having given the orbital elements of both disturbing and disturbed planets, S_0, T_0 and W_0 are functions of only the mean anomaly M of disturbed planet, and so are S_i, T_i and W_i. On the basis of (17), with suitable number of points well-distributed according to M, it is not difficult for using numerical method of harmonic analysis to solve and calculate (16) to obtain $[\dot{\sigma}]$.

With such method, the results of orbital computations for different conditions have been shown in Table 1. And now, we list an example of computation again in Table 2, which also contains the corresponding result obtained wholly by numerical method, to be a test for the entire method and formulae.

Taking

	mass	a	e	ω	Ω	i
disturbing planet	0.001	5.20	0.05	229°.18	114°.59	1°.15
disturbed planet	—	2.75	0.15	57.30	57.30	8.59

we have

Table 2 (for each day)

	$10^6[\dot{h}]$	$10^6[\dot{k}]$	$10^6[\dot{p}]$	$10^6[\dot{q}]$
Gauss' method	-0.054532696	-0.099315468	-0.092330152	+0.096448951
numerical integration	-0.054532701	-0.099315487	-0.092330147	+0.096448949

III. COMPUTATIONS, RESULTS AND DISCUSSION

In order to study the orbital evolution during a long period, by integrating (4) further, we have

$$\Delta\sigma = \int_{t_0}^{t} [\dot{\sigma}]\,dt = \frac{1}{4\pi^2}\int_{t_0}^{t}\int_{0}^{2\pi}\int_{0}^{2\pi}\frac{d\sigma}{dt}\,dM\,dM_1\,dt\,. \tag{18}$$

It has been mentioned to obtain the integrals for dM_1 and dM. To obtain the integral for dt, we use Adams' method; for the nine major planets of solar system, we solve jointly the equations of order 9×4×3=108 in all. We regard whether a_1/r or r/a_1 is smaller than 1/10 or not as a discriminating basis for taking (13) - (15) or (6) - (11). For harmonic analysis, we take 48 points a circle (as e≥0.15, i.e. the condition of Mercury and Pluto) and 24 points a circle (as e<0.15, for the other planets). We adopt 250,000 days as the step length of integration for dt. The initial epoch is adopted on 1941 1 6.0 E.T. (2430000.5). We take the plane of total moment of momentum of the nine major planets in heliocentric system as a fundamental reference plane (very close to the invariable plane), define the ascending node of ecliptic plane to the reference plane as the zero point of longitude and transform the orbital elements of planets to this system. Thus making integration backward, we calculate the orbital variation of the nine major planets during the past 2,100,000 years.

To check it, we have also made a computation during the same period by doubling the step length. Both results are consistent. It shows that the numbers of points and the step length adopted are permissible and the repeated computations are correct.

The main results of the whole computation can be listed in a simple

Table 3

	Eccentricity e		Inclination i	
	Max.	Min.	Max.	Min.
Mercury	0.2317 (0.2372)	0.1215 (0.1314)	9°.178 (9.968)	4°.740 (2.576)
Venus	0.0706 (0.0631)	0.0000 (0.0023)	3.272 (3.353)	* (0.097)
Earth	0.0677 (0.0571)	0.0000 (0.0024)	3.100 (2.948)	* (0.050)
Mars	0.1397 (0.1330)	0.0185 (0.0035)	5.933 (6.463)	* (0.126)
Jupiter	0.0608 (0.0625)	0.0255 (0.0266)	0.482 (0.470)	0.240 (0.251)
Saturn	0.0843 (0.0845)	0.0124 (0.0103)	1.010 (0.997)	0.788 (0.802)
Uranus	0.0780 (0.0905)	0.0118 (0.0082)	1.120 (1.119)	0.907 (0.906)
Neptune	0.0145 (0.0190)	0.0056 (0.0071)	0.788 (0.768)	0.562 (0.030)
Pluto	- (0.2525)	- (0.2172)	- (17.837)	- (15.288)

table (Table 3), in which are given the largest and the smallest values of the planetary eccentricities ane inclinations during the 2,100,000 years. To be compared, the corresponding extreme values used all along,[1] which were obtained by J.N.Stockwell in former times by making a tedious derivation and solving secular perturbation equations according to (3), are also given here. In Table 3, listed above in each row and column are Stockwell's results; those undecided for reason of the smallness of certain coefficients in the analysis are marked by asterisk. Because Pluto had not yet been discovered that time, no results for it are there in the corresponding places. The results we obtained are in parenthesis.

In Fig. 1 and Fig. 2, are given in detail the variations of eccentricities and inclinations of all the planetary orbits. It is another important contribution of this paper.

From Table 3 and the figures, many things are clear at a glance. We need not say any more except a few points as follows:

1. So-called secular variation is actually composed of many periodic variations. Except Mercury, Pluto and Neptune, most of planets present a continuous fluctuation with a period of tens of thousand years.

2. It is worth noticing that the curves of both eccentricity and inclination of Jupiter are exactly the same as that of Saturn's respectively. Actually both amplitudes are different, but they are drawn in a ordinate of equal length. The main characteristic is that the elements concerned of both planets have the same period of variation respectively but differ from each other exactly in one phase of variation: as one falls, another rises. This takes some explanation. Due to the immense mass of Jupiter and Saturn, the disturbing effect of the other planets is relatively insignificant. In mutual perturbation of only one couple of planets, based on the approximate formula (3), it is not difficult to obtain the following integral[1]:

$$m\sqrt{a}\,e^2 + m_1\sqrt{a_1}\,e_1^2 = \text{const},$$
$$m\sqrt{a}\,\sin^2 i + m_1\sqrt{a_1}\,\sin^2 i_1 = \text{const}. \qquad (19)$$

that just explains the repeated variation of the two planets' elements concerned, as discussed above.

3. Venus and Earth, for they are very close to each other, also have got similar circumstances in the main.

4. The orbital variation of Pluto differs from that of the most planets. Its eccentricity and inclination all present a period of variation of about 700,000 years and there is no small fluctuation. As seen from the curves, the peak value of eccentricity is getting higher while the valley value of inclination is getting lower.

5. As seen from recent data, the mass of Pluto (1/ 1812000) has been taken a little too large. It has little influence on secular variation of

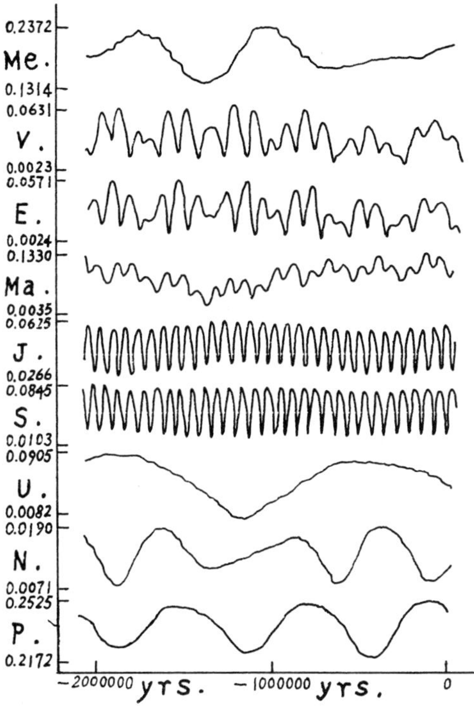

Fig. 1 *e*

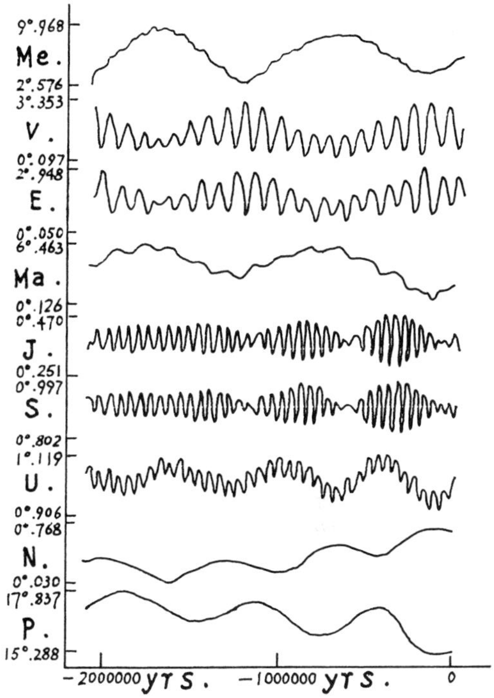

Fig. 2 *i*

Figures 1 and 2: Variations of eccentricities and inclinations of the planetary orbits.

---

the orbit of Pluto but there may be some influence on the adjacent planet, the Neptune.

All the computations of this paper are made by using the computers of our observatory. The author is deeply grateful to comrade Li Guang-yu and comrades in the computing laboratory for their constant help.

REFERENCES

1 Smart, W.M., Celestial Mechanics, 198-231.
2 Plummer, H.C., Dynamical Astronomy, 207-217.

PERTURBATIONS DUE TO THE ASTEROID BELT

Tong Fu, Chen Zhen
Purple Mountain Observatory, Nanjing, China

ABSTRACT: A new expansion for the gravitational potential of the asteroid belt is derived in this paper on the basis of binomial expansions. Its advantages are: (1)the unified form both for the inner and for the outer regions of the belt; (2)suitability for discussing the motions of the celestial bodies with perihelions within but apohelions beyond the belt; (3)rapidity of convergence.

The perturbations due to the asteroid belt are studied by using our expansion.

1. INTRODUCTION

According to the present knowledge about the solar system, the total mass of the asteroid belt is believed to be about o.1% of that of the Earth and, generally speaking, there is no need to consider its influence when the motions of celestial bodies in the solar system are studied. However, with the development of planetary exploration and the increase in the accuracy of observations, the perturbations due to the asteroid belt have become a factor not to be ignored. The effect of the asteroid belt on the orbit of Mars has been studied by A.P.Mayo,[1] who obtained the perturbations of the order of several kms within 700 days. They are big enough to be detected by modern observational techniques.

Similarly to Liu et al[3] and Mayo,[1] who both studied the influence of the asteroid belt on Mars, Plakhov[2] discussed the effect of Saturn's ring on the satellite orbits about the planet. In these studies, the disturbing function is expanded into a Laurent series in the heliocentric radius r of the disturbed body. The forms of the expansion at radii within the ring and at those beyond it are quite different. In fact, the disturbing function has the form of a power series in r in the first case, while it can only be expanded into a power series in 1/r in the latter case. When r is close to the inner or outer radius of the ring, the convergence of the expansion will be broken. The most serious defect of the expansion is that it cannot be used to discuss the motions in those orbits with

V. V. Markellos and Y. Kozai (eds.), Dynamical Trapping and Evolution in the Solar System, 73–79.

perihelions within but aphelions beyond the ring. But, as we know, that is precisely the form of motion in the case of the spacecraft exploring the outer solar system.

To avoid these drawbacks, another expansion of the disturbing function is provided in this paper on the basis of binomial expansion. Its advantages are: (1)the same form both for the inner and outer regions of the belt; (2)suitability for studying the motions of the celestial bodies with perihelions within but apohelions beyond the asteroid belt, such as Beira (1474) and Ganymed (1036); (3)rapidity of convergence.

Section 2 deals with the expansion of the disturbing function. Section 3 is devoted to decomposition of the disturbing function. In section 4 we will derive the short-period and secular (together with the long-period) perturbations respectively, being confined to the first nine terms of the expansion. Finally, some numerical results are given in the last section.

2. EXPANSION OF THE DISTURBING FUNCTION

For a point $P(r', \phi', \lambda')$, lying outside a belt with constant density and arbitrary thickness, the disturbing function by the belt can be written as

$$U = k^2 \rho \iiint \frac{1}{\Delta} r'^2 \cos\varphi' \, dr' d\varphi' d\lambda' . \tag{1}$$

Owing to the axisymmetry of the belt, the indirect part of the disturbing function vanishes.

According to the binomial formula, $1/\Delta$ can be expressed as

$$\frac{1}{\Delta} = (r^2 + r'^2 - 2rr'\cos H)^{-1/2} = \sum_{n=0}^{\infty} \frac{(2n)!}{2^n (n!)^2} (r+r')^{-(2n+1)} r^n r'^n (1+\cos H)^n . \tag{2}$$

By introducing a constant $C \geqslant (r + r')$, $(r + r')^{-(2n+1)}$ can be expanded as

$$(r+r')^{-(2n+1)} = C^{-(2n+1)} \sum_{m=0}^{\infty} \frac{(2n+1)_m}{m!} [1-(r+r')/C]^m . \tag{3}$$

In fact, for any concrete problem only a finite number of terms in (3) are needed. The number M is determined according to the magnitude of $1-(r+r')/C$ and the accuracy desired. On the other hand, $(1+\cos H)$ can be expanded according to the cosine law in spherical trigonometry and the addition theorem for the Legendre polynomials, and then, all the odd zonal harmonics and the tesseral harmonics will vanish after integration because of the belt symmetry.

After some calculations and arrangements we obtain finally

$$U = \varepsilon \sum_{N=0}^{\infty} \sum_{q=0}^{[N/2]} (r/C)^N P_{2q}(\sin\varphi) K_{Nq} , \tag{4}$$

where

$$\varepsilon = 4\pi k^2 \rho A^3 / C , \tag{5}$$

$$K_{Nq} = \frac{4q+1}{2^{2q}} \sum_{n=N_1}^{N} \sum_{m=N-n}^{M} \sum_{p=N-n}^{m} \frac{(-1)^p (2n+m)!}{(N-n)!(n-2q)!(n+2q+1)!(m-p)!(p-N+n)!} \times$$

$$\times \left(\frac{A}{C}\right)^{2n+p-N} K^*_{2n+p-N,q} , \tag{6}$$

$$N_1 = \max(2q, N-M) , \tag{7}$$

$$K^*_{\ell,q} = \sum_{j=0}^{q} (-1)^{q-j} \frac{(2q+2j)!}{(q+j)!(q-j)!(2j+1)!} \cdot \left(\frac{h}{2A}\right)^{2j+1} G_{\ell,j} , \tag{8}$$

$$G_{\ell,j} = \begin{cases} -\ln(B/A) , & \ell-2j+2=0 , \\ \dfrac{1-(B/A)^{\ell-2j+2}}{\ell-2j+2} , & \ell-2j+2 \neq 0 , \end{cases} \tag{9}$$

h, A, B denote the thickness, outer radius and inner radius of the belt, respectively.

The only limit during the process of expansion of the disturbing function is that P must be outside the belt.But there is no limit to its radius r. Therefore,(4) is a unified form of the disturbing function U for all the outside points, no matter $r>A$ or $r<B$.

To improve the convergence of (4), we arrange it as

$$U = \varepsilon \sum_{N=0}^{\infty} \sum_{\ell=0}^{[N/2]} F_{N\ell} (r/a)^N \sin^{2\ell} u , \tag{10}$$

where

$$F_{N\ell} = \sum_{q=\ell}^{[N/2]} \frac{(-1)^{q-\ell}}{2^{2q}} \binom{2q}{q+\ell} \binom{2q+2\ell}{2q} K_{Nq} \cdot \left(\frac{a}{C}\right)^N \sin^{2\ell} I , \tag{11}$$

$$\sin u = \sin\varphi / \sin I . \tag{12}$$

Expression (10) has a better feature of convergence than (4).

3. DECOMPOSITION OF THE DISTURBING FUNCTION

While using numerical method, we only need the partial derivatives $\partial U/\partial r$ and $\partial U/\partial \varphi$, both of which can easily be obtained from (10).

However, when the analytical method is used, the disturbing function must be expressed in terms of the orbital elements of the disturbed body and separated into secular, long- and short-periodic parts.

For example, we take (N,1) = (1,0),(2,0),(3,0),(4,0),(2,1),(3,1),(4,1),(5,1),(6,1) and use eccentric anomaly E as the independent variable, then U can be expressed in closed form

$$U = \varepsilon \sum_{i=0}^{6} (-1)^i \left[(P_i - Q_i \cos 2\omega) \cos iE + \sqrt{1-e^2}\, R_i \sin 2\omega \sin iE \right], \tag{13}$$

where

$$\begin{aligned} P_i &= e^i \left[\sum_{j=1}^{4} p_{j0}(e) F_{j0} + \sum_{k=2}^{6} p_{k1}(e) F_{k1} \right], \\ Q_i &= e^{|i-2|} \sum_{k=2}^{6} q_k(e) F_{k1}, \\ R_i &= e^{|i-2|} \sum_{k=2}^{6} r_k(e) F_{k1}, \end{aligned} \tag{14}$$

$p_{j0}(e)$, $p_{k1}(e)$, $q_k(e)$, $r_k(e)$ being polynomials in e.

Now, the Hamiltonian is

$$H = H_0 + \varepsilon H_1, \tag{15}$$

where

$$H_0 = \mu^2/2L^2, \qquad H_1 = U/\varepsilon. \tag{16}$$

The secular part of H_1 is given by

$$\langle H_1 \rangle = \frac{1}{2\pi} \int_0^{2\pi} H_1 \, d\ell = \left(P_0 + \frac{e}{2} P_1\right) - \left(Q_0 + \frac{e}{2} Q_1\right) \cos 2\omega, \tag{17}$$

and the corresponding first-order generating function by

$$S_1 = \frac{L^3}{\mu^2} \int (H_1 - \langle H_1 \rangle) \, d\ell = \frac{L^3}{\mu^2} S_1^*, \tag{18}$$

$$\begin{aligned} S_1^* = {} & \frac{e^2}{2} (P_1 - Q_1 \cos 2\omega) \sin E + \sum_{i=1}^{7} \frac{(-1)^i}{i} \left[\frac{e}{2} (P_{i-1} + P_{i+1}) + P_i \right] \sin iE - \\ & - \cos 2\omega \sum_{i=1}^{7} \frac{(-1)^i}{i} \left[\frac{e}{2} (Q_{i-1} + Q_{i+1}) + Q_i \right] \sin iE - \\ & - \sqrt{1-e^2} \sin 2\omega \sum_{i=1}^{7} \frac{(-1)^i}{i} \left[\frac{e}{2} (R_{i-1} + R_{i+1}) + R_i \right] \cos iE, \\ & (P_i = Q_i = R_i \text{ if } i \bar{\in} [1, 6]). \end{aligned} \tag{19}$$

Having eliminated the short-periodic terms, the new Hamiltonian can be written as

$$H^* = H_0 + \varepsilon \langle H_1 \rangle + O(\varepsilon^2). \tag{20}$$

4. PERTURBATIONS

(a) Short-periodic perturbations

The short-periodic perturbations of the orbital elements can be obtained immediately from the generating function S_1. But in many cases the perturbations of the coordinates are required.

Let δr, $r\delta u$ and δz denote the components of the displacement of the celestial body due to the belt along the directions of radial, cross and perpendicular to the orbital plane, respectively. Then

$$\delta r = \varepsilon \frac{a^3}{\mu r}\Big[-e\sin E\,\Big(3S_1^* + 2a\frac{\partial S_1^*}{\partial a} + \frac{1-e^2}{e}\Big) + \frac{\sqrt{1-e^2}}{e}(e-\cos E)\Big(\sqrt{1-e^2}\,\widetilde{H}_1 - \frac{\partial S_1^*}{\partial \omega}\Big) + 2\Big(\frac{r}{a}\Big)^2 \widetilde{H}_1\Big], \tag{21}$$

$$r\delta u = \varepsilon \frac{a\sqrt{1-e^2}}{\mu e}\Big[\Big(r - \frac{a^2(1-e^2)}{r}\Big)\frac{\partial S_1^*}{\partial e} + a\sin E\frac{\partial S_1^*}{\partial E} - \frac{a^2 e}{r}\Big(3S_1^* + 2a\frac{\partial S_1^*}{\partial a}\Big) + a\sin E\,(2+e\cos f)\Big(\widetilde{H}_1 - \frac{1}{\sqrt{1-e^2}}\frac{\partial S_1^*}{\partial \omega}\Big)\Big], \tag{22}$$

$$\delta z = \varepsilon \frac{ar}{\mu\sqrt{1-e^2}}\Big[\sin(f+\omega)\,\mathrm{ctg}\,I\,\frac{\partial S_1^*}{\partial \omega} - \cos(f+\omega)\sin 2I\,\frac{\partial S_1^*}{\partial(\sin^2 I)}, \tag{23}$$

where

$$\widetilde{H}_1 = H_1 - \langle H_1\rangle . \tag{24}$$

(b) Secular and long-periodic perturbations

Let $\bar{a}$, $\bar{e}$, $\bar{I}$, $\bar{\omega}$, $\bar{\Omega}$, $\bar{M}$ be the averaged keplerian elements. As H^* is independent of M and Ω, we have the integrals immediately

$$\bar{a} = \text{const}, \qquad \cos^2\bar{I}\,(1-\bar{e}^2) = \text{const}. \tag{25}$$

The second integral indicates that $\bar{I}$ well increases as $\bar{e}$ decreases, and vice versa.

If at the beginning we have

$$\bar{e} = \bar{I} = 0, \tag{26}$$

then the state will conserve forever.

The variations of the other elements are governed by the averaged equations of motion, especially we have

$$\frac{d\bar{e}}{dt} = -\varepsilon \bar{e}\sqrt{\frac{1-\bar{e}^2}{\mu\bar{a}}}\left(\sum_{k=2}^{6} C_k(\bar{e}) F_{k1}\right)\sin 2\bar{\omega} , \tag{27}$$

$$\frac{d\bar{\omega}}{dt} = \frac{\varepsilon}{\sqrt{\mu\bar{a}(1-\bar{e}^2)}}\left(\bar{A} - \bar{B}\cos 2\bar{\omega}\right) , \tag{28}$$

$$\bar{A} = \sum_{j=1}^{4} \bar{A}_{j0}(\bar{e}) F_{j0} + \sum_{k=2}^{6}\left[\bar{A}_{k1}(\bar{e}) - \frac{\bar{A}^*_{k1}(\bar{e})}{\sin^2 \bar{I}}\right] F_{k1} , \tag{29}$$

$$\bar{B} = \sum_{k=2}^{6}\left[\bar{B}_k(\bar{e}) - \frac{\bar{e}^2}{\sin^2\bar{I}}\bar{B}^*_k(\bar{e})\right] F_{k1} , \tag{30}$$

where $C_k(\bar{e})$, $\bar{A}_{j0}(\bar{e})$, $\bar{A}_{k1}(\bar{e})$, $\bar{A}^*_{k1}(\bar{e})$, $\bar{B}_k(\bar{e})$, $\bar{B}^*_k(\bar{e})$ are all polynomials in e.

Finally we conclude that (1)if $\bar{\omega} = \frac{1}{2}n\pi$, $\bar{A}=(-1)^n\bar{B}$, then equations (27),(28) have stationary solution; (2)if $\bar{A}=0$, $\bar{\omega}$ will oscillate around a libration point.

5. NUMERICAL RESULTS

Assuming

$a = 2.7$, $e = 0.5$, $I = 26^\circ.5$, $\omega = 5^\circ$,

$A = 4$, $B = 2$, $h = 2/3$, $m = 0.001m_\oplus$,

we have derived the following numerical results by taking $C = 11.23$:

$\varepsilon = 2.53\times10^{-12}$;

Coefficients $F_{N\ell}$ in the expansion (10) (see Sec. 2):

N	l	F ($\times10^{-2}$)
1	0	-1.06
2	0	0.26
3	0	-0.83
4	0	0.23
5	0	---
6	0	---

N	l	F ($\times10^{-2}$)
2	1	-0.72
3	1	0.45
4	1	-0.23
5	1	0.14
6	1	-0.05

Short-periodic perturbations (km):

E°	δr	rδu	δz	E°	δr	rδu	δz
0	-0.10	0.28	0.034	180	-0.41	-0.05	-0.080
10	-0.08	0.14	0.013	190	-0.42	0.03	-0.096
20	-0.11	0.00	-0.007	200	-0.43	0.11	-0.108
30	-0.15	-0.11	-0.024	210	-0.45	0.19	-0.114
40	-0.21	-0.20	-0.037	220	-0.46	0.26	-0.113
50	-0.26	-0.27	-0.044	230	-0.48	0.32	-0.104
60	-0.31	-0.32	-0.045	240	-0.50	0.38	-0.089
70	-0.34	-0.35	-0.042	250	-0.53	0.43	-0.067
80	-0.36	-0.37	-0.034	260	-0.55	0.47	-0.040
90	-0.37	-0.39	-0.026	270	-0.56	0.51	-0.012
100	-0.38	-0.39	-0.017	280	-0.56	0.54	0.017
110	-0.38	-0.39	-0.010	290	-0.55	0.57	0.042
120	-0.38	-0.38	-0.007	300	-0.53	0.60	0.063
130	-0.38	-0.35	-0.009	310	-0.48	0.61	0.076
140	-0.38	-0.32	-0.016	320	-0.41	0.61	0.082
150	-0.38	-0.27	-0.028	330	-0.32	0.58	0.079
160	-0.39	-0.20	-0.044	340	-0.23	0.51	0.069
170	-0.40	-0.13	-0.062	350	-0.15	0.41	0.053

Long-periodic perturbations:

$d\bar{e}/dt = 1.45 \times 10^{-13} \sin 2\bar{\omega}$ (day^{-1}),

$d\bar{\omega}/dt = -(7.60 + 3.16\cos 2\bar{\omega}) \times 10^{-8}$ ("/day).

REFERENCES

Mayo, A.P. 1979, Celest. Mech., Vol. 19, p. 317

Plakhov, Yu.V. 1968, Translation of Russian "Geodesy & Aerophotography", Vol. 1, p. 38

Liu, A.S. et al, 1969, JPL Space Programs Summary, Vol. 3 p. 37

PART II

COMETS AND METEOR STREAMS

PHYSICAL PROCESSES AFFECTING THE MOTION OF SMALL BODIES IN THE SOLAR SYSTEM AND THEIR APPLICATION TO THE EVOLUTION OF METEOR STREAMS

Iwan P. Williams
Applied Mathematics Department, Queen Mary College,
Mile End Road, London E1 4NS

ABSTRACT

In addition to planetary perturbations, the small particles which make up a meteor stream are subject to outward radiation pressure and the Poynting-Robertson effect. New particles can also be generated in a stream through being released from the nucleus of a comet. We summarise the main physical effects, discuss models for meteor stream evolution and give a brief account of the observational data available.

1. INTRODUCTION

One of the advantages (or disadvantages) of producing theoretical models for the dynamical evolution of meteor streams is the existence of a large body of observational evidence gathered in some cases over many centuries. There are in fact nearly twenty meteor streams which can currently be observed from Earth as the meteoroids from which they are composed burn up in the Earth's atmosphere. In some of these streams, the number of meteors seen per hour is very low, perhaps not exceeding 10 even at maximum. However streams such as the Quadrantids, Geminids and Perseids can give very impressive displays with upwards of 100 meteors per hour being observed. The Quadrantids and the Geminids streams appear to have become visible only fairly recently, the first recognition of the Quadrantids being in 1835 [Wartmann, 1841] while the Geminids were first mentioned with any certainty in 1862 [King, 1926]. On the other hand, the Perseids stream appears to be much older, dating back to 36 A.D., with at least a dozen recorded appearances between this date and 1451. Two other "old" streams are the Eta Aquarids (405 A.D.) and the Orionids (585 A.D.). These two streams are of considerable current interest since they are thought to be associated with Halley's comet.

Since a stream, in order to be detected, must interact with the Earth's atmosphere, it is safe to assume that the true population of meteor streams must be in excess of the twenty or so known streams. Meteors interacting with the atmosphere can in fact be detected in two ways, visually and be means of radar. Hughes (1978) deduced that the

V. V. Markellos and Y. Kozai (eds.), Dynamical Trapping and Evolution in the Solar System, 83–87.

radio meteors could have a mean mass of the order of 10^{-4}g and a mean density of 0.8 gcm^{-3} while the visual meteors have a mass of 10^{-1}g and a density of 0.3 gcm^{-3}. Their radii are thus around one third mm and a few millimeters respectively. They are thus small particles and it is worthwhile first considering the forces acting on such particles and the effect of such forces.

2. FORCES ACTING ON METEORITES

It is not our intention to give an exhaustive review of forces acting on small particles, but rather concentrates only on those parts which are relevant to meteroid motions. Clearly the dominant force must be the solar gravitational field, which at heliocentric distance r has magnitude $GM_\odot/r^2$, for if this were not the case, then the meteoroids would not move on even approximately regular orbits and no stream could develop. Donnison and Williams (1977) have shown that the effect of a solar wind is similar to that of the Poynting-Robertson effect and radiation pressure. We need not here therefore consider both phenomena.

Radiation pressure produces a force which opposes gravity on a body of radius a, of magnitude

$$\frac{a^2 Q L_\odot}{4r^2 c} ,$$

where $L_\odot$ is the solar luminosity, c the speed of light and Q is called the efficiency factor for radiation pressure. In general, Q is a factor of meteoroid radius, composition and wavelength of the incident light and a general discussion is given for example by Wickramasinghe (1967). For meteoroids of the dimensions under consideration it will be satisfactory to regard Q as a constant with a value close to unity. As can be seen, both the radiation pressure and gravity have the inverse square dependence on heliocentric distance and it is therefore convenient to write the radial force as

$$\frac{GM_\odot m}{r^2} (1 - \beta) ,$$

where $\beta = 3QL_\odot/16\pi c G a \sigma M_\odot$,
σ being the density of a meteoroid. β is a nondimensional constant for a given meteoroid, but as can be seen, is smaller for large bodies than for small ones. For object of 1 km radius like a cometary nucleus, β is effectively zero, while it increases to near unity for micron sized particles. Of course once $\beta \sim 1$ there is no attractive force and so we would expect sub-micron sized particles to be lost from a stream very quickly. One important consequence of the above, pointed out by Kresak (1974), is that small meteoroids on being released from a comet will start pursuing an orbit of larger seminajor axis than the comet because of the β factor.

In addition to providing an outward push, radiation also removes angular momentum from a meteoroid. If h is the specific angular momentum, then

$$\frac{dh}{d\theta} = -\frac{GM_{\odot}\beta}{c} .$$

If, in addition, it is assumed that the meteoroid is on or near eliptic orbit, then this equation can be converted into an equation which gives the time taken for orbital decay as

$$\frac{dl}{dt} = -\frac{8GM_{\odot}}{lc}(1 - e^2)^{3/2} .$$

e and l being the eccentricity and semi-latus rectum of the orbit. Similarly the eccentricity changes are given by

$$\frac{de}{dt} = \frac{10GM_{\odot}e(1 - e^2)^{3/2}}{l^2c} .$$

Of more interest perhaps is the rate of change of aphelion, Q. By combining the two above equations, we have

$$\frac{dQ}{dt} \doteq \frac{1.5 \times 10^{-7}}{a\sigma} \text{AUy}^{-1}$$

for a meteoroid on an orbit similar to the Quadrantid stream [See Burns *et al* 1979].

Thus the duration in the stream of particles less than about .01 mm is getting rather short and they may well not exist in streams.

In addition to the above forces, meteor streams are also perturbed by the gravitational field of the planets. Because of the general dimensions and eccentricity of their orbits, meteor streams are in fact more susceptible than most other solar system bodies to such perturbations as they can pass close to more than one planet. It is possible to include planetary perturbations, either by calculating the secular changes in an orbit, as for example Babadzhanov and Obrubov (1979) do, or alternatively by going into a full numerical treatment which we shall mention later.

3. OTHER POSSIBLE EFFECTS

Collisions do not in fact play an important role in the general evolution of the stream. Though they do of course play a role in causing meteoroids to be lost from the stream and become sporadic meteors. For an average meteor shower, the number density of meteoroids, n, can be calculated in two ways. Either we can proceed from the mean hourly rate of observation of meteors during a shower E_n say, and assuming the relative speed of stream and Earth is of the order of the Earth's orbital velocity V_E we have roughly

$$\pi R_E^2 V_E n = E_N ,$$

R_E being the earth's radius.

Secondly, one estimate could be obtained on the assumption that a stream represents the break-up of a significant fraction of a cometary nucleus and so estimating the volume of a stream and the mean meteorite mass, again allows a determination of n. Both these approximations yield similar values of about 10^{-25} cm^{-3} for n. With such a number density, any specific meteoroid could expect to be involved in a collision every 10^{12} orbits, thus the majority of a stream is in essence collision-free. However there are some 10^{17} meteoroids in a stream and so 10^5 of them can expect a collision on an orbit. This is a very significant number in terms of populating the sporadic family, but is insignificant in the context of stream evolution.

A final effect which warrants mention is the possibility that a cometary nucleus is feeding particles into the stream. Particles may be expected to be fed in with small velocities in random directions relative to the nucleus. Once released these new meteoroids become subject to all the effect mentioned above and must be included in any calculation. A recent model incorporating this effect is given by Fox *et al* (1982).

The above describes the main forces and effects that operate on meteor streams. I shall now briefly describe our model for meteor stream evolution and summarize our results.

4. A COMPUTER MODEL

In all the models we have produced, the meteor stream is represented by a set of independent particles, each moving on a given initial orbit (not necessarily the same orbit for all particles) and having a given position on the orbit at $t = 0$. Each particle is then subject to the forces mentioned above. The positions of the planets, required to evaluate planetary perturbations, are obtained by integrating their motion backwards in time from the most convenient ephemeris point. The first models were concerned with the Quadrantid meteor stream and Hughes *et al* (1979) showed that the variations in mean orbital parameters between A.D. 1830 and the present agreed well with observations while Murray *et al* (1980) showed that the influx of meteoroids into Earth crossing orbits from the model was consistent with the first sighting of the stream in the 1830's. In these early models, numerical integration was carried out using the fourth order Runge-Kutta technique with self-adjusting step length. It was later found that the Gauss-Jackson method was more convenient [See Herrick 1971, Fox 1982].

The apparent different behaviour of the radio and visual meteors in the Quadrantid stream was explained in terms of the effect mentioned earlier (Radiation-pressure and the Poynting-Robertson effect) placing the smaller set in the vicinity of the 2:1 resonance with Jupiter by Hughes *et al* (1981).

Fox *et al* (1982) have included the effect of insertion of particles from a cometary nucleus into the model and produce a pictorial representation of the cross section of the stream as it crosses the ecliptic, thus giving a visual representation of its evolution. This was applied to the Geminid meteor stream and helped to explain the anomaly where the stream appears to have no retrogression of the ascending node despite all calculations suggesting that it should.

More recent work using the same model will be described by Fox (1983).

5. CONCLUSIONS

The evolution of meteor streams can be studied be means of computer models. These models give results which are in excellent agreement with the observed results from meteor streams which indicates that the physics inserted into the models could also be used beneficially to study areas such as the Trojan asteroid family and ring systems around planets.

6. REFERENCES

Babadzhanov, P.B. and Obrubov, Yu.V., 1979, *Dokl. Akad. Sci. USSR*, 22, 466.

Burns, J.A., Lamy, P.L. and Soter, S., 1979, *Icarus*, 40, 1.

Donnison, J.R. and Williams, I.P., 1977, *Mon. Not. R. astr. Soc.*, 180, 289.

Fox, K., 1982, Ph.D. Thesis, London University.

Fox, K., 1983, This volume, p. 89.

Fox, K., Williams, I.P. and Hughes, D.W., 1982, *Mon. Not. R. astr. Soc.*, 199, 313.

Herrick, S., 1971, *Astrodynamics*, Van Nostrand Reinhold, London.

Hughes, D.W., 1978, In *Cosmic Dust*, ed. Mc Donnell, John Wiley, New York.

Hughes, D.W., Williams, I.P. and Murray, C.D., 1979, *Mon. Not. R. astr. Soc.*, 189, 493.

Hughes, D.W., Williams, I.P. and Fox, K., 1981, *Mon. Not. R. astr. Soc.*, 195, 625.

King, A., 1926, *Mon. Not. R. astr. Soc.*, 86, 638.

Kresak, L., 1974, *Bull. astr. Inst. Csl.*, 13, 176.

Murray, C.D., Hughes, D.W. and Williams, I.P., 1980, *Mon. Not. R. astr. Soc.*, 190, 733.

Wartman, M., 1841, *Bull. Acad. R. Brux.*, 8, 226.

Wickramasinghe, N.C., 1967, *Interstellar Grains*, Chapman-Hall, London.

THE ORBITAL EVOLUTION OF THE PERSEID AND QUADRANTID METEOR STREAMS.

Ken Fox,
Department of Applied Mathematics,
Queen Mary College,
Mile End Road,
London. E1 4NS.

ABSTRACT

Some mathematical models of the formation of meteor streams are developed. Some of the testable predictions of these models are compared with observations.

1. THE PERSEID METEOR STREAM

In 1861 Daniel Kirkwood wondered; "may not our periodic meteors be the debris of ancient but now disintegrated comets, whose matter has become distributed round their orbits." (see Brandt and Chapman (1982)). In July of the following year a new comet was discovered, on an Earth crossing orbit, which was designated 1862 III P/Swift-Tuttle. Between 1864 and 1866 Giovanni Schiaparelli was able to compute the orbits of some of the meteoroids in the Perseid meteor stream. It soon became apparent that these particles had very similar orbits to this new comet, thus confirming the correctness of Kirkwood's conjecture.

Starting from the orbit determined from the 1862 observations, Marsden (1973) calculated the orbital evolution of Swift-Tuttle in an unsuccessful attempt to link the 1862 orbit with earlier recorded cometary apparitions. A forward integration by Yeomans (1972), which included the effects of all nine planets, predicted that the comet should return to perihelion on June 30th 1981. It has yet to be spotted despite intensive searches. Non-gravitational forces, which are observed in many comets, could be responsible for these anomalies. Whatever has happened, the parent comet should be in the vicinity of perihelion at sometime close to the present time.

The Perseid / Swift-Tuttle orbit does not evolve rapidly in time, indeed it is an incredibly stable orbit. Hughes and Emerson (1982) have estimated from observations that the ascending node of the Perseid stream is progressing by only $(38 \mp 27) \times 10^{-5}\ ^\circ\ yr^{-1}$, an order of magnitude less than any other stream. The antiquity of the Perseid shower, it was first

V. V. Markellos and Y. Kozai (eds.), Dynamical Trapping and Evolution in the Solar System, 89–95.

observed in AD 36, confirms that the stream has remained on an Earth crossing orbit for a considerable time and hence is very stable. Celestial mechanics also predicts slow evolution. The high inclination of the stream, about 114°, means that it can only be in the vicinity of perturbing planets at its nodes, but there are no major planets near these points. Therefore it can be concluded that there is a comet with a poorly determined future on a very stable Earth crossing orbit. This might be slightly worrying as the chances of a collision between this comet and Earth must be quite high. If the Tunguska event is typical of such collisions then this must be very worrying indeed. (see Turco et al (1982))

Many authors have predicted that the activity of the Perseid shower should increase as its parent comet returns to perihelion surrounded by a dense swarm of recently ejected particles. By observing changes in the flux rates of the Perseid shower it might be possible to get a fix on the approximate whereabouts of the parent comet. Due to its extreme stability, the early lives of particles ejected from Swift-Tuttle into the Perseid stream can be determined ignoring planetary perturbations.

Whipple (1951) gives the following expression for the ejection speed of dust from the nucleus of a comet

$$c = \left[\frac{1}{n\, s\, \rho\, r^{9/4}} - 0.013\, R_c \right]^{1/2} R_c^{1/2} \times 656 \text{ cms}^{-1} \quad (1)$$

where R_c is the radius of the nucleus of the comet in km, 1/n is the fraction of solar radiation utilised for sublimation and s and ρ are the radius and density of the spherical dust particles in cgs units. Whipple (1978), states that a typical value for the radius of a cometary nucleus is about 1km. A value of 1 for n will be used here. Following Plavec (1955), if a particle is ejected with velocity c from a comet while it is in the vicinity of perihelion r_o, then the resulting overall particle velocity v can be split into components given by

$$v_r = c_r$$
$$v_u = v_o + c_u$$
$$v_b = c_b$$

where c_r is in the direction of the comet's radius vector, positive away from the Sun, c_u is tangential to the orbit, positive in the direction of motion and c_b completes the right handed set. The orbital speed of the comet at perihelion being v_o, therefore

$$v^2 = v_o^2 + 2\, v_o\, c_u + c^2.$$

At this point Plavec, ignorant of the effects of radiation, related the semi-major axis of the parent comet's orbit to that of the new orbit taken up by the ejected particles. Therefore correcting this omission one gets

$$a = \frac{\mu'}{v^2 - 2\mu'/r_o}$$

where a is the semi-major axis of the new orbit taken up by the ejected particle and $\mu' = GM_o\,(1 - \beta)$ with β given by

$$\beta = 5.74 \times 10^{-5}\, Q_{PR}\, s^{-1}\, \rho^{-1}$$

where Q_{PR}; the radiation pressure efficiency factor, is a number that in most cases is close to unity. However c is given for any β by equation (1) and depending upon the direction of ejection c_u can take on any value in the range (−c,+c). The period of the ejected particle is given by $P = 2\pi\,(a^3/\mu')^{1/2}$.

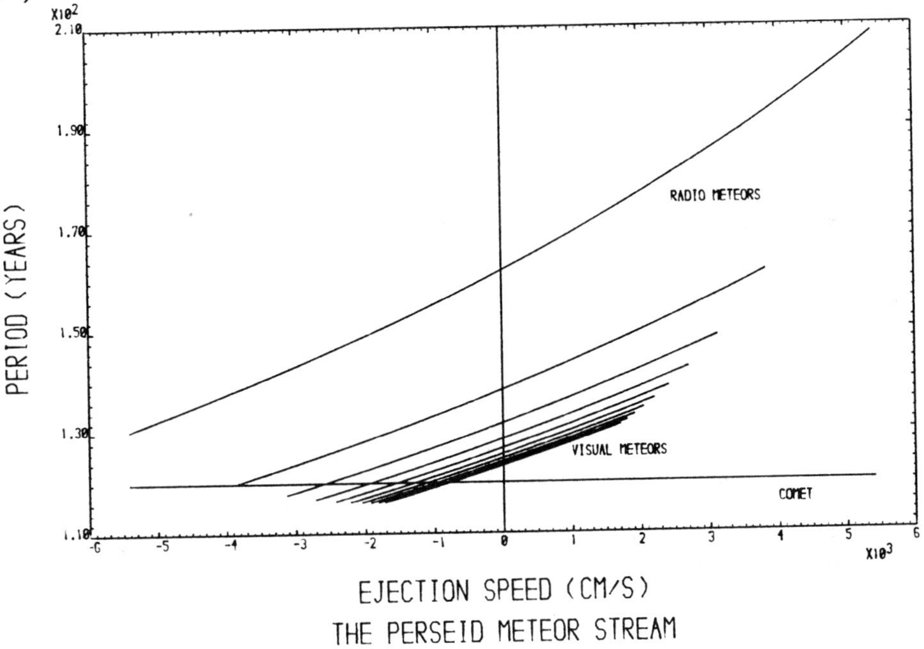

Figure 1 is a plot of ejection speed against the subsequent periods of particles ejected from comet Swift-Tuttle. The 120 year period of this comet also being indicated on this figure. The particles are taken as having ten evenly spaced β values between those corresponding to the approximate observed spread of the small radio meteors and the larger visual meteors. The curves are asymmetrical about the parent comet's period. This is because radiation pressure, which always increases the period of the ejected particles, is competing with the ejection velocity effects which can either increase or decrease the period. If all directions of ejection are equally likely then there will always be more particles behind the comet than ahead of it. There are always some particles that have been ejected so that their period differs from that of the parent comet by at least one year. Therefore Earth will always encounter a dense swarm at least once while the comet is close to perihelion. This is substantiated by the increased flux of Perseid meteors observered the last time Swift-Tuttle was at perihelion; in 1862.

Russel (1982) reckons that the flux of Perseid meteors peaked in 1980, as his photographic observations since 1977 show double the flux of meteors in 1980 compared with the years either side of this date. This means that it is possible that Swift-Tuttle has been missed altogether this time around and will not be seen again for a further 120 years.

2. THE QUADRANTID METEOR STREAM

The Quadrantid meteor stream is undergoing substantial evolution due to its close proximity to Jupiter's orbit. This means that the model just used to describe early evolution of particles ejected from a comet is no longer applicable as Jovian perturbations will now considerably effect such evolution.

Unfortunately there is no known comet associated with this stream. This could be due to two things; either the parent comet had a close approach to Jupiter and was ejected from the solar system leaving behind just its trail of debris or the comet disintegrated entirely to leave no visible trace. This fact makes modelling the early evolution of the stream rather awkward! However, by calculating the mean behavior of the stream in the past, a suitable type of orbit for such a comet can be determined and used for such modelling. Starting from the observed mean orbit as given by Poole et al (1972), Williams et al (1979) placed ten particles at equally spaced intervals of eccentric anomaly around this orbit and from this starting configuration, then integrated the equations of motion of these particles moving through the gravitational fields of the Sun, Earth and Jupiter, over the time period from 300 BC to 3780 AD. Therefore to model the formation and subsequent evolution of the stream some time has to be chosen to place a comet on this calculated mean orbit. By considering the observed mass-segregation in the stream, Fox (1982) has deduced that the stream is about 1000 years old. However computing restraints have limited the period of integration to just 500 years. Starting from the mean orbit of 500 years ago, a theoretical comet placed at the perihelion of this orbit ejected twenty particles in random directions but with ejection speeds given by Whipple's formula; equation (1). Ten of these particles had $s = 0.22$ cm and $\rho = 0.3$ g/cm^3 corresponding to visual meteors, while the other ten had $s = 0.035$ cm and $\rho = 0.8$ g/cm^3 corresponding to radio meteors. The equations of motion of these twenty particles were integrated numerically back to the present time. Radiation pressure and drag as described by Burns et al (1979) were included in the model. The initial comet orbit is illustrated in figure 2 and the final stream orbit in figure 3.

An observer on Earth only sees a meteor when a meteoroid burns out in Earth's atmosphere. This can only happen at the appropriate node of the meteoroid's orbit as these are the only points where the meteoroid crosses the ecliptic. For the Quadrantid stream, the descending node is the one in the vicinity of Earth's orbit, so it is instructive to look at the evolution of this node with time. This evolution is shown in figure 4.

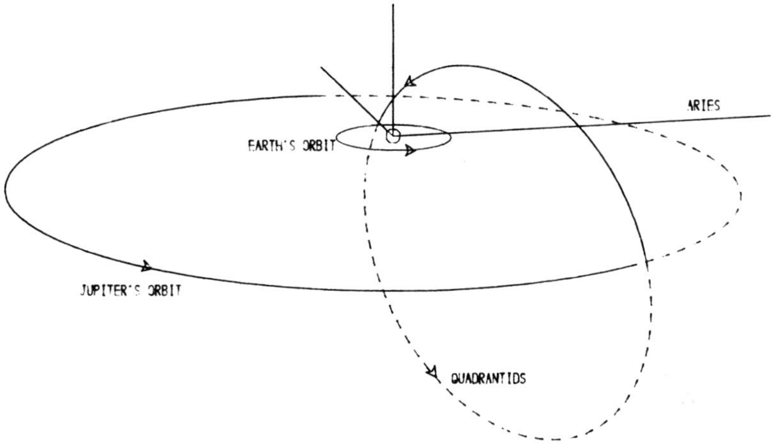

Figure 2. The initial Quadrantid orbit.

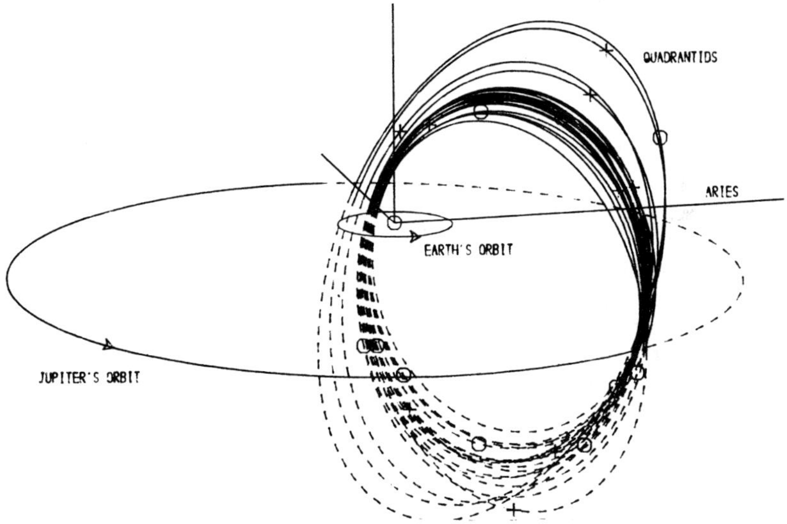

Figure 3. The final Quadrantid orbits.

It is immediately seen that a long, thin stream cross-section develops. The stream does not spread out much along Earth's orbit but it does spread out considerably in the solar direction. This situation is necessary in order that theory and observations should match. The

Quadrantid shower is an extremely short shower, it only lasts for about one day. This vindicates the shortness of the theoretical stream along Earth's orbit. Jovian perturbations are causing the stream to be swept across Earth's orbit at a great rate and as the shower has been observed since 1835 the observations imply that the stream should be spread out by a much greater amount in the solar direction than it is along Earth's orbit Earth would pass through this theoretical shower in a time of about 18 hours. It would reach the centre of the radio shower about four hours before it reached the centre of the visual shower. Poole et al give three values for the observed visual stream width of 8 hours, 17 hours and 24 hours, while for the radio stream they give a width of between 3 hours and 29 hours. Hughes et al (1981) give a value of fourteen hours as the time between the peaks of the visual and radio meteors.

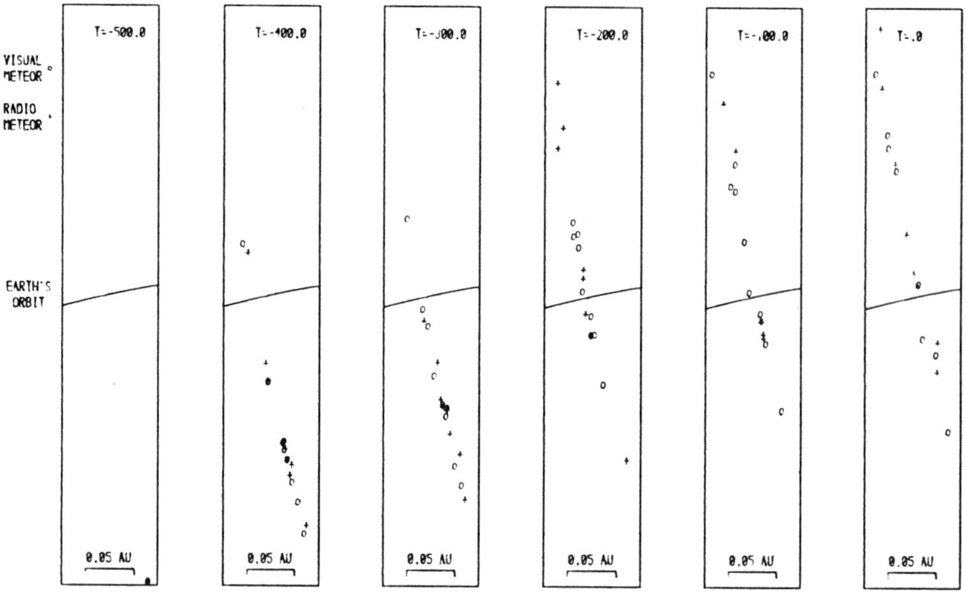

Figure 4. The evolution of the Quadrantid ascending node.

It is also noticable that in the short time that the integration has been carried out that a continuous loop of meteoroids has formed around the stream. This is in contradiction to early theories that only took into account the effects of ejection velocities and ignored radiation effects. Both radiation and Jovian perturbations are important in spreading particles around the Quadrantid stream. Thus it is no longer correct to assume that a regular meteor shower is produced by a very old stream, it can in fact be produced by a relatively young one.

REFERENCES.

Brandt, J. C. and Chapman, R. D. : 1982, Introduction to Comets, CUP.

Burns, J. A., Lamy, P. L. and Soter, S. : 1979, Icarus, 40, 1.
Fox, K. : 1982, PhD Thesis, London.
Hughes, D. W. and Emerson, B. : 1982, The Observatory, 102, 39.
Hughes, D. W., Williams, I. P. and Fox, K. : 1981, Mon. Not. R. astr. Soc., 195, 625.
Marsden, B. G. : 1973, Astron. J., 78, 654.
Plavec, M. : 1955, Bull. astr. Inst. Csl., 6, 20.
Poole, L. M. G., Hughes, D. W. and Kaiser, T. R. : 1972, Mon. Not. R. astr. Soc., 156, 223.
Russel, J. A. : 1982, Sky and Telescope, 63, 10.
Turco, R. P., Toon, O. B., Park, C.,Whitten, R. C., Pollack, J. B. and Noerdlinger, P. : 1982, Icarus, 50, 1.
Whipple, F. L. : 1951, Ap. J., 113, 464.
Whipple, F. L. : 1978, 'Comets' in 'Cosmic Dust' ed McDonnell, J. A. M., Wiley, New York.
Williams, I. P., Murray, C. D. and Hughes, D. W. : 1979, Mon. Not. R. astr. Soc., 189, 483.
Yeomans, D. L. : 1972, Computer Sciences Corporation Report, 9101-11200 -01TR.

STEADY STATE NUMBER OF THE EXTINCT COMETS IN HIGH-INCLINATION ORBITS

Tsuko Nakamura
Dodaira Station, Tokyo Astronomical Observatory
Tokigawa, Hiki-gun, Saitama 355-05 Japan

Abstract

Steady state number of the short-period(SP) comets captured from nearly parabolic(NP) orbits with $60° \leqslant$ i(inclination) $\leqslant 120°$ and $0.5 \leqslant$ q (perihelion distance) $\leqslant 1.5$ AU is calculated. Due to smallness of the q and slowness of the capture process, almost all these SP comets become completely extinct. Combining annual flux of the observed NP comets of high inclination with the capture probability from NP to SP orbits and the ejection rate in SP orbits obtained by Monte Carlo simulations, we find that the steady state number of the extinct SP comets in high-inclination orbits is at least 1-2 hundred. If this number can be considered to correspond to the observed one of Apollo-type objects with $i \geqslant 60°$, it is concluded that only less than a few percent of extinct comets leaves sizable non-volatile cores or shells. The number of asteroid-like bodies deduced from the extinct comets with small perihelion distance and high inclination has less ambiguity than that of low inclination, because contribution from asteroid belt is negligible and the source comets are visible(free from observational selection) through the whole course of orbital evolution.

It is of primary importance to properly estimate the fraction of the comets which leave sizable core or shell of stony material after losing volatile components(extinct comets), both in order to determine the dominant source of Apollo-Amor(AA) objects and to get information on internal structure of cometary nuclei. Major contributions to this problem have been made by Öpik(1963), Wetherill(1979), Kresák(1980), Rickman and Froeschlé(1980), Levin and Simonenko(1981) and others.

The purpose of this paper is to estimate an upper limit of the fraction of asteroid-like survivors among extinct(EX) comets without ambiguity of the source as far as possible. We also intend to understand the population of EX comets within the framework of orbital evolution of long-period(LP) comets. For those purposes, we restrict our treatment to the LP comets with $0.5 \leqslant q \leqslant 1.5$ AU and $60° \leqslant i \leqslant 120°$. The reasons and advantage for this restriction are :

V. V. Markellos and Y. Kozai (eds.), Dynamical Trapping and Evolution in the Solar System, 97–104.

1) contribution from asteroid belt to high-inclination AA objects is negligible,
2) the effects of observational selection with respect to q are minimum among various LP comets,
3) physical disintegration is active and at a nearly constant rate during the whole course of orbital evolution, and
4) the ejection rate is comparatively small in spite of slowness of orbital change.

As is shown later, the observed number of the AA objects of high inclination is extremely small. This prevents us from defining observationally the orbital region of the EX comets of high inclination. Therefore we regarded somewhat arbitrarily the region of a(semi-major axis) $\lesssim 10$ AU, $0 < q \lesssim 1.5$ AU and $60° \lesssim i \lesssim 120°$ as the place where EX comets are expected to be found. This selection, however, is justified from the viewpoint of orbital evolution of LP comets(Nakamura 1981).

The steady state number of the EX comets is given by n/λ (e.g., Wetherill 1979), where n is the annual injection rate into this region and λ is the expulsion rate per comet per year by Jovian perturbations. Since the value of n cannot be known from observation, it must be estimated from n_0 the observed annual rate of Oort-cloud(NP) comets and the dynamical capture efficiency of NP comets via LP comets into short-period(SP) comets. n_0 was found to be 0.195 objects/yr, by picking up 14 objects(/135 yr) of the NP comets with $1/a_{orig} \leqslant 50 \times 10^{-6}$ AU^{-1}, $0.5 \leqslant q \leqslant 1.5$ AU and $60° \leqslant i \leqslant 120°$ from the list of 200 original orbits of LP comets (Marsden et al., 1978), and by correcting the number by a factor of 375/200, the ratio of the total number of the LP comets which appeared past 135 years to the 200 original LP comets.

The capture efficiency of the NP comets was calculated by a means of Monte Carlo simulation of the orbital evolution due to perturbations of Jupiter. The method adopted here is similar to that described in Nakamura(1981): orbital evolution is traced as a random walk in the 3-dimensional space of Kepler energy, total angular momentum and its z-component, where each step was chosen following the trivariate distributions of Jovian perturbations calculated beforehand. About 2.3 % of the comets which start from the inner Oort cloud region is found to reach the region of $a \lesssim 5.2$ AU after the mean revolution of nearly 8000. Then we have $n = 0.023n_0 = 4.5 \times 10^{-3}$ objects/yr.

As for high-inclination comets, there are good reasons to believe that argument of perihelion ω may play an important role in their orbital evolution. This situation does not apply to low-inclination case. Geometrical considerations suggest that the LP comets of $q \lesssim 1$ AU with $\omega \sim 90°$ or $\sim 270°$ may not interact with Jupiter strongly enough to be brought into SP orbits. Dynamical considerations, on the other hand, indicate that secular perturbation on ω may gradually change its value, resulting in close approaches to Jupiter within an appropriate period of time. Only detailed numerical simulations will be able to decide which is true. However, an orbital statistics of high-inclination LP comets seems to support the former. Fig.1 represents the distributions

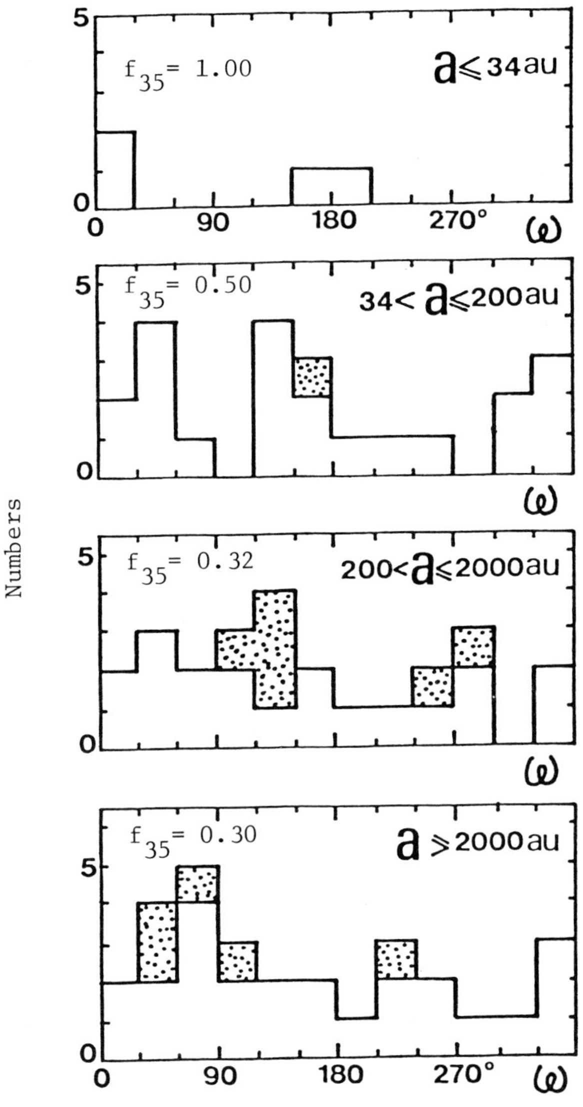

Fig.1. Distributions of argument of perihelion for LP comets with $60° \leq i \leq 120°$ and $0 < q \leq 2$ AU(dotted blocks are for $1.5 < q \leq 2.0$ AU).

of ω for the LP comets with $0 < q \leq 2$ AU and $60° \leq i \leq 120°$, in order of a_{orig}, which can be regarded as an indicator of degree of orbital evolution. The data are from Marsden(1979) and Marsden et al.(1978). It is clearly seen that along with the decrease of a_{orig}, the comets with $\omega \sim 90°$ or $\sim 270°$ are removed and finally only the comets with $\omega \sim 0°$ or $\sim 180°$ are left. Adoption of a factor f_{35} will make the situation clearer. f_{35} is the ratio of the comets whose ω lie both between $-35°$ and $35°$ and between 145° and 215° to all the comets considered. Fig.1 will be interpreted as: the NP comets other than those with $\omega \sim 0°$ and $\sim 180°$ cannot evolve into shorter-period ones by Jovian perturbations. Then, we should modify the value of n by f_{35}; namely, $n = 4.5 \times 10^{-3} f_{35} = 1.75 \times 10^{-3}$ objects/yr, in which a flat distribution of ω for the LP comets is assumed.

In Fig.2 are shown the mean paths of orbital evolution of the LP comets with high inclination. On an average the increases of i were less than 20-30°.

N	100	500	1000	2000	5000	(rev)
$\bar{a}$	156	56	34	21	11	(AU)

This is a relation of mean semi-major axis($\bar{a}$) versus revolution(N) for the initial q of 1 AU.

According to sublimation theory of H_2O ice nucleus, about 1000 revolutions are a typical lifetime for a nucleus of 1 km radius and for q = 1 AU(Cowan and A'Hearn 1979, Weissman 1980). This figure will be slightly reduced if the effects of outburst and splitting are taken into account. Since the average diameter of LP comets is estimated to be 5 km or so(Whipple 1978, Rahe 1981), it is understood that, in combination with the result of the above table, most of the LP comets considered here will become completely extinct due to physical disintegration before they reach the orbital region of EX comets. This seems to be also the reason why we do not observe SP comets of high inclination (Fig.2).

Next we will discuss briefly non-gravitational(NG) effects on orbital elements. Table 1 is the net orbital changes per revolution due

q(AU)	$\Delta(1/a)$ (AU^{-1})	Δ e	Δ q (AU)
0.2	1.91×10^{-3}	-3.82×10^{-4}	-3.48×10^{-5}
0.5	2.56 10^{-4}	-1.28	-1.85
0.7	1.09	-7.6 10^{-5}	-1.21
1.0	3.74 10^{-5}	-3.7	-6.2 10^{-6}
1.5	7.5 10^{-6}	-1.1	-1.8
2.0	1.5	-3.0 10^{-6}	-4.5 10^{-7}

Table 1.

to jet reaction caused by sublimating gases on a nucleus for various q. These values were obtained by integrating numerically Lagrange equations of Gauss's form along parabolic orbits. $1/r^3 \cdot \exp(-r^2/2)$ is assumed as dependency of reactive force on heliocentric distance (Marsden et al., 1973). In this calculation $-0.98 \times 10^{-8} AU^4/day^2$ (retrograde rotation) is assigned to A_2 the tangential component of NG parameters, which is an average of the absolute values of nine A_2 of the LP comets tabulated in Marsden et al.(1973). $\Delta(1/a)$ in Table 1 shows that if NG effects act every return constantly in one sense --- this is no doubt improbable, it takes more than 5000 revolutions for a NP comet of q = 1 AU to

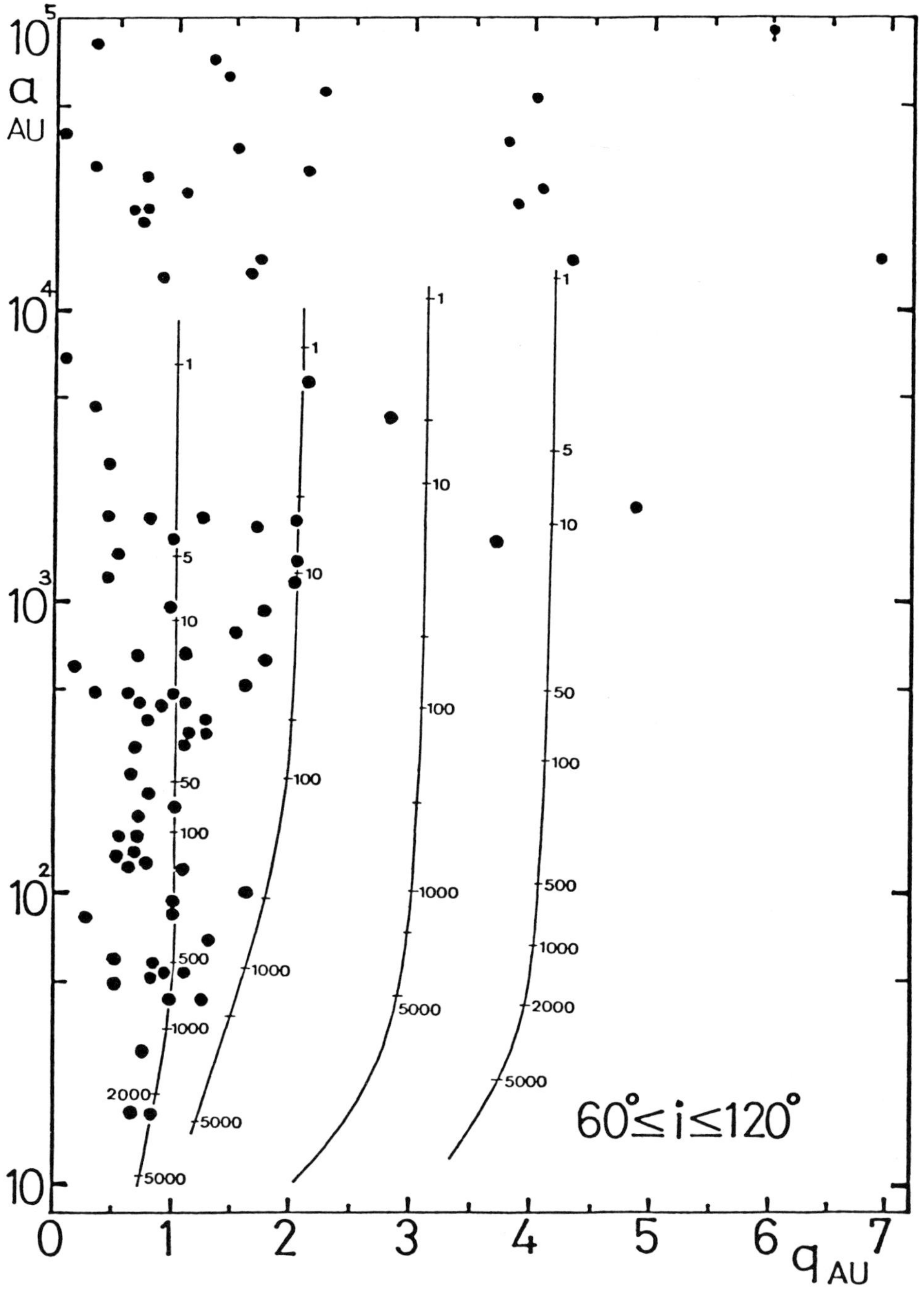

Fig.2. Orbital evolution of high-inclination LP comets. Figures attached to each curve represent mean revolutions. Filled circles are the observed comets(Marsden et al., 1978, Marsden 1979).

reach near the orbit of Jupiter. This suggests that inclusion of NG effects on 1/a does not change sharply the lifetime obtained in consideration of orbital evolution and physical disintegration.

The calculation of λ was carried out as follows. Many hypothetical EX comets were first scattered uniformly in the domain of $5.5 \leqslant a \leqslant 8.1$ AU, $0.5 \leqslant q \leqslant 1.5$ AU and $60° \leqslant i \leqslant 120°$ and their subsequent orbital evolution was traced by another Monte Carlo simulation. λ (/yr/comet) is expressed as

$$\lambda = -d\nu \;/\; \nu \;/(N_2 - N_1)/\; \bar{P},$$

where ν is the initial number of EX comets, $d\nu$ the number ejected from the region of $0.5 \leqslant q \leqslant 1.5$ AU between N_2-th and N_1-th returns, and $\bar{P}$ the mean orbital period(yr) of EX comets. For $\nu = 210$, $N_2 - N_1 = 100$ (this simulation was traced up to 500 revolutions) and $\bar{P} = 17$ yrs, an averaged $d\nu$ was 3.6; thus $\lambda = 1.0 \times 10^{-5}$/yr. This value might be a considerable overestimate, since the bodies once ejected from the region of $0.5 \leqslant q \leqslant 1.5$ AU are implicitly assumed to never come back to the same region. The loss rate due to planetary collision is shown to be about 10^4 times smaller than this λ (Wetherill 1979). Therefore, using the obtained values of n and λ, a steady state number of EX comets of 175 immediately results for the orbital range assigned above.

Now we are in a position to deduce the fraction of asteroidal survivors by comparing with the observed number of AA objects of high inclination. As the initial i of LP comets are limited to more than 60° and the orbital evolution has a trend to increase i, it would be then reasonable to regard the Apollos with $i \geqslant 60°$ as the candidates. Only two of such Apollos have been known so far:

No.	q(AU)	Q(AU)	i	ω
1973NA	0.88	3.98	68°	118°
1975YA	0.91	1.69	64	61 .

If we receive this number at its face value, the fraction of asteroidal survivors is about 1 %. However, as for 1975YA, it seems improbable that this is an EX comet, because its aphelion distance Q is too small to be accounted for by jet effects.

Attention must be called to the point that discovery condition of high-inclination bodies may be different from that of low inclination. Intuition predicts that the relative speed of high-inclination bodies to the earth will be larger than that of low inclination. Fig. 3 is a magnitude-observed daily motion diagram for the asteroids designated as "fast-moving objects" near their discovery in IAU Circulars past ten years. This diagram indicates that the intuition is correct to some extent, though exceptional cases are not rare. Large relative speed leads to failure of detection of faint objects, resulting in the raise of their limiting magnitude. We assume the size distribution of $dn \propto D^{-b}dD$ (D: asteroid's diameter and b = 3 - 3.5 for typical asteroids) and that the number density of silver grains in an asteroidal

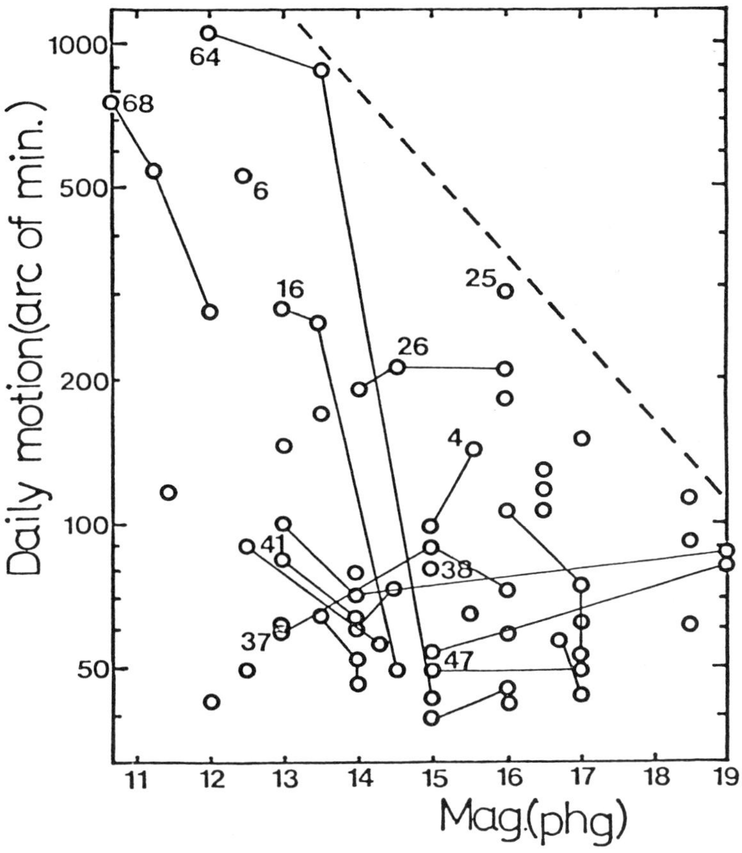

Fig.3. Observed daily motions of "fast-moving objects" near their discovery(1972-81) plotted against apparent magnitude. Attached numbers are inclinations. The circles connected by straight lines belong to the same object.

image on a photographic plate is proportional to 1/v (v: mean relative speed of high-inclination objects measured in the unit of low-inclination ones'). Unobservable number N_u due to large relative speed and still observable number N_v are given respectively as follows,

$$N_u \propto (\sqrt{v}\, D_L)^{1-b} - D_L^{1-b},$$

$$N_v \propto D_{max}^{1-b} - (\sqrt{v}\, D_L)^{1-b},$$

in which D_L is the diameter corresponding to the limiting magnitude for low-inclination bodies and D_{max} is the diameter of the brightest bodies. Simple calculation brings about a result that the ratio of N_u to N_v is roughly equal to v - 1. This expression holds exactly for the case b = 3.0.

Mean observed daily motions deduced from the data of Fig.3 for high- and low-inclination Apollos are 435' and 91' respectively; thus v = 4.8. Then the number of high-inclination Apollos corrected for this selection effect will be 7.6 or so. In this case, the fraction of asteroidal survivors is 4.3 %. Even if ambiguity of the adopted constants involved in our analysis is taken into account, the resultant fraction will be raised by a factor of two at most, whereas it is very probable that the fraction may be lowered, say, by one order of magnitude. In conclusion, the fraction of the cometary nuclei which survive as sizable asteroid-like bodies is less than 2-8 %. This result coincides well with the estimate by Rickman and Froeschlé(1980), though quite different data are analysed.

At present the origin of SP comets is understood, at least qualitatively, as a result of orbital evolution of NP or LP comets(Everhart, 1972). If the existence of the comet Encke is a probable event among SP comets and the fraction obtained above is also applicable to low-inclination comets, the number of Apollo objects deduced from Encke-type comets will have to be reduced by the same fraction. As far as the problem of Apollo objects is considered in the framework of orbital evolution of LP comets, this seems to be an inevitable outcome. The origin of AA objects is obviously still an unsettled problem.

REFERENCES

Cowan,J.J. and A'Hearn,M.F.: 1979, The Moon and the Planets, 21, 155.

Everhart,E.: 1972, Astrophys. Letter, 10, 131.

Kresák,L.: 1980, The Moon and the Planets, 22, 83.

Levin,B.J. and Simonenko,A.N.: 1981, Icarus, 47, 487.

Marsden,B.G., Sekanina,Z. and Yeomans,D.K.: 1973, Astron. J., 78, 211.

Marsden,B.G., Sekanina,Z. and Everhart,E.: 1978, Astron. J., 83, 64.

Marsden,B.G.: Catalogue of Cometary Orbits, 3rd Ed.(1979), Central Bureau of Astronomical Telegram, IAU.

Nakamura,T.: 1981, Icarus, 45, 529.

Öpik,E.J.: 1963, Advances Astron. Astrophys., 2, 219.

Rahe,J.: Landoldt-Bornstein, New Series, Vol.2, Subvol.a, Methods, Contants, Solar System(Ed. by K. Schaifers and H.H. Voight), Chap.3, Springer-Verlag, Berlin(1981).

Rickman,H. and Froeschlé,C.: 1980, The Moon and the Planets, 22, 125.

Weissman,P.R.: 1980, Astron. Astrophys., 85, 191.

Wetherill,G.W.: 1979, Icarus, 37, 96.

Whipple,F.L.: Comets, in Cosmic Dust(Ed. by J.A.M. McDonnell), Chap.1, John Wiley & Sons, New York(1978).

EJECTION OF PARTICLES FROM COMET LEXELL: THE GRAVITATIONAL INFLUENCE OF JUPITER

M. Kresáková[1], A. Carusi[2] and G.B. Valsecchi[2]
[1]Astronomical Institute - S.A.V., Bratislava
[2]I.A.S. - Reparto planetologia, Roma

The orbital evolution of comet P/Lexell has been characterized by a sequence of three close planetary encounters within 12 years; the first of these encounters was with Jupiter, 1767, and lead to a reduction of the perihelion distance from 2.9 AU to 0.67 AU. Then the comet encountered the Earth at a distance of 0.015 AU, in 1770, and was discovered. The final, very close encounter with Jupiter, in 1779, removed the comet from observability, sending it into an orbit of perihelion distance in excess of 5 AU (Kazimirchak-Polonskaya, 1967).

We have investigated the orbital evolution of the comet, and of an associated stream of particles, between 14.0 August 1770 and 20.0 March 1782 (JD 2367764.5 - JD 2372000.5), thus covering the second encounter with Jupiter (Carusi, Kresáková, and Valsecchi, Astron. Astrophys., in press). The computations have taken into account the perturbations of the planets from Venus to Neptune, and the n-body integrator is based on Greenspan's Discrete Mechanics.

The aim of the work has been to check if a sufficient fraction of the particles ejected from P/Lexell at the two perihelion passages in 1770 and 1776 could escape far enough from the parent body to avoid being removed from their Earth crossing orbits by jovian perturbations in 1779. Note that P/Lexell is the only known case for which we know the exact time of particle release, and that this time coincided with the only period of appreciable cometary activity (Kresáková, 1980).

Our fictitious stream is composed by 44 particles, eje-

V. V. Markellos and Y. Kozai (eds.), Dynamical Trapping and Evolution in the Solar System, 105–106.

cted at the starting date at tangential velocities V between -50 and +50 m/s relative to the comet. More specifically, the spacing in V is 5 m/s between -50 and -10, 1 m/s between -10 and +20, and again 5 m/s between +20 and +50. The motion of the comet has also been integrated for comparison.

The trajectories of the particles at the close encounter turn out to be very interesting and rather surprising. A great part of the stream is very strongly perturbed, and only the particles farthest from the comet remain in orbits resembling the initial ones. The central part of the stream is practically unchanged until immediately before the encounter; then the particles are deflected in all the directions, so that, looking at them in a frame centred on Jupiter, the resulting stream looks like an expanding circle. Plotting the individual trajectories in three dimensions reveals that the paths of the most perturbed particles are nearly coplanar, and that the plane of the motion contains the position of Jupiter at the time of the encounter, but not that of the Sun.

The spatial density of particles in the vicinity of the comet drops dramatically after the encounter. Jupiter removes from Earth crossing orbits not only the comet, but also the particles ejected in 1770 with V comprised between -3.5 and 14.5 m/s; moreover, an analogous stream ejected at the perihelion passage of 1776 would suffer much stronger depletions, due to the shorter time available for the dispersion along the orbit.

REFERENCES

Kazimirchak-Polonskaya, E.I.: 1967, Sov. Astron. - A.J. 11, pp. 349-365.

Kresáková, M.: 1980, Bull. Astron. Inst. Czechosl. 31, pp. 193-206.

CAPTURE OF THE COMET P/BOETHIN BY JUPITER.

BENEST Daniel
Observatoire de Nice, France

BIEN Reinhold
Astronomisches Rechen Institut, Heidelberg, Fed. Rep. Germany

RICKMAN Hans
Astronomiska Observatoriet, Uppsala, Sweden

ABSTRACT

The possible effects of non-gravitationnal forces on the motion of the comet P/Boethin are investigated for various values of the orbital period. A time interval of 2000 years backward and forward is treated. The authors find in all cases that the comet librates temporarily around the 1/1 resonance with Jupiter as a remote jovian satellite during at least two centuries.

1. INTRODUCTION

Comet P/Boethin (1975 I) has been previously shown to librate around the 1/1 resonance with Jupiter during at least a few centuries, and to have a very remote satellite motion around this planet (Benest et al., 1980; see also Benest et al., 1981). More recently, the authors began to study the influence of non-gravitationnal forces upon this behaviour, first using a restricted four-body model (Sun-Jupiter-Saturn-comet; Benest et al., 1982).

We present here the results obtained for a more complete set of conditions with the restricted three-body model, integrated with a Runge-Kutta method with variable step, where we combine the influence of varying orbital period with non-gravitationnal forces.

2. CALCULATION

In a rotating-pulsating frame centered on Jupiter, the motion of a remote satellite is composed of a fast motion along a bean-shaped curve

V. V. Markellos and Y. Kozai (eds.), Dynamical Trapping and Evolution in the Solar System, 107–114.

which librates slowly around the planet (see figure 1a). Before capture and after escape, the comet can be still in 1/1 resonance with Jupiter, but the bean-shaped curve either librates around a point approximately symmetric to Jupiter with respect to the Sun (which we call an "anti-satellite libration"; see figure 1b), or circulates continuously around the Sun.

We have computed 25 orbits for the comet over 2000 years backward and forward. One of these, hereafter referred to as the "central orbit", is based on the elements in Marsden's (1979) catalogue and corresponds to purely gravitationnal motion. Besides, we have treated four different gravitationnal orbits varying the orbital period P at the starting epoch (1974 Dec. 19) by $\Delta P = \pm 8$ d and ± 16 d. We have also treated 20 non-gravitationnal orbits, 4 for each starting period. As for the four-body calculation, we adopted the standard expression for the non-gravitationnal force (Marsden, 1974) with the same set of values for A_2: ± 1.5 and $\pm 3.\ 10^{-9}$ a.u./day^2.

For each orbit, we have plotted $\underline{a}$ (semi-major axis of the orbit of the comet), $l-l_J$ (cometary mean longitude minus Jupiter's mean longitude) and d_{CJ} (minimum distance between Jupiter and the comet over one period of the comet) versus time (see figure 2, here for the central orbit). On these plots, the libration motions correspond to oscillations of $\underline{a}$ around the value of a_J (Jupiter's semi-major axis) and of $l-l_J$ around 0 (satellite) or $\pm 180°$ (anti-satellite); during a circulation, $\underline{a}$ stays always greater or less than a_J and $l-l_J$ varies monotonously.

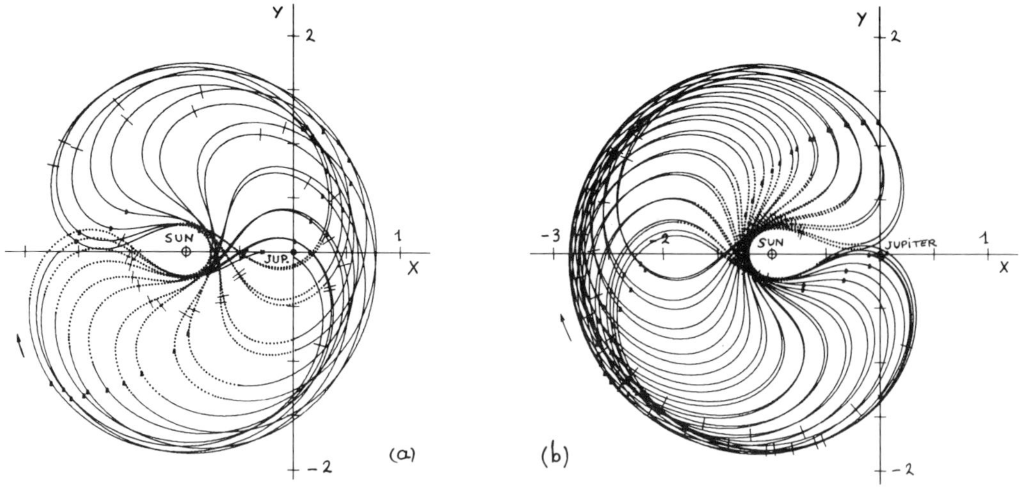

Figure 1. Satellite (a) and anti-satellite (b) motion for P/Boethin in a rotating-pulsating frame. Full (resp. dotted) line: part of the orbit above (resp. below) Jupiter's orbital plane; the perpendicular slashes indicate when the comet is at a maximum distance from this plane; the little open marks indicate when Jupiter is at perihelion or aphelion. The arrow on the left side indicates the direction of motion.

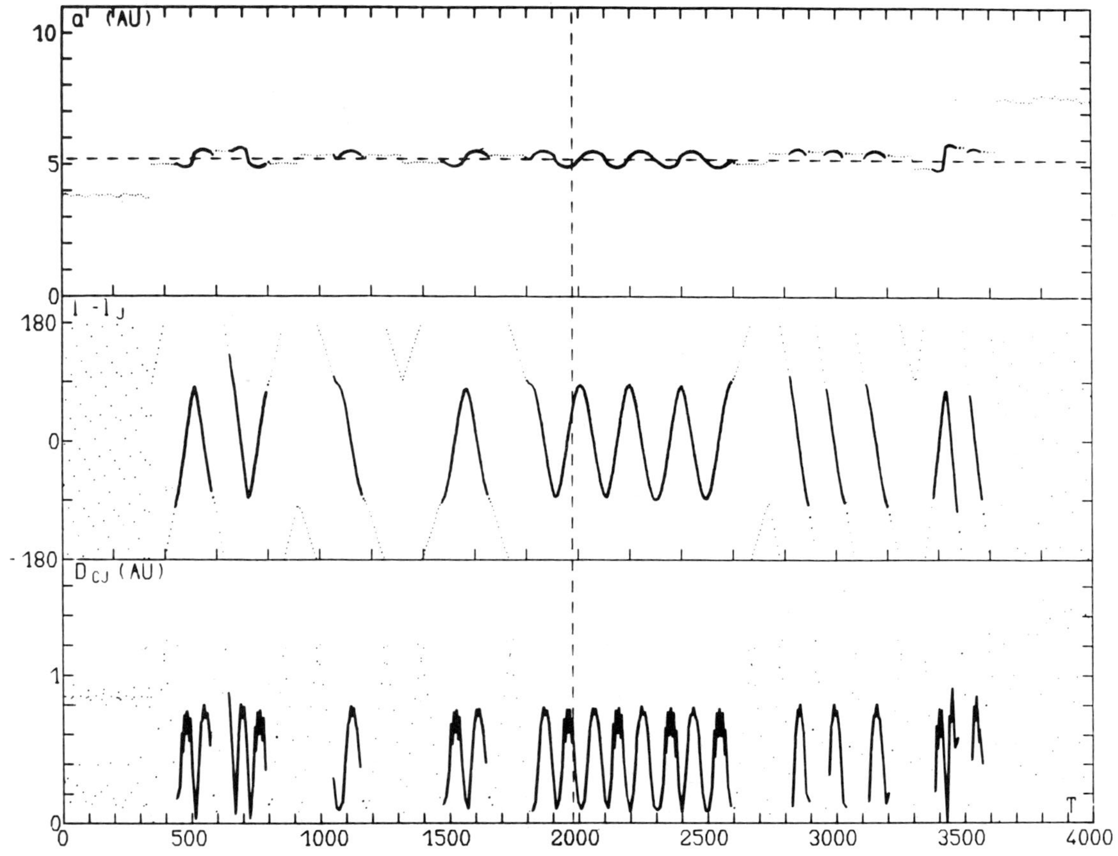

Figure 2. Semi-major axis $\underline{a}$ (in a.u.), mean longitude of the comet $\underline{l}$ minus Jupiter's mean longitude l_J (in degrees) of the orbit of P/Boethin and the minimum distance D_{CJ} between Jupiter and the comet over one period of the comet, from 0 to 4000 A.D. for the central orbit. Full line: satellite libration; dotted line: other types of motion (anti-satellite libration or circulation).

3. RESULTS

All our 25 orbits show temporary satellite librations enclosing the starting date (see Table 1). The minimum duration is about 200 years (for $\Delta P < 0$ and $A_2 < 0$), slightly less than one complete libration period; the maximum of stability is obtained when ΔP and A_2 increase.

A_2 ($\times 10^{-9}$ a.u./d^2) \ ΔP(d)	−16	−8	0	8	16
3.	622–2007 (1400)	1152–1318 1473–1660 1818–> 4000 (> 2200)	1473–1648 1807–> 4000 (> 2200)	283–463 583–761 1013–1164 1401–> 4000 (> 2600)	< 0–3474 3723–3901 (> 3500)
1.5	1710–2005 (300)	1509–1660 1818–2125 (300)	1473–1648 1807–2501 2647–2834 (700)	29–3544 3651–3795 (3500)	965–2975 (2000)
0.	1710–2005 (300)	774–1056 1485–1660 1818–2137 3679–3821 (300)	442–581 652–796 1473–1648 1807–2597 (700)	1236–3366 3521–3688 (2100)	< 0–2975 (> 3000)
−1.5	1485–1662 1807–2005 (200)	1509–1660 1818–2017 2172–2335 (200)	133–452 784–973 1461–1648 1807–2311 (500)	823–962 1332–3462 (2100)	< 0–2975 (> 3000)
−3.	1497–1660 1807–2005 2910–3073 (200)	349–629 905–1660 1818–2017 2160–2326 (200)	345–511 784–962 1461–1648 1807–2299 (500)	1343–2798 3074–3237 (1400)	478–639 752–2975 (2200)

Table 1. Periods of jovian satellite motion for comet P/Boethin, when ΔP and A_2 vary. Underline: periods enclosing the starting date (1974 Dec. 19); parenthesis indicate the duration of these periods.

A_2 ($\times 10^{-9}$ a.u./d^2) \ ΔP(d)	−16	−8	0	8	16
3.	516–2434 3640–3806 (1100)	<0–>4000 (>4000)	1343–>4000 (>2700)	<0–>4000 (>4000)	<0–>4000 (>4000)
1.5	1569–2434 (900)	1104–>4000 (>2900)	1233–>4000 (>2800)	<0–3866 (>3800)	785–3117 (2300)
0.	107–465 831–1281 1569–2375 (1400)	466–2350 3679–3821 (1100)	348–3381 (3000)	1018–>4000 (>3000)	<0–3192 (>3200)
−1.5	1367–2446 (1100)	1317–2446 (1100)	<0–2597 (>2600)	823–3747 (2900)	<0–3105 (>3100)
−3.	1379–3217 (1900)	<0–2422 (>2400)	<0–2540 3735–>4000 (<2800)	620–3521 (2900)	382–3093 (2700)

Table 2. Periods of 1/1 resonance with Jupiter for comet P/Boethin. Same notations as for Table 1.

Moreover, before and after this satellite libration, the comet stays in 1/1 resonance over a longer interval, at least 900 years (see Table 2); during this time, there can be one or several satellite and anti-satellite librations with sometimes intervals of circulation. Generally, the transition between these three types of motion correspond to close encounters with Jupiter, as do the definitive departures from the 1/1 resonance.

As examples, figures 3 a and b show two orbits with a very short interval of 1/1 resonance ($\Delta P=-16$, $A_2=-1.5\ 10^{-9}$) and a very long one ($\Delta P=16$, $A_2=3\ 10^{-9}$).

4. CONCLUSION

We have here confirmed the main result of the previous calculation. We must now determine the interval of time during which the three-body model is sufficient, that is to compare these results with same calculations by four- and nine-body models. In fact, we have as yet compared

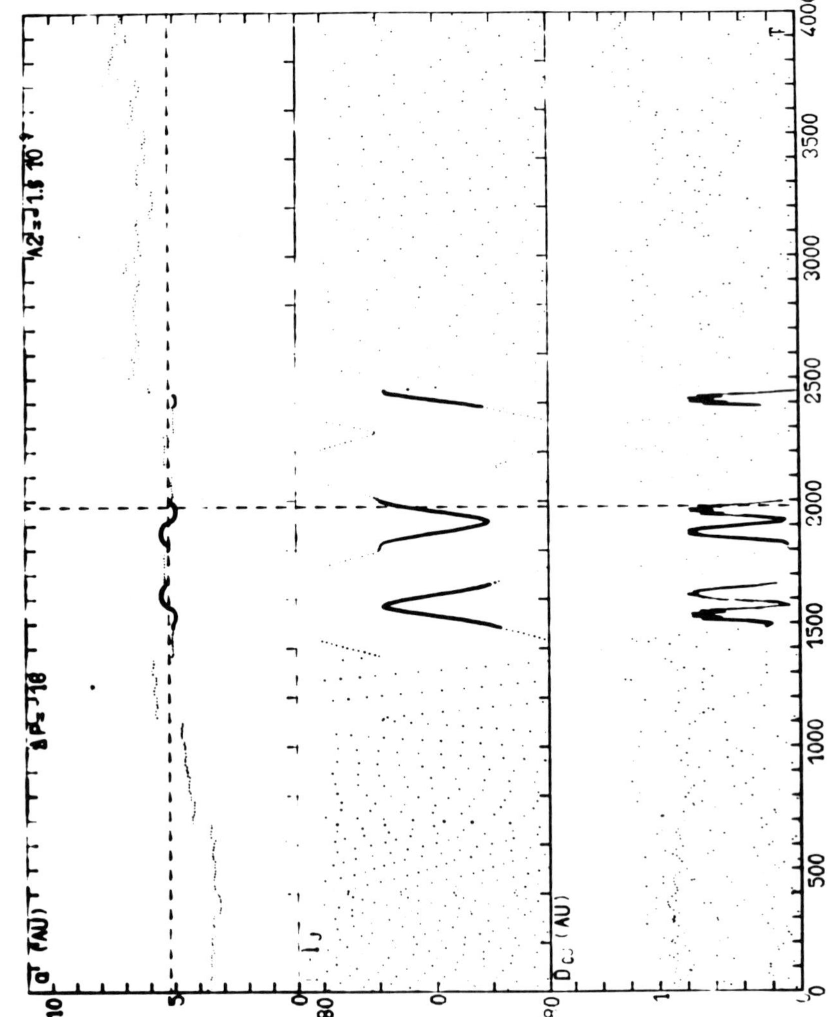

Figure 3a. Elements $\underline{a}$, $l-l_J$ and D_{CJ} for P/Boethin when $\Delta P=-16$ and $A_2= -1.5\ 10^{-9}$, showing a very short interval of 1/1 resonance with Jupiter. Same notations as for Figure 2.

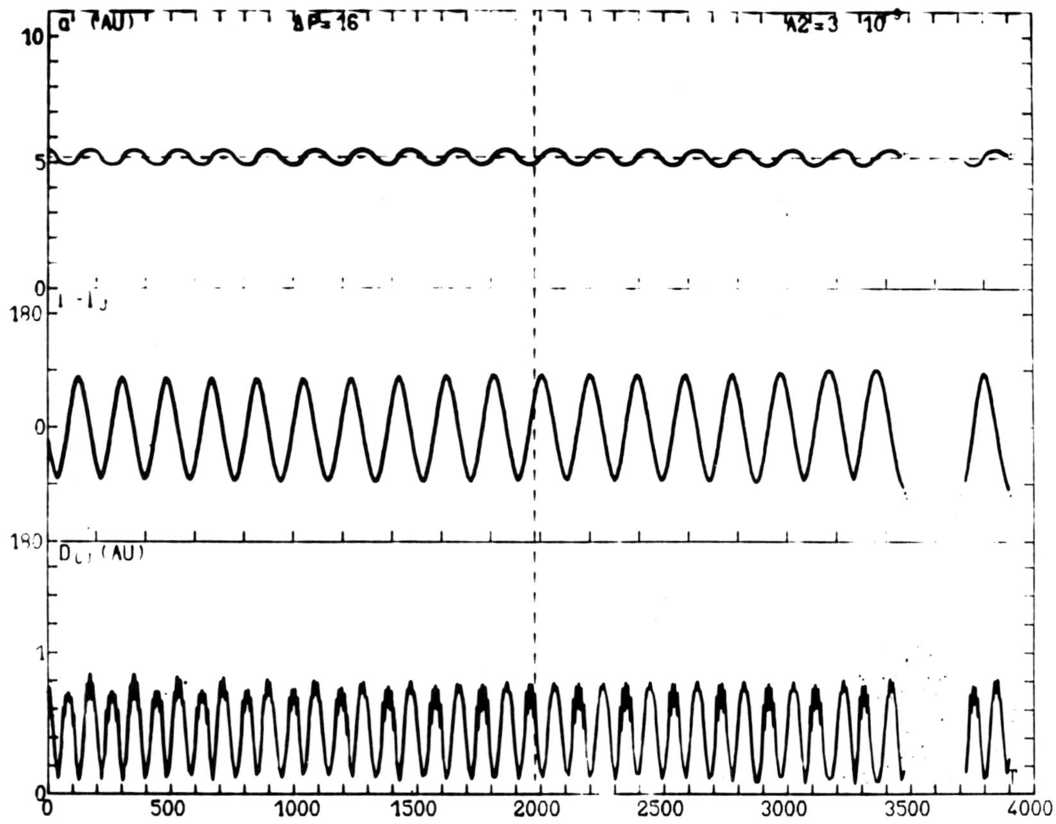

Figure 3b. Elements $\underline{a}$, $l-l_J$ and D_{CJ} for P/Boethin when $\Delta P=16$ and $A_2 = 3\ 10^{-9}$, showing a very long interval of 1/1 resonance with Jupiter. Same notations as for Figure 2.

the results for three- and four-body models for the 9 orbits where ΔP and A_2 are varied separately, and we have found that a moderately close encounter with Saturn in 1836-37 (Benest et al., 1982) causes an appreciable divergence between the orbits before this date; however, for the future, the two models seem in good agreement during a longer interval of time.

Finally, we must wait for the next perihelion passage of the comet in 1986 to establish the real value of ΔP and for yet another perihelion passage ($\sim$ 1997) to determine the value of A_2.

REFERENCES

Benest, D., Bien, R., Rickman, H.: 1980, Astron. Astrophys. 84, pp.L11-L12.
Benest, D., Bien, R., Rickman, H.: 1981, Astr. Ges. Mitt. 52, pp.27-28.
Benest, D., Bien, R., Rickman, H.: 1982, in "Sun and Planetary System", eds. W.Fricke and G.Teleki, Reidel pp.397-400.
Marsden, B.G.: 1974, Ann. Rev. Astron. Astrophys. 12, pp.1-21.
Marsden, B.G.: 1979, "Catalogue of Cometary Orbits", third edition, I.A.U. Central Bureau for Astronomical Telegrams, Cambridge (Mass.).

PART III

ASTEROIDS

FAMILIES OF ASTEROIDS

Yoshihide Kozai
Tokyo Astronomical Observatory, Mitaka, Tokyo 181, Japan

The numbered asteroids are classified into families by a new method proposed by the author(Kozai, 1979a) and their characteristics are studied by several aspects. It seems to the author that there are two kinds of the families, one being very compact in the phase space and containing many faint as well as a few bright asteroids and one being rather loose and bounded by secular and mean motion commensurable regions.

Families of asteroids were discovered by Hirayama(1918) by finding several clusterings of the asteroids with similar osculating values of the semi-major axis, a, the eccentricity, e, and the inclination, i. He then showed that the proper eccentricity and inclination of each member in the same clustering take values more similar to each other. As the semi-major axis as well as the proper elements are stable quantities in the theory of the secular perturbations of the asteroids, he thought that the asteroids in the same clustering have a common origin, and, therefore, the name of the family was given to the clusterings. Later Brouwer(1951) and others(Arnold, 1969; Williams, 1971 and 1979; Lindblad and Southworth, 1971) restudied the family of the asteroids. Except for Williams they tried to group the asteroids into families by using the proper elements which are based on the classical theory of the secular perturbations with the semi-major axis. Williams used his own theory to compute the proper elements.

In the classical theory the eccentricity and the inclination of any of the asteroids as well as those of the disturbing planets are assumed to be small quantities of the order of the square root of the disturbing mass, namely the mass of Jupiter and the terms with factors of the squares of the disturbing masses are neglected in the disturbing function and the orbital elements of the disturbing planets are assumed to be known functions of time, namely the expressions obtained by the classical secular perturbation theory of the major planets. Then the equations of motion are reduced to two independent sets of linear differential equations, one for the eccentricity and the longitude of the perihelion and one for the inclination and the longitude of the node, and the solutions are expressed by the sums of free oscillation and forced oscillation terms which are

V. V. Markellos and Y. Kozai (eds.), Dynamical Trapping and Evolution in the Solar System, 117–122.

produced by the secular perturbation terms in the expressions of the orbital elements of the disturbing planets. The frequencies of the free oscillations for the two sets are the same in the absolute value but opposite in the sense, the sum of the two frequencies vanishing. The proper frequencies represent for most of the cases the secular motions of the longitudes of the perihelion and of the node, respectively, and depend on the semi-major axis of the asteroid only for the linear theory. The amplitudes of the free oscillations are called the proper eccentricity and inclination which are constant in the linear theory.

However, it is evident that when higher order and degree terms are included in the disturbing function the equations of motion for the secular perturbations are no more linear and the sum of the two secular motions does not generally vanish. Since there are many asteroids, for which the eccentricity and/or inclination are as large as 0.1 to 0.2 the effects of the neglected terms in the linear theory are not so small that they cannot be regarded as small corrections for some cases.

The author(Kozai, 1962) showed that when the eccentricity and/or inclination are not small they change with the argument of perihelion appreciably due to the neglected terms in the disturbing function in the linear theory and the amplitudes of the variations are for some cases larger than those of the forced oscillations. The secular motions depend also strongly on the squares of the eccentricity and the inclination. There are some asteroids, for which the arguments of perihelion do not complete revolutions but librate around 90° or 270°, which is not possible in the classical linear theory.

When it can be assumed that the disturbing planets move along circular orbits on the same plane, the value of $[a(1 - e^2)]^{1/2}\cos i$, the z-component of the angular momentum density, is constant. After averaging the equations of motion or the Hamiltonian with respect to the mean longitudes of the asteroid and the disturbing planets the energy integral can be obtained, namely, the Hamiltonian is constant then since the time does not appear explicitly there. The averaging with respect to the longitudes can be done when there is no commensurable relation easily. Since the semimajor axis is constant in the averaged system, the system of the equations is now reduced to that of one degree of freedom with the energy integral, and, therefore, it can be solved by a quadrature mathematically.

The equations of motion show that the value of $(1 - e^2)^{1/2}$ is a periodic function of twice the argument of perihelion. Since $\Theta = (1 - e^2)^{1/2}\cos i$ is constant both the eccentricity and the inclination are also periodic functions of twice the argument of perihelion. The eccentricity is minimum and the inclination is maximum when the argument of perihelion is 0° and 180° and vice versa when it is 90° and 270°. However, when the argument of perihelion librates around 90° or 270°, both the maximum and the minimum of the eccentricity and the inclination take place at 90° or 270° which the argument of perihelion takes twice in one period. The inclination in this paper is referred to the orbital plane of Jupiter which is also assumed to be the common plane for all the disturbing planets.

For all the numbered asteroids the maximum and the minimum values of the eccentricity and the inclination as functions of the argument of perihelion are computed and stored in a computor file. In one of the previous papers(Kozai, 1979a) it was proposed that the semi-major axis, the minimum value of the inclination, i_{min}, and Θ be used as the three parameters to classify the asteroids into families. They are more adequate than the semi-major axis and the two proper elements based on the linear theory, particularly,when the eccentricity and/or the inclination are not small. In fact there are several such families.

It is, again, tried to find several clusterings in the three-dimensional phase space of a, i_{min} and Θ for 2 700 numbered asteroids. In Figure 1 distributions of i_{min} with respect to the semi-major axis are shown. In the figure several clusterings clearly show up. In Table 1 data for nine major families which were found by Hirayama(1918 and 1923) are given. The names of the families were also given by Hirayama. However, Flora family which is originally one family(Hirayama, 1923) is divided here into three families as the original one by Hirayama is very wide.

Table 1. Data for Nine Major Families

Family	Number	a(AU)	Θ	i_{min}	Hirayama	Brouwer
Flora I	66	2.16 - 2.28	0.992 - 0.999	0°.5 - 5°.5		
II	155	2.16 - 2.30	0.983 - 0.992	0.6 - 7.5	63	125
III	143	2.18 - 2.30	0.954 - 0.983	2.0 - 9.6		
Phocaea	52	2.30 - 2.43	0.850 - 0.916	20.3 - 26.0	11	21
Maria	42	2.52 - 2.58	0.938 - 0.975	12.1 - 15.9	14	17
Pallas	9	2.62 - 2.79	0.794 - 0.827	27.9 - 30.6	3	6
Coronis	76	2.83 - 2.91	0.994 - 1.000	1.7 - 2.5	20	33
Eos	160	2.97 - 3.07	0.973 - 0.989	8.0 - 11.3	27	38
Themis	157	3.07 - 3.22	0.976 - 1.000	0.6 - 2.5	32	53

In the column under the heading "Number" the numbers of the member asteroids are given, whereas under the headings "Hirayama" and "Brouwer" the numbers in their works are written. It is clear that the numbers of the members have been increased very much as Hirayama and Brouwer used, respectively, about 950 and 1540 asteroids for their family works.

In the Table 1 the ranges of the semi-major axis, i_{min} and Θ are given for each family and it is clear that the families of Coronis, Eos and Themis are very compact clusterings in the phase space with defined boundaries whereas those of Flora, Phocaea, Maria and Pallas are not. However, for Pallas family, for an example, the density in the phase space is much higher in the family than that in the surroundings since there are only few asteroids with so large inclinations and eccentricities as those of Pallas family asteroids generally. Therefore, the clustering for Pallas family is very remarkable in the phase space. The situations for Flora families are a little different, as the inclinations and the eccentricities for the asteroids in Flora families are rather small. However,there are so many asteroids in Flora families to make very big clusterings.

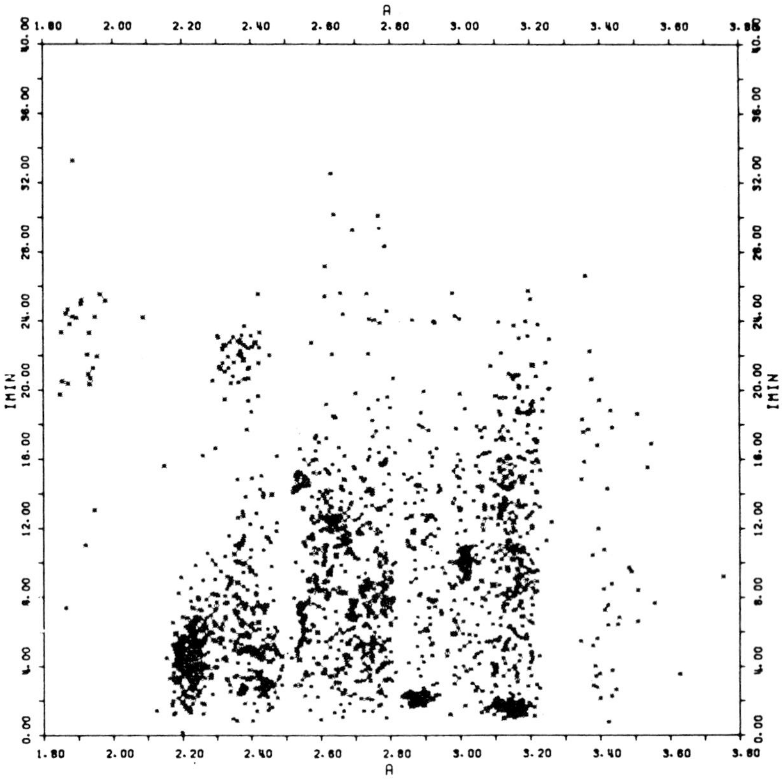

Figure 1. Distribution of i_{min} with respect to the semi-major axis for the numbered asteroids.

In Table 2 the numbers of the member asteroids are given for the seven groups according to their serial registration numbers, the number regions being 1-450, 451-900, 901-1350, 1351-1800, 1801-2250 and 2251-2700.

Table 2. Numbers of the Member Asteroids in Each Family according to the Serial Registration Numbers.

Family	1 - 450	451 - 900	901 -1350	1351-1800	1801-2250	2251-2700
Flora I	6	9	2	15	12	20
II	12	17	23	36	26	40
III	8	14	30	35	32	24
Phocaea	7	3	9	10	17	6
Maria	6	15	7	5	3	6
Pallas	1	2	3	1	1	1
Coronis	8	8	11	16	22	11
Eos	10	26	25	32	31	35
Themis	15	16	23	24	35	43
Total	73	110	133	174	179	186

Table 2 shows that compact and/or large families contain much more faint and small asteroids rather than bright and large ones as asteroids with large serial registration numbers are generally fainter as they could not be discovered by using small telescopes in old days. This tendency is remarkable for Eos, Themis and Flora families as about two fifths of the asteroids numbered in 2251-2700 belong to these families. This may suggest that in such compact and/or large families certainly some kinds of fragmentations of asteroids took place very frequently while in small families like Pallas, Phocaea and Maria fragmentations did not take place so frequently as there are not many faint asteroids there.

It is also true that every major family is near resonant regions. In fact it is possible to find some commensurable ratios for the mean motions with Jupiter corresponding to the boundaries of the families as it is shown in Table 3.

Table 3. Relations between Ratios of the Mean Motions with Jupiter and Boundaries of Families.

Ratio	15:4	17:5	19:6	3:1	17:6	5:2	12:5	7:3	11:5	2:1
a(AU)	2.16	2.30	2.41	2.50	2.60	2.82	2.90	2.96	3.07	3.28
Family	Flora		Phocaea		Maria	Pallas	Coronis		Eos	Themis

Also some of the boundaries in the phase space correspond to secular commensurable regions(Williams and Faukner, 1981). It seems to the author, therefore, that such commensurable regions did have some effects for the fragmentations of asteroids in the places of the families as it is suspected that the motions might be unstable there. In fact an idea of the author on the origin of the families is that originally several regions with high number density of asteroids were created by the resonances of the mean motions and secular motions and since there were much more asteroids collisions to create families took place more frequently than other regions.

Besides the nine families shown here many other families can be found as there are other clusterings in Figure 1. However, for other possible families clear boundaries such as those for the compact families cannot be found. According to criteria assigned to classify the asteroids different families may be found. Also it is necessary to depend on more exact theory of the secular perturbations as well as periodic ones to identify more families. As it was mentioned earlier many families are situated near commensurable regions, either for the mean or secular motions. Therefore, it is very difficult to develop any adequate theory which is valid for such regions. In fact the discussions about the origin of the families need more advanced theories on the motions of asteroids which will be the next important target of celestial mechanics. And any study on the families will contribute much to researches on origins of the asteroids and the solar system.

The computations in this paper were made by using the UNIVAC 1100/80B computer at the Tokyo Astronomical Observatory.

REFERENCES.

Arnold, J.R.: 1969, "Astron. J." 74, p.1235.

Brouwer, D.: 1951, "Astron. J." 56, p.9.

Hirayama, K.: 1918, "Astron. J." 31, p.185.

Hirayama, K.: 1923, "Japan J. Astron. Geophys." 1, p.55.

Hirayama, K.: 1928, "Japan J. Astron. Geophys." 5, p.137.

Kozai, Y. : 1962, "Astron. J." 67, p.591.

Kozai, Y. : 1979a, In "Dynamics of the Solar System" ed. by R.L.Duncombe, Reidel, p.231.

Kozai, Y. : 1979b, In "Minor Planets" ed. by T. Gehrels, University of Arizona Press, p.334.

Lindblad, B.A. and Southworth, R.B.: 1971, In "Physical Studies of Minor Planets" ed. by T. Gehrels(NASA SP-267), p.337.

Williams, J. G. : 1971, In "Physical Studies of Minor Planets" ed. by T. Gehrels(NASA SP-267), p.177.

Williams, J.G.: 1979, In "Minor Planets" ed. by T. Gehrels, University of Arizona Press, p.1040.

Williams, J.G. and Faukner, J.: 1981, "Icarus" 46, p.390.

REGIONS OF STABILITY OF ASTEROIDS

[1]Victor Szebehely, [2]Raimundo Vicente, [3]John Lundberg
[1]L.B. Meaders Professor, University of Texas, Austin
[2]University of Lisbon, Professor
[3]University of Texas, Austin

ABSTRACT. Using Hill's modified stability criterium, regions of orbital elements are established for conditions of stability. The model of the three-dimensional restricted problem of three bodies is used with the Sun and Jupiter as the primaries. Four different cases are studied: direct and retrograde, outside and inside asteroidal orbits. The directions of the asteroidal orbits refer to the synodical reference frame and the positions refer to Jupiter's orbit. The orbital parameters of the asteroids are the semi-major axis (a), the eccentricity (e), and the inclination from Jupiter's orbital plane (i). The effects of the other orbital elements are not investigated in this paper. The argument of the perihelion and the longitude of the ascending node are fixed at $\Omega = \omega = 90^\circ$ and the time of perihelion passage is T = 0 for all orbits.

The aim of this paper is to give quantitative evaluation of the stability of asteroids, the results being also applicable to comets and meteor streams. The evolution of the solar system may be studied using planets, satellites or smaller bodies like asteroids. The unquestionable advantage of approaching the problem via the investigation of asteroids is that there are a very large number of asteroids with well defined orbital elements while the number of planets and natural satellites in the solar system is much smaller.

Establishing regions of stability enhances the location and discovery of additional minor planets. On the other hand, bodies with unstable orbits might be, under certain conditions, available for capture or for significant orbital changes without large artificial perturbations. Furthermore, changes in the observed orbital parameters may change the character of the motion from stability to instability and various evolutionary trends could be observed concerning the solar system.

ANALYSIS

For simplicity's sake the first results are derived for the

V. V. Markellos and Y. Kozai (eds.), Dynamical Trapping and Evolution in the Solar System, 123–136.

two-dimensional case to clarify the underlying ideas. This corresponds to the physical simplification of assuming that the asteroids' orbits are coplanar with Jupiter's orbit. Our final results do include three-dimensional effects.

The key quantity in the analysis is known as the Jacobian constant, given by

$$C = 2\Omega - v^2, \tag{1}$$

where Ω is the dimensionless potential function of the restricted problem in the synodic system and v is the dimensionless velocity of the third particle relative to this system. The potential function is a combination of the gravitational and centrifugal effects and is given by

$$\Omega = \frac{1}{2}(x^2 + y^2) + \frac{1-\mu}{r_1} + \frac{\mu}{r_2} + \frac{1}{2}\mu(1-\mu), \tag{2}$$

where μ is the mass parameter obtained from the masses of the primaries (m_1 and m_2) as follows

$$\mu = \frac{m_2}{m_1 + m_2}, \text{ with } m_2 \leqq m_1 . \tag{3}$$

The distances between the primaries and the third body are r_1 and r_2 . The primaries are located on the axis of syzygies which rotates with the unit angular velocity around the center of mass. The unit of length is the distance between the primaries and the dimensionless time is $t = t^*n$ where t^* the actual time and n is the mean motion of the primaries.

The above-mentioned distances are computed from

$$r_1^2 = (x-\mu)^2 + y^2$$

and

$$r_2^2 = (x-\mu+1)^2 + y^2 .$$

When computing Jupiter's effect on an asteroid, we have

$$\mu = \frac{m_j}{m_s + m_j} = 9.53875 \times 10^{-4} ,$$

using for m_j and m_s the recent values given by Circular No. 163 of the

U.S. Naval Observatory.

The basic idea of establishing regions of stability is to find regions in the orbital plane where motion may take place. For a given asteroid, the value of C is computed using Eqn (1). This requires knowledge of its position and velocity, which might be obtained from its orbital elements. Once the actual value of the Jacobian constant C_{ac} of the asteroid is known, this is compared with the critical value C_{cr} at which the forbidden regions of motion change. These critical values are described in considerable detail in the literature (see for instance, Szebehely 1967) and tabulated values are available. If the actual value of the Jacobian constant is much higher than the critical value, then the motion is confined to a well defined region of the plane, that is, no exchanges or escapes are possible. The forbidden and allowable regions of motion are separated by the so-called curves of zero velocity which close or open up at the critical values of the Jacobian constant. These curves intersect the axis of syzygies at the collinear equilibrium points located on both sides of Jupiter at a dimensionless distance of approximately $(\mu/3)^{1/3} \simeq 0.06825$. The third collinear equilibrium point is located approximately at a unit distance from the sun on the opposite side from Jupiter. For large values of C_{ac} we have three distinct and separate regions for possible motion. One is outside the Sun-Jupiter system which region can not be penetrated by asteroids originally moving in this outside regions. Such asteroids cannot be captured by either the Sun or by Jupiter. The second region encloses the Sun and an asteroid in this region can not be captured by Jupiter nor can it leave the system. The third region is around Jupiter and an asteroid (or a satellite of Jupiter) can not leave Jupiter's neighborhood, cannot become a minor planet governed by the Sun and cannot escape the system. This is the case when $C_{ac} > C^1_{cr} \simeq 3.03971$. If $C_{ac} < C^1_{cr}$ the inside two regions (around the Sun and around Jupiter join forming a region where the body might join or leave the Sun and/or Jupiter. This happens when $C^1_{cr} > C_{ac} > C^2_{cr} \simeq 3.03844$. The outer region still remains separated from the inner region and neither penetration nor escape is possible. When $C^2_{cr} > C_{ac}$ the outer region joins the inner region and outside particles may penetrate and inside particles may escape. Further reduction of the value of $C_{ac} \leq 3$ eliminates all forbidden regions and in Hill's sense there is no stability. Since the values of C^1_{cr} and C^2_{cr} are close and anytime $C_{ac} < C^2_{cr}$ we have the possibility of exchange, of communication, of penetration or of escape we select $C_{cr} = C^2_{cr}$ as our critical value to determine stability and introduce a measure of stability by the equation

$$S = \frac{C_{ac} - C_{cr}}{C_{cr}} , \qquad (4)$$

where $C_{cr} = 3.03844$ corresponds to the collinear equilibrium point usually denoted by L_1 and located close to Jupiter on the opposite side from the Sun.

The next step is to evaluate the actual value of the Jacobian constant for an asteroid. Given the semi-major axis and the eccentricity of the orbit, the relative velocities are computed at the perigee-distance, $a(1-e)$ or at the apogee-distance, $a(1+e)$ with respect to the Sun, assuming, once again for simplicity's sake that these points are on the axis of syzygies. The Jacobian constant becomes

$$C = \frac{1}{a} \pm 2\,[a(1-e^2)]^{\frac{1}{2}} \quad , \tag{5}$$

once the appropriate substitutions are made into Eqn (1). Note that the above equation applies at the perigee and uses two-body approximations because of the small value of μ. The + sign refers to direct and the - sign to retrograde orbits. For the straightforward but tedious derivation see for instance, Szebehely (1967). For circular orbits the Jacobian constant is

$$C = \frac{1}{a} \pm 2a^{\frac{1}{2}} \quad . \tag{6}$$

If the limiting critical value, C_{cr} is substituted for C and the resulting, essentially cubic equation is solved for a we obtain the limiting value(s) of a for direct and for retrograde orbits. These values are $a = 0.80438$ for direct and $a = 0.24786$ for retrograde orbits. If $C_{ac} > C_{cr}$ the orbit is stable, consequently, the next step is to investigate the effect of changes in the quantity a on the Jacobian constant. Since

$$\frac{dC}{da} = -\frac{1}{a^2} \pm \frac{1}{a^{\frac{1}{2}}} \tag{7}$$

we see that for retrograde orbits $dC/da < 0$, for any value of a. Therefore if a is increased above the previously given value ($a = 0.24786$) C_{ac} will be smaller than C_{cr} and instability will set in. For example, if $a = 0.3$, the actual value of the Jacobian constant becomes $C_{ac} = 2.2379$ which results in a negative value for the measure of stability and according to Eqn (4) it becomes $S = -0.2635$, indicating instability. Similar analysis may be performed for direct orbits.

Taking the partial derivative of Eqn (5) with respect to the eccentricity we may once again establish its role. This is left to the reader since our figures and tables given in the next chapter, clearly demonstrate the effects of changes in the eccentricity. The only remark to be made is that $\partial C/\partial e > 0$ for retrograde and negative for direct orbits.

The three-dimensional effects are brought into the picture by using the zero-velocity surfaces instead of the previously mentioned zero-velocity curves of the restricted problem of three bodies. The Jacobian integral is identical in form in the two and three-dimensional cases, that is, Eqn (1) is still applicable with

$$v^2 = \dot{x}^2 + \dot{y}^2 + \dot{z}^2 ,$$

$$r_1^2 = (x-\mu)^2 + y^2 + z^2 ,$$

and

$$r_2^2 = (x-\mu+1)^2 + y^2 + z^2 .$$

The method used is identical to the two-dimensional case excepting the complications of the topology of the three-dimensional limiting surfaces.

RESULTS

The results of the computations are represented in Figures 1 to 3 and in the corresponding Tables 1 to 3. Figure 1 represents direct inner orbits. The parameter is the semi-major axis, the value of which is shown on the curves, using astronomical units (A.U.). Note that the unit of distance in the discussion of the previous chapter was the Sun-Jupiter distance, but the Figures and Tables give results in A.U.-s to comply with astronomical conventions.

Consider Figure 1 and the corresponding Table 1. For zero eccentricity and inclination, the previous section gave for the limiting semi-major axis $a = 0.80438$. Since we now use A.U.-s we multiply this value by the Sun-Jupiter distance or by 5.2028 A.U. and obtain 4.185 A.U., corresponding to the origin of Figure 1, i.e., to the point $e = i = 0$. The same value is shown as the first entry in Table 1.

As an example consider asteroid No. 25 (Phocaea) with $i = 21^\circ\!.59$, $e = 0.254$ and $a = 2.4$ A.U. Locating the point corresponding to the given i and e values our Table 1 or Figure 1 gives $a \simeq 3.2$ (see point A). Consequently, if Phocaea's semi-major axis would be larger than 3.2 its orbit would show instability. In fact, its semi-major axis is 2.4, consequently, its orbit is stable, and its measure of stability as defined by Eqn (4) is $S = 0.119$. The charts, of course may be used for finding limiting values of i and e. Moving to the curve corresponding to $a = 2.4$ on Fig. 1 we see that if the eccentricity of Phocaea would be as high as 0.5 (see point B), then its limiting inclination should be less than $42^\circ\!.50$. In other words, for a given semi-major axis higher eccentricity and higher inclination reduce the stability, as expected. In fact, the non-existing asteroid mentioned above and marked by B on Fig. 1 has a measure of stability $S = 0$, while asteroid #25 is stable.

In general, therefore, the use of Fig. 1 (or Table 1) is to locate the point for a given (e,i) set and read off the corresponding value of the limiting semi-major axis, a^*. If the minor planet's actual

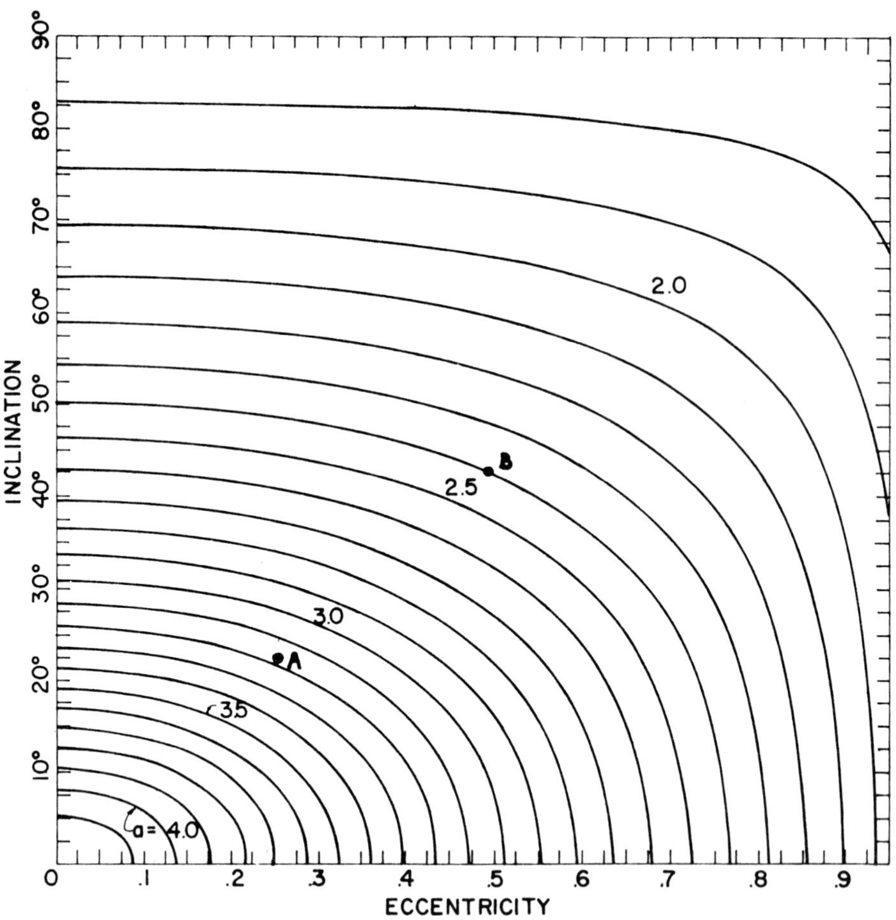

FIGURE 1. DIRECT INSIDE ORBITS

TABLE 1. LIMITING VALUES OF SEMI-MAJOR AXES FOR DIRECT INNER ORBITS

	i° = 0	10	20	30	40	50	60	70	80	90
e = 0.0	4.185	3.919	3.455	3.034	2.686	2.404	2.176	1.989	1.837	1.712
.1	4.083	3.850	3.420	3.015	2.675	2.398	2.172	1.988	1.836	1.712
.2	3.849	3.677	3.322	2.960	2.644	2.380	2.162	1.982	1.834	1.712
.3	3.572	3.453	3.180	2.875	2.594	2.350	2.145	1.973	1.831	1.712
.4	3.296	3.214	3.013	2.767	2.526	2.309	2.121	1.961	1.826	1.712
.5	3.033	2.978	2.832	2.643	2.444	2.257	2.090	1.944	1.819	1.712
.6	2.787	2.749	2.647	2.506	2.350	2.195	2.052	1.923	1.810	1.712
.7	2.554	2.529	2.460	2.359	2.243	2.122	2.005	1.896	1.799	1.712
.8	2.327	2.312	2.268	2.203	2.123	2.036	1.948	1.863	1.784	1.712
.9	2.094	2.086	2.063	2.027	1.981	1.929	1.874	1.818	1.763	1.712

semi-major axis, a is smaller we have stability and if $a > a^*$ we have instability. A new measure of stability could be established by evaluating the ratio $S' = (a^* - a)/a(0,0)$, where $a(0,0) = 4.185$. This measure for Phocaea becomes $(3.2 - 2.4)/4.185 = 0.167$. It is not recommended that S' should replace S as given by Eqn (4) since this latter has more general applications.

Another observation of general validity may be made from Fig. 1 (or from Table 1). Considering a fixed value for the semi-major axis we see that small eccentricity allows higher inclination, while high eccentricity requires low inclination for stability. Furthermore, as the value of the semi-major axis decreases larger eccentricities and larger inclinations are allowed.

Finally, we note that some of the constant semi-major axis curves may be approximated by elliptic arcs, especially, in the middle range of the chart.

Figure 2 and Table 2 represent retrograde orbits inside the limiting zero-velocity surface. The topology of the zero velocity surfaces is identical to the case discussed in connection with direct orbits but the evaluation of the actual values of the Jacobian constants is significantly different, as shown by the ± signs in Eqn (5). The origin of Figure 2 corresponding to $e = i = 0$ is $a = 0.24786 \times 5.2028$ A.U. = 1.2896 A.U. and the semi-major axis increases with increasing eccentricity and inclination. The reason for this is that, as mentioned before, $\partial C/\partial e > 0$ for retrograde orbits.

Comparing Figs. 1 and 2 we observe the low values of the semi-major axes for retrograde orbits indicating that asteroids in retrograde orbits must be closer to the Sun than those in direct orbits. This is the consequence of using Hill's criterion for stability and it should be kept in mind that this criterion is a necessary but not sufficient condition. In previous papers the difference between linearized stability investigations and Hill's method have been subjected to analysis (Szebehely, 1978). The linear analysis shows considerably higher stability for retrograde orbits than Hill's method.

Figure 3 and Table 3 refer to outer direct orbits. These are asteroids or meteoroids outside of Jupiter's orbit and the limiting value for the semi-major axis may be computed from Eqn (6). We use the plus sign and find the solution to the equation

$$C_{cr} = \frac{1}{a} + 2a^{\frac{1}{2}} \tag{8}$$

for $a > 1$. Note that the solution of this equation for $a < 1$ was already established before ($a = 0.80438$ or $a = 4.185$ A.U.). The solution of Eqn (8), using once again, $C_{cr} = 3.03844$ is $a = 1.25$ or $a = 6.504$ A.U. If the semi-major axis (for $i = e = 0$) is larger than this value we have $C_{ac} > C_{cr}$ and stability occurs. It is important to

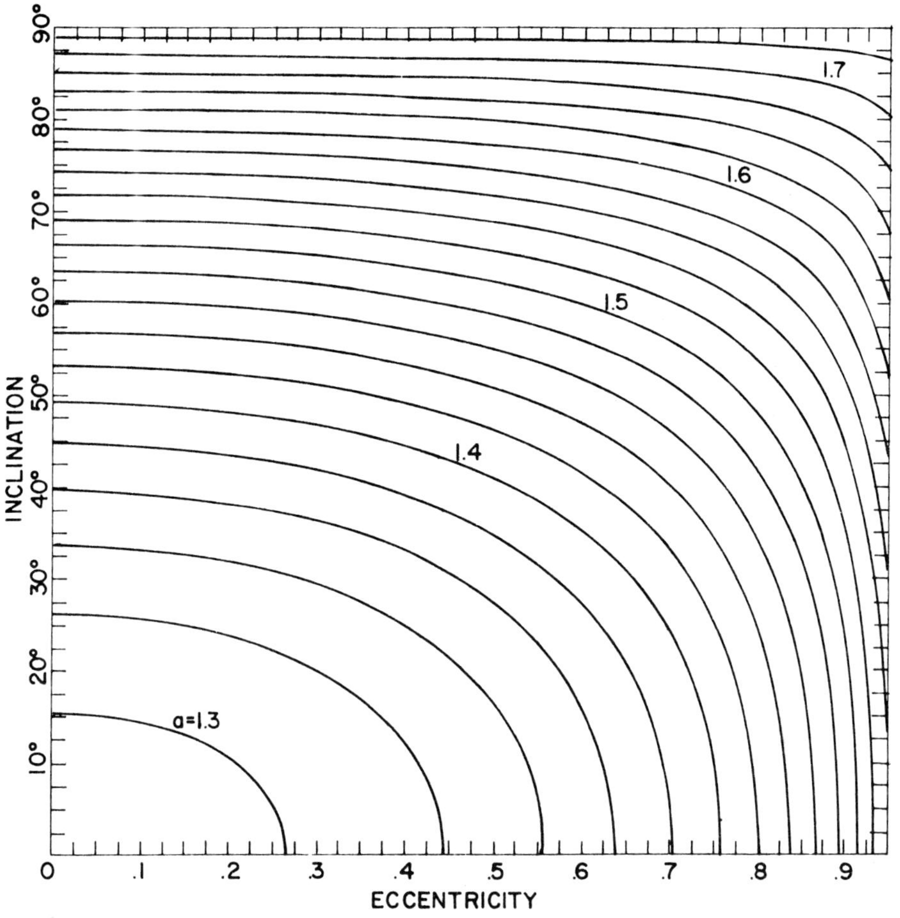

FIGURE 2. RETROGRADE INSIDE ORBITS

TABLE 2. LIMITING VALUES OF SEMI-MAJOR AXES FOR RETROGRADE INNER ORBITS

	i° = 0	10	20	30	40	50	60	70	80	90
e = 0.0	1.290	1.294	1.307	1.329	1.361	1.404	1.458	1.526	1.610	1.712
.1	1.291	1.295	1.308	1.331	1.362	1.405	1.459	1.527	1.610	1.712
.2	1.295	1.300	1.313	1.335	1.366	1.408	1.462	1.529	1.612	1.712
.3	1.303	1.307	1.320	1.342	1.373	1.414	1.468	1.533	1.614	1.712
.4	1.314	1.318	1.331	1.352	1.383	1.423	1.475	1.539	1.618	1.712
.5	1.329	1.333	1.346	1.366	1.396	1.436	1.486	1.548	1.623	1.712
.6	1.350	1.354	1.366	1.386	1.415	1.452	1.500	1.559	1.629	1.712
.7	1.379	1.382	1.394	1.413	1.440	1.475	1.519	1.573	1.637	1.712
.8	1.419	1.423	1.433	1.450	1.475	1.506	1.546	1.593	1.649	1.712
.9	1.485	1.487	1.496	1.510	1.530	1.555	1.587	1.623	1.665	1.712

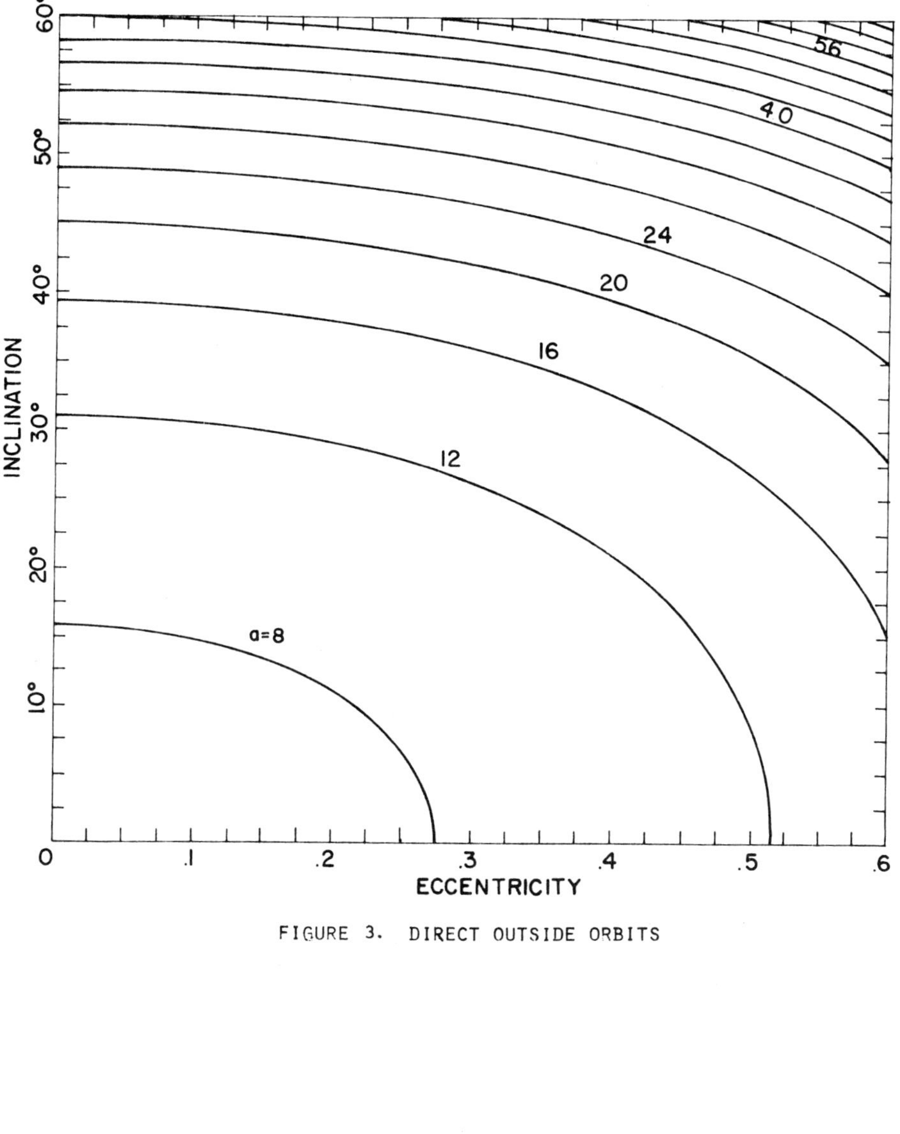

FIGURE 3. DIRECT OUTSIDE ORBITS

TABLE 3. LIMITING VALUES OF SEMI-MAJOR AXES FOR DIRECT OUTER ORBITS

i° =	0	10	20	30	40	50	60
e = 0.0	6.504	7.138	8.800	11.601	16.347	25.137	44.142
.1	6.727	7.332	8.974	11.784	16.568	25.445	44.650
.2	7.338	7.893	9.498	12.342	17.245	26.382	46.186
.3	8.294	8.818	10.411	13.336	18.461	28.071	48.952
.4	9.658	10.179	11.813	14.899	20.391	30.763	53.365
.5	11.619	12.167	13.918	17.294	23.379	34.949	60.236
.6	14.592	15.207	17.195	21.074	28.136	41.642	71.243
.7	19.571	20.323	22.762	27.560	36.353	53.244	90.352

clarify the meaning of Hill's stability criteria for inside and for outside orbits. If a direct inside orbit has larger semi-major axis than 4.185 A.U. its Jacobian constant is below C_{cr} and the separation of the outside and inside regions ceases to exist. In other words, the asteroid will not be confined to a closed region including the Sun and Jupiter. On the other hand, if an outside direct orbit has smaller semi-major axis than 6.504 A.U. then, once again the inside and outside regions may communicate and the orbit of the outside asteroid may enter the region including the Sun and Jupiter, in fact it may be captured by Jupiter. So Hill's criteria for "inside" and "outside" are quite different. Stability for inside asteroids means they cannot leave a region including the Sun and Jupiter, while stability for outside asteroid means they cannot enter the same region. This explains the different trends shown on Figs. (1) and (3). The larger the semi-major axis is of an outside asteroid, the larger its inclination and eccentricity may be without instability setting in (see Fig. 3). The opposite is the situation for inside orbits as we have seen on Fig. 1. (Note that other perturbations such as Saturn's for outside and Mars' for inside orbits are not included in these considerations.)

Finally, a discussion is offered of retrograde outside orbits. These orbits are all unstable according to Hill's definition since Eqn (6) with a negative sign does not have real solutions for $a > 1$. In other words, all outside retrograde orbits may enter the inner region since their Jacobian constant is smaller than C_{cr}. (Note that this does not apply to inside retrograde orbits since Eqn (6) has a real solution (using the negative sign) for $a < 1$.

Further consequences (and in some respect disadvantages) of Hill's method are that stability for inside orbits means that they cannot escape the Sun-Jupiter system. Outside orbits may escape but are not allowed to enter the inside regions. In terms of instability this means that if an inside orbit is unstable it may in fact escape the system and an unstable outside orbit may enter the Jupiter-Sun region. Stable inside bodies may have chaotic orbits, may be captured by the Sun or by Jupiter, may have figure-eight orbits, etc., but cannot leave the region. Stable outside orbits, once again may take any shape (including escape) as long as they do not enter the region of the inside orbits.

CONCLUSIONS

Well defined regions of stability and instability of minor planetary orbits are established using Hill's stability criterion. For direct inside orbits the maximum value of the semi-major axis is 4.185 A.U. when the eccentricity and inclination are zero. For retrograde inside orbits the minimum value of the semi-major axis is 1.290 A.U., again for $e = i = 0$. For outside direct orbits, the minimum value of the semi-major axis is 6.504 A.U. for $e = i = 0$. There are no stable outside retrograde orbits.

For given orbital parameters (a,e,i) and direction of motion the Hill-type stability of minor planets or of meteor-streams can be determined by the tables and charts given.

ACKNOWLEDGEMENTS

Support of the University of Texas Research Institute, of the National Aeronautics and Space Administration and of the Scientific Affairs Division of NATO are gratefully acknowledged. The authors wish to express their appreciation to Drs. Y. Kozai and H. Scholl for their comments.

REFERENCES

Hill, G. (1978), American Journal of Mathematics, Vol.1, p.5, p.129 and p. 245.

Szebehely, V. (1967), "Theory of Orbits", Academic Press, New York.

Szebehely, V. (1978), Proceedings of the National Academy of Sciences, Vol. 75, p.5743.

THE STABILITY OF SOME ASTEROIDS

R.O. Vicente
Faculty of Sciences, University of Lisbon, Portugal

ABSTRACT

The utilization of two different stability criteria, namely, Hill´s modified stability criterium and the method of surface of section, has been employed for asteroid orbits. The idea is to compute different criteria of stability for the same asteroids in order to compare the results and see the practical interest of the computations for researches about evolutionary trends of individual asteroids, groups and families of asteroids.

1. INTRODUCTION

This paper is an application of the criteria of stability defined and employed in the papers by Szebehely *et al* (1983),and Hadjidemetriou and Ichtiaroglou (1983); it is the result of discussions among the authors previously mentioned.

It is well known there are several definitions of stability in celestial mechanics but, unfortunately, some of them are not easy to apply to practical cases. Also, it would be convenient to try to verify the conclusions of different criteria of stability when applied to the same bodies of the solar system.

As pointed out by Vicente (1979) it is advantageous to apply different criteria of stability to the asteroids because they are far more numerous than planets and natural satellites, and, therefore, there is greater likelihood to find cases of instability among the asteroids. Unfortunately, there are only about 2,700 asteroids with reliable computed orbits.

We shall consider some well known asteroids, computing the measure of stability S, introduced by Szebehely (1978) and corresponding to the criteria defined by Hill (1878). The results obtained are confirmed by the computations of Hadjidemetriou and Ichtiaroglou (1983) whenever possible.

V. V. Markellos and Y. Kozai (eds.), Dynamical Trapping and Evolution in the Solar System, 137–139.

2. PRACTICAL APPLICATIONS

The Trojans group, east or west of Jupiter, present S slightly negative ($S < 0$); they correspond to a resonance of 1/1 with libration cycles of about 150 years. Considering this result they might escape from their present positions.

In the Hilda group some (Hilda, Ismene, Bononia) show $S < 0$ while others (Chicago, Normannia) show $S > 0$; they correspond to a resonance of 2/3 with libration cycles of 250 to 300 years. The resonance 2/3 is represented by branches B_1 and B_2 of Fig. 1 of Hadjidemetriou and Ichtiaroglou (1983).

The Hecuba group shows $S > 0$ corresponding to a resonance 1/2, and can be identified on the stable branch A_1 of Fig. 1 (Hadjidemetriou and Ichtiaroglou, 1983). The Hestia type of asteroids also present $S > 0$ and correspond to stable orbits with non-zero eccentricity at resonance 1/3.

The Hirayama families of Themis, Eos and Koronis show positive values of S and correspond to stable resonant orbits. There is therefore agreement between the two criteria of stability and we can interpret that as corresponding to stable families of asteroids.

There are different opinions not only about the existence of certain families of asteroids but also on the advantages or disadvantages of grouping asteroids into families. There is no doubt that, at least certain families, do have dynamical and evolutionary meaning, and the criteria of stability employed in the present paper corroborate the existence of such families, showing stable values.

The regions exemplified by the Hungaria and Phocaea asteroids present positive values of S, having non-zero eccentricities, and they are isolated by resonances as it was shown by Brouwer and van Woerkom (1950).

At the resonance 1/3, asteroid Alinda shows $S > 0$, having an eccentricity of 0.543 and appearing on the stable family of periodic orbits. It crosses the orbit of Mars and is a known librator.

The regions corresponding to the asteroids Maria and Flora, sometimes considered as families and other times not considered as families, present positive values of S. Following our criteria these regions are stable and, for that reason, there are numerous asteroids in those regions. Kozai (1983) has proposed better parameters, among the orbital elements of the asteroids, for characterising asteroid families, for instance the Flora and Maria families.

Among the group of Earth crossing asteroids, we mention some of the Apollo group, for instance, Icarus and Geographus showing $S < 0$. The groups represented by Amor and Betulia also show $S < 0$. The case of Betulia is an interesting one because it might be one of the candidates for a cometary origin. Eros, also approaching the Earth, and which was extensively observed for astrometric purposes, has $S < 0$ but with smaller value than the Apollo and Amor groups.

Considering the asteroids with orbits going outside Jupiter´s orbit, we mention Hidalgo and Telamon, both having $S < 0$, but Hidalgo presenting a larger negative value because its inclination is $42°$ while Telamon´s is only $6°$. This, again, justifies the conclusions presented by Szebehely *et al* (1983) that greater inclinations correspond to greater degrees of instability. The same conclusions can be inferred for greater values of the eccentricity. Chiron is another interesting case because $a = 13.69$ and the computed value of S is negative. It has been shown that it corresponds to a chaotic orbit, and, therefore, our conclusions are in agreement.

3. CONCLUSIONS

Laplace (1798) was impressed by the regularities of the solar system, that is, small inclinations and eccentricities, and, about 200 years later, we can prove, thanks to the asteroids, that these features correspond to greater stability and, therefore, are more prominent in the solar system.

We must remember that present day theories of the motions of the asteroids are not as well developed as for other bodies of the solar system, namely, the planets. One reason for that is the comparative neglect of the systematic study of the orbits of asteroids which is unfortunate because they show far greater variability in their motions than other bodies of the solar system. As a consequence, some families of asteroids are not yet so well defined and we cannot infer many conclusions about possible evolutionary trends of the solar system.

REFERENCES

Brouwer, D. and van Woerkom, A.J.J.: 1950 Astron. Papers Am. Ephem. 13, part II, pp. 81-107

Hadjidemetriou, J.D. and Ichtiaroglou, S. : 1983, This volume , p. 141

Hill, G.W.: 1878 Am. J. Math. 1, pp. 5-26, 129-147, 245-260

Kozai, Y.: 1983, This volume , p. 117

Laplace, P.S.: 1798,Exposition du système du monde, 2nd ed., Paris

Szebehely, V., Vicente, R and Lundberg, J.: 1983, This volume , p. 123

Vicente, R.O.: 1979 in V.G. Szebehely (ed.) "Instabilities in Dynamical Systems", D. Reidel Publ. Co., Dordrecht, Holland, pp. 211-225

ON THE STABILITY OF RESONANT ASTEROID ORBITS

J. D. Hadjidemetriou and S. Ichtiaroglou
University of Thessaloniki, Thessaloniki, Greece

ABSTRACT. The stability of the asteroid orbits has been studied by the method of surface of section. Families of simple symmetric periodic orbits of the asteroid and their stability have been computed and this served as a guide for the selection of the energy levels for the surface of section. In this way all possible cases for the structure of phase space have been obtained. It was found that the region in phase space around the resonant orbits at the resonances 1/3, 3/5, 5/7,.... is unstable, but small stability regions of doubly symmetric periodic orbits near the above resonances are also present. At the resonances 1/2, 2/3, 3/4, ... it was found that there exist two separate regions in phase space at about the same resonance 1/2, 2/3, 3/4,...., respectively, one being stable and the other unstable. At certain energy levels only the stable region appears. The above results are consistent with the observed distribution of the asteroids.

1. INTRODUCTION

The purpose of this paper is to study the stability of the asteroid orbits and in particular those orbits whose mean motion is in resonance with that of Jupiter. It is well known that the distribution of the asteroids is not smooth but gaps exist at some resonances, the most conspicuous being at 1/3, 1/2, 3/5 and also, to a lesser extend, at 2/5, 3/7. These are the well known Kirkwood gaps whose explanation is not yet clear despite the fact that much work has been done. Several mechanisms have been proposed for the explanation of the Kirkwood gaps, but recent work based on statistical analysis (Dermott and Murray, 1981) supports the gravitational hypothesis, i.e. that the gaps are due to instabilities in the asteroid orbits produced by the gravitational perturbation of Jupiter.

Several papers have been published, in which the gravitational hypothesis is studied, both from the analytical and the numerical point of view. A review of this work is made by Hagihara (1972). We note in this respect that the appearance of small divisors in the resonant cases

V. V. Markellos and Y. Kozai (eds.), Dynamical Trapping and Evolution in the Solar System, 141–152.

cannot explain the gaps since there do exist groups of resonant asteroids, for example the Hecuba group and the Hilda group at the 1/2 and 2/3 resonances, respectively. On the other hand, numerical integrations at or near the resonant cases do not always produce the gaps that we would expect (Froeschlé and Scholl, 1974, 1975, 1979, Lecar and Franklin, 1973, Sinclair, 1969, 1970). So, the explanation of the Kirkwood gaps is still an open problem.

In this paper we present a global view for the totality of asteroid orbits. The orbit of Jupiter will be considered as circular and the orbits of the asteroids will be considered as coplanar with Jupiter, moving in the same direction. This will be done by giving the structure of the phase space by the method of surface of section. In this way the stable and unstable regions will be presented clearly and the relation between the various resonant cases with stability or instability and the corresponding generation of gaps will become evident.

A different approach to the study of the stability of the asteroid orbits can be made by using Hill's stability criterium. This has been done by Szebehely, Vicente and Lundberg, 1983.

We present here the qualitative aspects and the main results of this work. The complete work with all the numerical results will be presented elsewhere.

2. FAMILIES OF PERIODIC ORBITS

The periodic orbits and their stability characteristics determine criticaly the structure of the phase space. For this reason we present here families of periodic orbits of asteroids, moving under the gravitational attraction of the Sun and Jupiter. The mass of the asteroid will be considered negligible and Jupiter will be assumed to describe a circular orbit around the Sun. The orbit of the asteroid will be considered with respect to a rotating frame whose origin is at the center of mass of Sun-Jupiter and the x-axis is the line from Sun to Jupiter. This is the well known circular restriced 3-body problem. We shall study planar motion only and the orbit of the asteroid will be considered inside the orbit of Jupiter and moving in the same direction.

Families of periodic orbits for the asteroid, in the planar circular restricted 3-body problem have been computed by Colombo, Franklin and Munford (1968) and Broucke (1968). We have recomputed these families to a high accuracy in order to have reliable results for the stability, especially at some critical cases, which have been predicted analytically (Hadjidemetriou, 1982b). The value of the small parameter $\mu=m_j/(m_s+m_j)$ is taken equal to $\mu=0.001$, where m_s, m_j are the masses of the Sun and Jupiter, respectively.

The above families are the continuation, for $\mu\neq 0$, from families of periodic orbits of the asteroid, in the rotating frame defined above, for the case $\mu=0$ (no perturbation from Jupiter). We describe now briefly the families for $\mu=0$:

a). There exists a family of circular orbits, in the rotating frame, symmetric with respect to the x-axis. At t=0 we have $y=\dot{x}=0$, so the initial conditions of such a periodic orbit are the values of x and $\dot{y}$ at t=0. Instead of $\dot{y}$ we shall use the value C of the Jacobi constant (e.g. Szebehely 1967). This family is shown in Fig. 1. The normalization of the variables is such that the mean motion of Jupiter is n=1. Along this family the value n/n' varies, and several resonant (unperturbed) orbits exist, where n' is the mean motion of the asteroid.

b). There exist families of elliptic orbits for the asteroid, which bifurcate from the circular resonant orbits 1/2, 2/3, 3/4, ... These are simple periodic orbits, i.e. they close after the first intersection with the x-axis. From Keplerian theory it can be proved that these elliptic families (Fig.1) are symmetric with respect to a line normal to the x-axis, passing through the corresponding circular resonant orbit. All along such a family, the resonance is constant, equal to the resonance of the generating circular orbit.

c). There exist also families of elliptic multiple periodic orbits (for $\mu=0$) which bifurcate from resonant circular orbits at the resonance p/q ($q-p\neq1$), in the same way as the families described in (b) above. The multiplicity of such an orbit is equal to q-p. It is clear that a dense set of such resonant orbits exists, though of measure zero.

When the perturbation from Jupiter comes into effect, i.e. for $\mu\neq0$, the above families are continued to a set of families of periodic orbits of the restricted circular 3-body problem, as shown in Fig. 1. The existence proof for the continuation of the circular orbits is given by Birkhoff (1927) and for the elliptic orbits by Arenstorf (1963) and Schmidt (1972a). Also, the form of the continued orbits at the vicinity of the resonant circular orbits 1/2, 2/3, 3/4, ... has been studied by Guillaume (1969) and Schmidt (1972b).

We note that there exist several distinct families of periodic orbits for $\mu\neq0$, three of them shown in Fig. 1. Each family has a part which is the continuation of the circular orbits for $\mu=0$ (periodic orbits of the first kind) and two branches (one for the first family, to the lower left of the Figure) of resonant elliptic orbits at the resonances 1/2, 2/3, 3/4, ... Along these latter branches the eccentricity increases as we move away from the resonant circular orbit 1/2, 2/3,... .

In particular, we have two resonant elliptic branches at the 1/2 resonance, denoted by A_1 and A_2 respectively. The branch A_1 corresponds to a mean motion n' of the asteroid such that $n/n'\simeq1/2$ but $n/n'<1/2$ and the branch A_2 corresponds to $n/n'\simeq1/2$ but $n/n'>1/2$.

Apart from the above mentioned families of periodic orbits, which are all simple periodic orbits, there also exist families of multiple

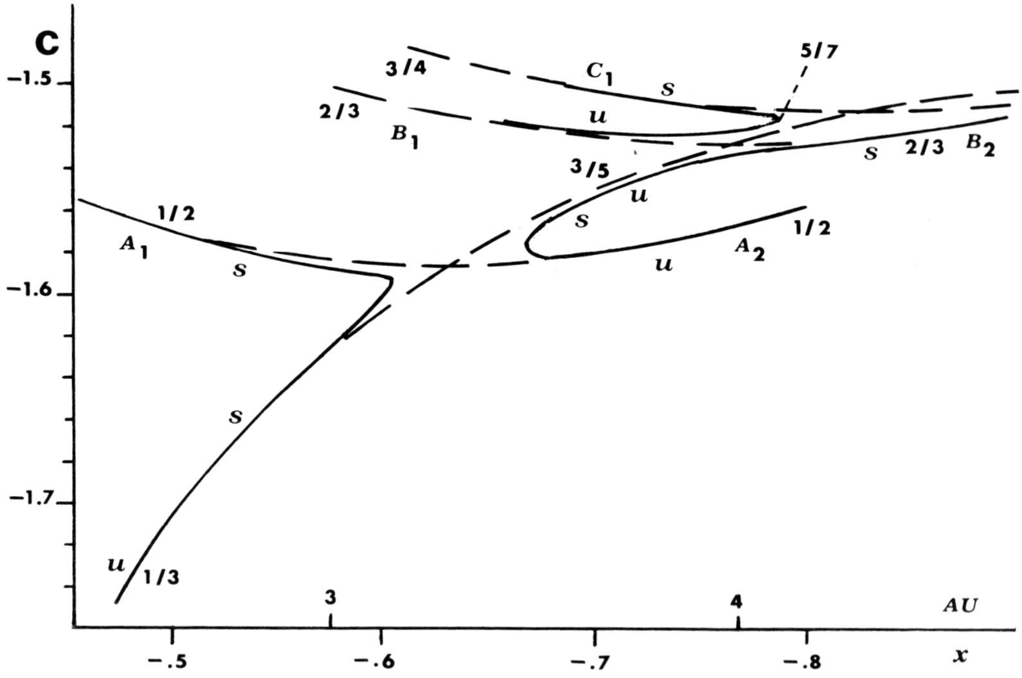

Figure 1. Families of simple periodic orbits for the asteroid (μ=0.001). The dotted lines represent the unperturbed families (μ=0). The resonance on the elliptic branches A_1, A_2, B_1, B_2, C_1, and the stability (s) or instability (u) is indicated.

resonant periodic orbits of the form p/q (q-p$\neq$1) which fill the whole space densely. They are not shown in Fig. 1 but some of them will appear on the surface of section, as we shall see in the next section.

The resonant branches 1/2 and 2/3 can be identified with the Hecuba and the Hilda group of the asteroids, respectively. This will become clear in the next section when the stability will be studied.

3. STABILITY

It can be proved analytically (Hadjidemetriou 1982b) that all the resonant circular orbits at the resonances 1/3, 3/5, 5/7, ... are continued, for $\mu\neq0$, to periodic orbits of the first kind which are unstable. Note from Fig. 1 that to each family for $\mu\neq0$ there belongs only one such unstable resonant orbit. This implies that on the extended family for $\mu\neq0$ there exists a small region, around the corresponding resonant orbit, which is unstable. It was also found that this unstable region extends as the value of μ (i.e. the perturbation) increases, and also the magnitude of the unstable eigenvalue increases.

All the other simple periodic orbits of the first kind are stable and no Hamiltonian perturbation exists that could make them unstable.

The resonant branches A_1, A_2, B_1, B_2, C_1,.... can be proved to be unstable, in the sense that there always exists a Hamiltonian perturbation on the unperturbed elliptic orbit which generates instability. This however does not mean that all the above branches are unstable when $\mu \neq 0$. In fact, it is found by numerical integrations that the branches A_1, B_2 and C_1, at the resonances 1/2, 2/3, and 3/4, respectively, are stable.

A similar situation holds for the resonant branches p/q ($q-p \neq 1$) of multiple elliptic periodic orbits.

A measure of the instability can be provided by the magnitude of the unstable eigenvalue, $|\lambda|>1$. We obtained that for the same value of μ the magnitude of $|\lambda|$ increases, for the resonant orbits $n/n'= =(2\nu-1)/(2\nu+1)$ as ν increases. This is shown in the Table I below (the max. value of $|\lambda|$ in the corresponding unstable area is given):

Table I

n/n'	λ
1/3	-1.005
3/5	-1.076
5/7	-1.257

As far as the unstable resonant branches A_2 and B_1 are concerned, we found that $|\lambda|$ increases as we proceed outwards to higher eccentricities, for the same branch, and also $|\lambda|$ is larger as ν in $n/n'=\nu/(\nu+1)$ increases. The value of $|\lambda|$ is much larger on these brances than at the resonant orbits 1/3, 3/5, 5/7, in Table I. This is shown in Table II (each orbit on a branch is identified by the initial value x, as can be seen from Fig. 1):

Table II

Branch A_2: $n/n' \simeq 1/2$		Branch B_1: $n/n' \simeq 2/3$	
x	λ	x	λ
-.681	1.101	-.727	1.876
-.691	1.202	-.716	2.745
-.737	1.517	-.697	4.841
-.767	1.823	-.666	29.6

4. THE STRUCTURE OF PHASE SPACE

The best way to obtain a global view of the totality of orbits and the stable or unstable regions in phase space is to consider a mapping on the surface of section. We have in our case two degrees of freedom and consequently a 4-dimensional phase space. The surface of section is now defined by C=constant, y=0, which is the 2-dimensional space, on which we use the cartesian coordinates x, $\dot{x}$. The periodic orbits are the fixed points of this mapping.

In order to understand the structure of the phase space on the surface of section, we start with the unperturbed case $\mu=0$ (Keplerian, circular or elliptic orbit of the asteroid, referred to the rotating frame). From Fig. 1 we can see that for each energy level C we have only one circular orbit, which corresponds to a central fixed point on the surface of section. It can be proved that this is a central fixed point which is surrounded by smooth invariant curves, topologically equivalent to circles (Hadjidemetriou 1982a). A dense subset of them corresponds to resonant orbits p/q, though in general p and q are large integers. These resonant invariant curves correspond to the intersections of the line C=constant in Fig. 1 with the resonant elliptic branches p/q mentioned in section 2. As we go outward, starting from the central fixed point, the ratio n/n' of the mean motions decreases. The mapping around the central fixed point is a twist mapping.

When the perturbation from Jupiter is applied, $\mu\neq 0$, the nonresonant invariant curves, p/q=irrational, survive as smooth invariant curves, as expected by the Kolmogorof-Arnold-Moser theorem. The resonant invariant curves evolve to a set of stable and unstable fixed points, and thus dissolution occurs, (e.g. Arnold and Avez, 1968). This dissolution however is in most cases negligible, except at the low order resonances. Chaotic behavior appears near the unstable fixed points with large value of the eigenvalue λ, as is indeed the case with the orbits of the branch B_1 (see Table II).

We present now some representative cases: In all the figures, together with the invariant curves we have plotted the curve $\dot{y}^2=0$ (dotted line), which is the boundary of the motion on the surface of section.

(a). Energy level C=-1.7367295694

The value of C is so selected that the central fixed point is an unstable fixed point at the resonance 1/3. In order to save computer time, we used the value $\mu=0.01$ instead of $\mu=0.001$, so that the eigenvalue λ is larger at the central fixed point and its effects appear in a shorter time. The value of λ is in this case $\lambda=-1.056$. As expected, the mapping around the central fixed point is a hyperbolic twist mapping (Fig. 2). One doubly symmetric stable periodic orbit appears. Also, resonant orbits at the 2/7 and 3/11 resonances are clearly seen. The stable fixed points are surrounded by islands, shown in Fig. 2 and the unstable fixed points are indicated (schematically) at the place we should expect them.

From this diagram we can deduce that the asteroids cannot stay near the unstable resonant orbit 1/3, and thus a gap is expected at that resonance. A few multiple periodic orbits can still exist at about that resonance, trapped around the doubly symmetric periodic orbit at 1/3, as shown in Fig. 2.

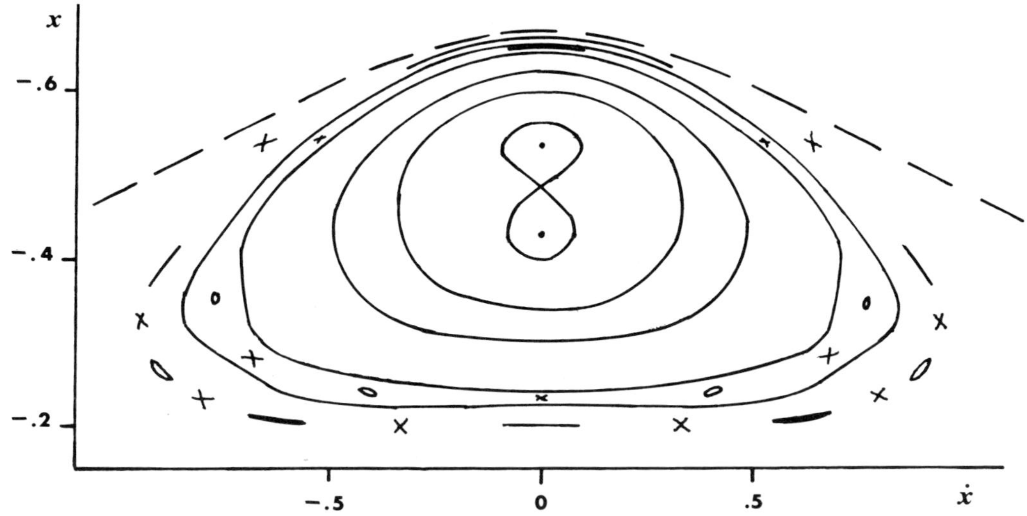

Figure 2. Invariant curves at C=−1.7367295694 for μ=0.01. The central unstable fixed point is at the resonance 1/3. Resonant orbits at 2/7 and 3/11 are also present.

At the 2/7 and 3/11 resonance stable regions exist, trapped around the stable periodic orbits at that resonance, but unstable regions also exist. Thus a smaller density of asteroids is expected at the above resonance cases, which results in minor gaps in the distribution of the asteroids.

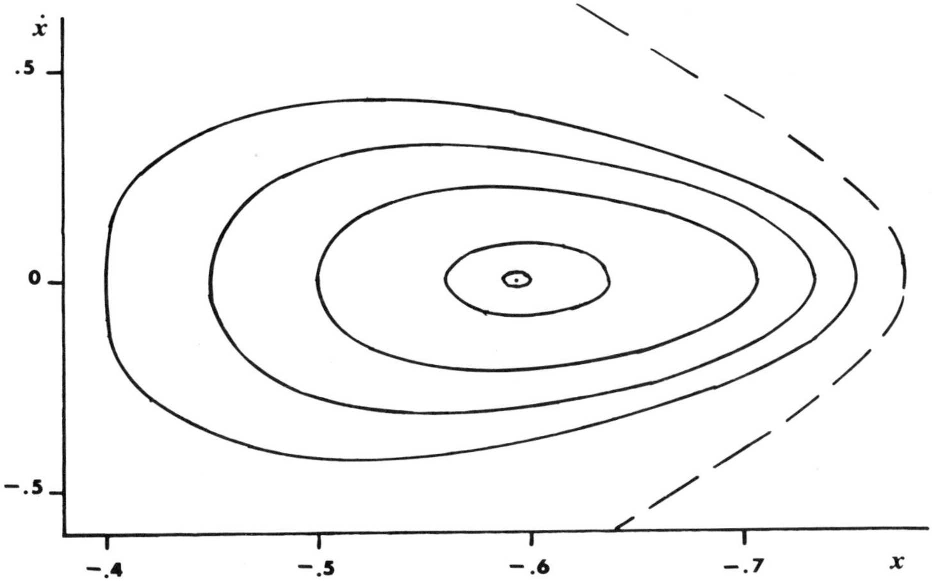

Figure 3. Invariant curves at C=−1.59. Only the stable resonant orbit 1/2 appears.

(b). Energy level C=-1.59

At this value of the energy the line C=-1.59 intersects only the stable branch A_1 at the resonance 1/2 and not the unstable branch A_2 at about the same resonance. This happens because the perturbed branch A_1 evolves to a position "lower" than that of the unperturbed one (Fig.1) while the unstable branch A_2 evolves to a position "above" the corresponding unperturbed branch. As a consequence, only the stable fixed point at the resonance 1/2 appears on the surface of section. Clearly, trapping at this resonance is possible, and this corresponds to the Hecuba group of the asteroids. A similar situation also appears at the 2/3 resonance where the line C=-1.525 intersects only the stable branch B_2 (Hilda group of the asteroids).

(c). Energy level C=-1.574982425

The central fixed point corresponds to a nonresonand periodic orbit of the first kind, which belongs to the second family of periodic orbits in Fig. 1. The value of n/n' at this central fixed point is larger than 1/2 and consequently, as we go outwards, there exists an invariant curve at the resonance 1/2 (for $\mu=0$) which dissolves when the perturbation $\mu \neq 0$ is applied into a stable and an unstable fixed point (Fig.4). This can be clearly seen from Fig.1 where the line C=-1.574982425 intersects both resonant elliptic branches A_1 and A_2. The intersection with the stable resonant branch A_1 corresponds to the stable fixed point and the

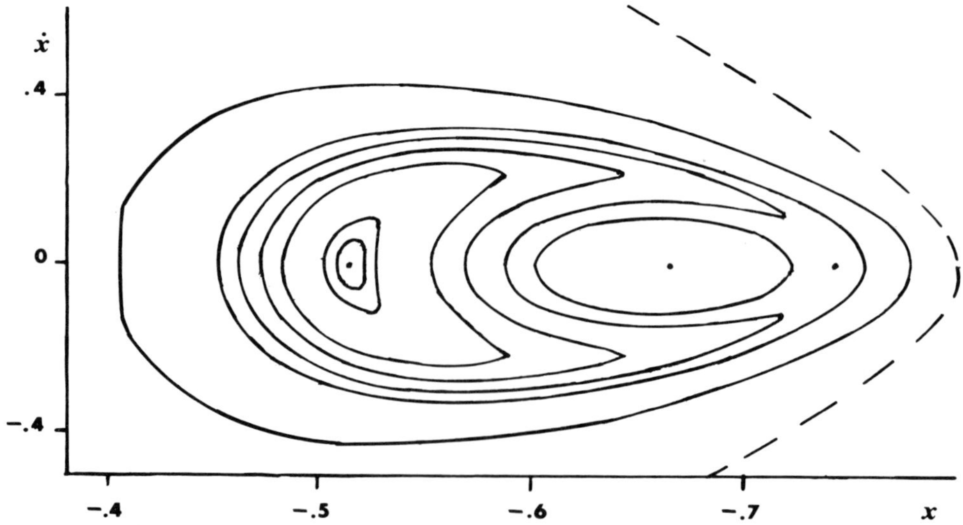

Figure 4. Invariant curves at C=-1.574982425. The central fixed point is a nonresonant periodic orbit of the first kind. Two resonant orbits near 1/2 appear, one stable and the other unstable.

intersection with the unstable resonant branch A_2 corresponds to the unstable fixed point. Both these latter points are at a resonance $n/n' \simeq 1/2$. The stable point is however at a resonance slightly smaller than 1/2, as already mentioned in section 2.

As a consequence of the existence of the above mentioned stable fixed point at the resonance $\simeq 1/2$, asteroids are expected there. This is the Hecuba group. On the other hand, at a resonance n/n slightly larger than 1/2 a quite extended unstable region exists, which corresponds to the 1/2 Kirkwood gap.

(d). Energy level C=-1.54696142.

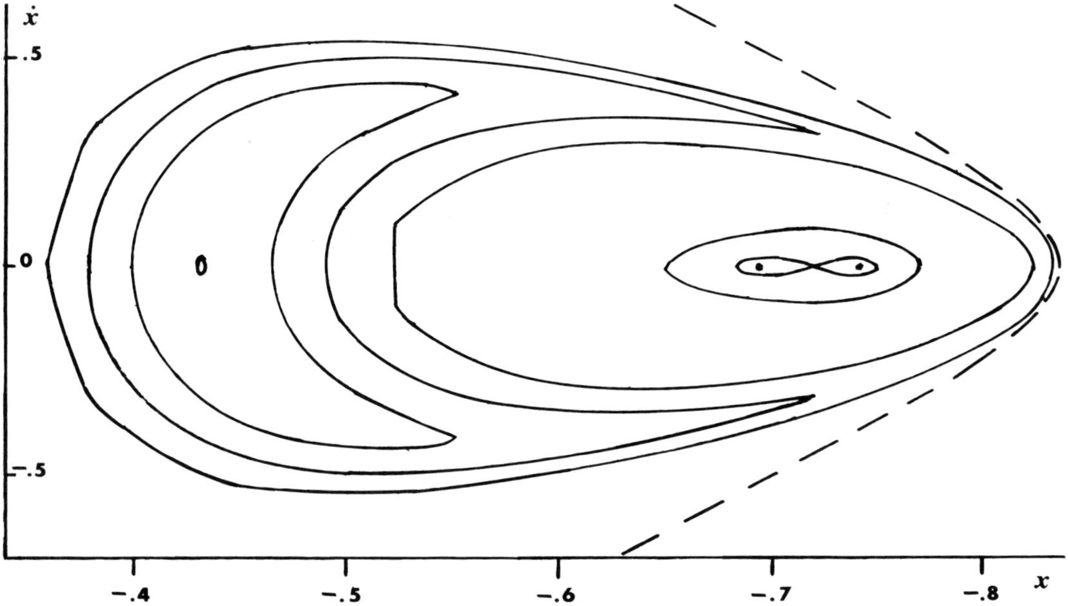

Figure 5. Similar to Fig.4, at C=-1.54696142. The central fixed point is the resonant periodic orbit at 3/5. The stable and unstable resonances at 1/2 are present.

This case, shown in Fig.5, is completely similar to the case (c) above. The only difference is that the central fixed point is now in the unstable area at the resonance 3/5 (see Fig.1). As a consequence, a hyperbolic twist mapping appears, which prevents the concentration of asteroid orbits at this resonance. This is indeed observed in the distribution of the asteroids. This situation is similar to the resonant gap at $n/n' \simeq 1/3$, shown in Fig.2, but in this case the gap is expected to be wider, as the unstable eigenvalue λ is, absolutely, larger (see Table I). Also, the instability area at the resonance $n/n' \simeq 1/2$ is more prominent than that in Fig. 4. This is so because the unstable eigenvalue λ increases along the unstable resonant branch A_2 (i.e. as the eccentricity increases).

(e). Energy level C=-1.5201681

This is shown in Fig. 6. The central fixed point is in the unstable area at the resonance 5/7 and consequently we have a hyperbolic twist mapping, as in the cases 1/3 and 3/5. We have also two fixed points at the resonance 2/3, which correspond to the intersection of the line C=-1.5201681 with the two resonant elliptic branches B_1 and B_2. The intersection with the stable branch B_2 corresponds to the stable fixed

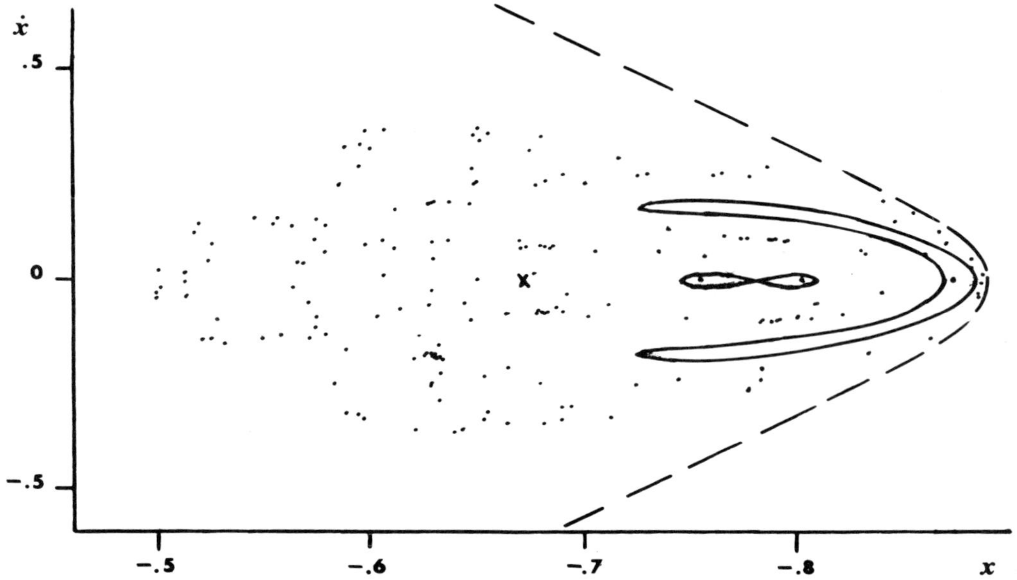

Figure 6. Invariant curves at C=-1.5201681. The central fixed point is at the resonance 5/7. Two resonant fixed points appear near the resonance 2/3. The stable one is surrounded by smooth invariant curves while dissolution appears at the unstable fixed point (indicated by x).

point and the intersection with the unstable branch B_1 corresponds to the unstable fixed point. Contrary however to the previous cases where smooth invariant curves appeared, even in the vicinity of the unstable fixed point, in this case we have chaotic behavior generated by the unstable fixed point. This is so because the unstable eigenvalue λ is large, as shown in Table II. Note that all scattered points belong to the same orbit. But around the stable fixed point a stable area clearly exists, which can explain a trapping at the resonance 2/3. This is the Hilda group of the asteroids, in accordance with the observations.

4. DISCUSSION

From the above study we see that all the observed Kirkwood gaps can be explained, at least qualitatively, on the assumption that the gaps are due to instabilities generated by the gravitational attraction of Jupiter.

Note that no other gaps appear in this study than those observed in the actual motion of the asteroids.

It is also clear that not all resonances are of equal importance, and this is not directly related to the order of the resonance. The 1/3 and 3/5 resonances are more important than the 1/2 or 2/3 resonances. The latter type of resonance has been studied extensively, but little work has been done for the 1/3, 3/5 resonances, especially the 3/5 resonance. Recent work on the 1/3 and 3/5 resonances is by Colombo and Franklin (1982). From the present study we find that both these resonances are of the same nature qualitatively, but the gap expected at the 3/5 resonance is wider, because the corresponding unstable eigenvalue is larger. This is indeed the case in the actual situation. Note however that stable areas at about the same resonances, 1/3 and 3/5, do exist, though small. So one would expect a few such resonant orbits of asteroids, librating around the above mentioned stable doubly symmetric periodic orbits at the corresponding resonances. These stable areas are clearly seen in Figs 2,5 and 6.

The 1/2 and 2/3 resonances are of different nature than the 1/3 and 3/5 resonances in that the stable and unstable areas in the 1/2, 2/3 case are comparable, contrary to the 1/3, 3/5 resonances where the instability character on the surface of section dominates. Compare for example Fig.5 where both the 1/2 and 3/5 resonance show clearly. As a consequence, stable areas at the resonances 1/2, 2/3 do exist. In particular, the stable region at the 1/2 resonance corresponds to mean motions such that n/n' is slightly smaller than 1/2 (corresponding to the Hecuba group) and the unstable region at 1/2 corresponds to n/n' slightly larger than 1/2, (corresponding to the observed gap at 1/2). The same situation, more pronounced, occurs at the resonance 2/3 (Hilda group.

Finally, we note that all resonances of the form $(2\nu-1)/(2\nu+1)$ are unstable (Hadjidemetriou, 1982a), i.e. the resonances 1/3, 3/5, 5/7,... These resonances have an accumulation point at the orbit of Jupiter, and for this reason the whole space near Jupiter is dominated by the instabilities generated by the above orbits. Thus, no asteroids are expected near Jupiter, as is indeed the case. Note that the continuation from the above periodic orbits, for $\mu=0$, to $\mu\neq0$ is always possible, but the max. value of μ for which the continuation can be extended decreases as ν increases. This means that no periodic orbits for $\mu=0.001$ exist beyond a certain value of ν. This however seems to enhance the chaotic situation at the area near Jupiter, but we do not have detailed numerical examples.

Note: The energy constant has been computed from the formula

$$C= \frac{1}{2}(\dot{x}^2+\dot{y}^2)- \frac{1}{2}[\mu r_1^2 + (1-\mu)r_2^2] - \frac{\mu}{r_1} - \frac{1-\mu}{r_2}$$

where

$$r_1 = [(x-1+\mu)^2+y^2]^{\frac{1}{2}}, \qquad r_2 = [(x+\mu)^2+y^2]^{\frac{1}{2}}.$$

REFERENCES

Arenstorf , R. F., 1963, Amer. J. Math. 83, 27.

Arnold V. I. and Avez, A., 1968, "Ergodic Problems in Classical Mechanics", Benjamin

Birkhoff, G. D., 1927, "Dynamical Systems", Amer, Math. Soc. New York.

Broucke, R. A., 1968, NASA-JPL Technical Report, 32-1168.

Colombo G., Franklin, F. A., 1982, in V. Szebehely (ed.) "Applications of Modern Dynamics to Celestial Mechanics and Astrodynamics", p.339, D. Reidel Publ. Co.

Colombo G., Franklin, F. A. and Munford, C. M., 1968, Astron. J. 73,111.

Dermott, S. F. and Murray, C.D., 1981, Nature 290, 664.

Froeschlé, C. and Scholl H., 1979, in V. Szebehely (ed.), "Instabilities in Dynamical Systems", p.115, Reidel Publ. Co.

Guillaume, P.: 1969, Astron. Astrophys. 3, 155.

Hadjidemetriou, J. D., 1982a in V. Szebehely (ed.), "Applications of Modern Dynamics to Celestial Mechanics and Astrodynamics", p.25, Reidel Publ. Co.

Hadjidemetriou, J. D., 1982b, Celes. Mech. 27, 305.

Hagihara, Y., 1972,"Celestial Mechanics", Vol.2, part 1, MIT Press.

Lecar, M. and Franklin, F. A., 1973, Icarus 20, 422.

Schmidt, D., 1972a, in L. Weiss (ed.) "Ordinary Differential equations", p.553, A.P.

Schmidt, D., 1972b, SIAM J. Appl. Math. 22, No 1.

Scholl, H. and Froeschlé, C.: 1974, Astron. Astrophys. 33, 455.

Scholl, H. and Froeschlé, C.: 1975, Astron. Astrophys. 42, 457.

Sinclair, A. T., 1969, Monthly Not. Roy. Astron. Soc. 142, 289.

Sinclair, A. T., 1970, Monthly Not. Roy. Astron. Soc. 148, 325.

Szebehely, V., 1967, "Theory of Orbits", Academic Press.

Szebehely, V., Vicente, R. and Lundberg, J.B.: 1983, this volume, p.123.

LONG PERIODS IN THE THREE-DIMENSIONAL MOTION OF TROJAN ASTEROIDS

R. Bien and J. Schubart
Astronomisches Rechen-Institut, Heidelberg, Germany, Federal Republic

ABSTRACT. The long periods caused by Jupiter, or Jupiter and Saturn in the motion of a Trojan are studied by numerical integration. Orbital inclinations between 0° and 50° are considered. Results on the length of the periods and on their main effects in the osculating elements are presented. It turns out that secular resonances can be important for the evolution of Trojan orbits.

1. INTRODUCTION

E.W. Brown's theory of the Trojan group of asteroids (Brown and Shook, 1933) describes an analytical way of approximation of three-dimensional motion of Trojans. Recently Érdi (1981) has applied a different theoretical method. Numerical results by Chebotarev et al. (1974) refer to real asteroids and cover an interval of 400 yr. One of us (Bien, 1978, 1980a) has studied by numerical integration the motion of real and fictitious Trojans on the basis of a simplified version of the planar elliptic three-body problem, and then (Bien, 1980b) on the basis of the rigorous problem Sun-Jupiter-Saturn-asteroid in three dimensions. Here we present a continuation of this work that refers to orbits of both small and high inclination, and consists in a study of the long periods by means of the main effects caused by them. Especially, we are interested in the long-period effects shown by eccentricity and inclination of an orbit and by the longitudes of perihelion and node. The following is a preliminary discussion of these effects. Our basic numerical integrations cover intervals between 1.5×10^3 and 2×10^5 yr.

2. MODELS AND EXAMPLES

For our studies of the three-dimensional motion of massless Trojan asteroids we have used four models. The following abbreviations will refer to them in the text:
"circ.3b." = circular restricted three-body problem with sun and Jupiter.
"ell.3b." = elliptic restricted three-body problem.

V. V. Markellos and Y. Kozai (eds.), Dynamical Trapping and Evolution in the Solar System, 153–161.

"av.ell.3b." = the same problem, but the short-period terms are eliminated in the equations of motion by a procedure of numerical averaging (Schubart, 1978).
"simp.4b." = four-body problem with sun, Jupiter and Saturn. The two major planets move on elliptic orbits, but in the same plane as a simplification (compare Schubart, 1979, model 3).
With the exception of "av.ell.3b." we realize these models by numerical integration of the rigorous equations of motion, using the n-body program by Schubart and Stumpff (1966). The masses of Jupiter and Saturn correspond to the IAU (1976) System of Constants. All other perturbing masses are entirely neglected.

We use the following symbols for the osculating orbital elements of a Trojan: a, e, i, Ω, $\tilde{\omega} = \omega + \Omega$, ℓ = mean longitude. The subscript J relates a symbol to Jupiter, so that $\mu = \ell - \ell_J$ represents the mean heliocentric angular distance between asteroid and Jupiter. In the four models mentioned above the orbital plane of Jupiter is the plane of reference for ω, Ω, and i ($i_J = 0°$). At the start of an integration we put $a_J = 1$, $e_J = 0.048$ (except in "circ.3b."), and $\tilde{\omega}_J = 0°$. The time T is counted from a moment that is comparatively close to the present. The longitudes count from the fixed direction or starting value of $\tilde{\omega}_J$. We prefer to replace e, $\tilde{\omega}$ by

$$\psi_1 = e\cos(\tilde{\omega} - \tilde{\omega}_J) , \qquad \psi_2 = e\sin(\tilde{\omega} - \tilde{\omega}_J) .$$

Variations of e, $\tilde{\omega}$ are studied by curves in rectangular coordinates ψ_1, ψ_2 (compare Bien, 1978). If the attraction of Saturn is neglected, the simpler relations $\psi_1 = e\cos\tilde{\omega}$, $\psi_2 = e\sin\tilde{\omega}$ hold. In studies of the planar subcase of "av.ell.3b." Bien (1978, 1980a) has found the two long periods P_L (period of libration) $\approx$ 150 yr and $P_{\tilde{\omega}}$ (period of perihelion) $\approx$ 3 600 yr. μ librates around +60° or -60° for preceding and following Trojans, respectively, while a oscillates around 1, and both variations correspond to P_L as the main period. In this case the "ψ_1, ψ_2 curve" is approximately a circle around a point $\psi_1 = e_J\cos(\pm 60°)$, $\psi_2 = e_J\sin(\pm 60°)$, where the sign of 60° corresponds to that of μ. For nonplanar cases with large i the ψ_1, ψ_2 curve becomes an ellipse, and the nodal rate can be positive or negative (Bien, 1980b). In general, the ψ_1, ψ_2 curve does not always surround the point $\psi_1 = \psi_2 = 0$. Thus $P_{\tilde{\omega}}$ is given by the mean period of revolution along this curve.

Starting values will be designated by the subscript 0 and similarly extremes by max or min. We have chosen a standard set of starting values: $a_0 = 1$, $\mu_0 = 53°$ (then $\mu_0 \approx \mu_{min}$ because μ reaches extremes near a = 1), $e_0 = 0.11$, $\tilde{\omega}_0 = 60°$ (then e is not very far from its maximum), $\Omega_0 = 70°$. i is varied from 2° to 50°. The choice of $\mu_0 > 0$ is not essential, because we expect similar periods and effects for $\mu_0 < 0$ according to a property of symmetry of "av.ell.3b.". For the three values $i_0 = 4°$, 30°, 40° we have integrated forward over 150 000 yr in both "ell.3b." and "simp.4b.". The two following Trojans ($\mu < 0$), (1208) Troilus and (1873) Agenor have been integrated in "av.ell.3b." over more than 10^5 yr.

3. RESULTS FROM THE THREE-BODY CASES

We have integrated the standard set of starting values with $i_0 = 2°$, $4°$, ... , $50°$ over 1 500 yr in "av.ell.3b.", and derived from this approximate values for P_L and μ_{max} (with respect to P_L), see Fig.1. We find comparatively large P_L values for high-inclination orbits, whereas μ_{max} decreases with increasing i_0 such that μ oscillates around mean values of less than 60° for large i_0. We can approximately identify i with i_0 in case of Trojans, because the variations of i are small. Besides the dominant influence of P_L on μ and a, our more extended calculations mentioned above reveal additional periodicities. An integration of high-inclination orbits in "circ.3b." shows already a superposition of effects by a period of $\frac{1}{2} P_\omega$, where in this model P_ω can be defined as the period of revolution of ω. Using "ell.3b.", an additional superposition of effects due to $P_{\tilde{\omega}}$ appears. The superimposed effects cause oscillations of about 0°5 in μ_{max} and μ_{min} with respect to mean values.

The influence of $\frac{1}{2} P_\omega$ appears clearly in all the curves of Fig.2, which are smoothed with respect to P_L and shorter periods, and which correspond to "circ.3b.". Strong effects appear in e and $\dot{\tilde{\omega}}$, and smaller effects in i and $\dot{\Omega}$. In this standard example with $i_0 = 30°$ the four curves coincide roughly in the position of the extremes and can be

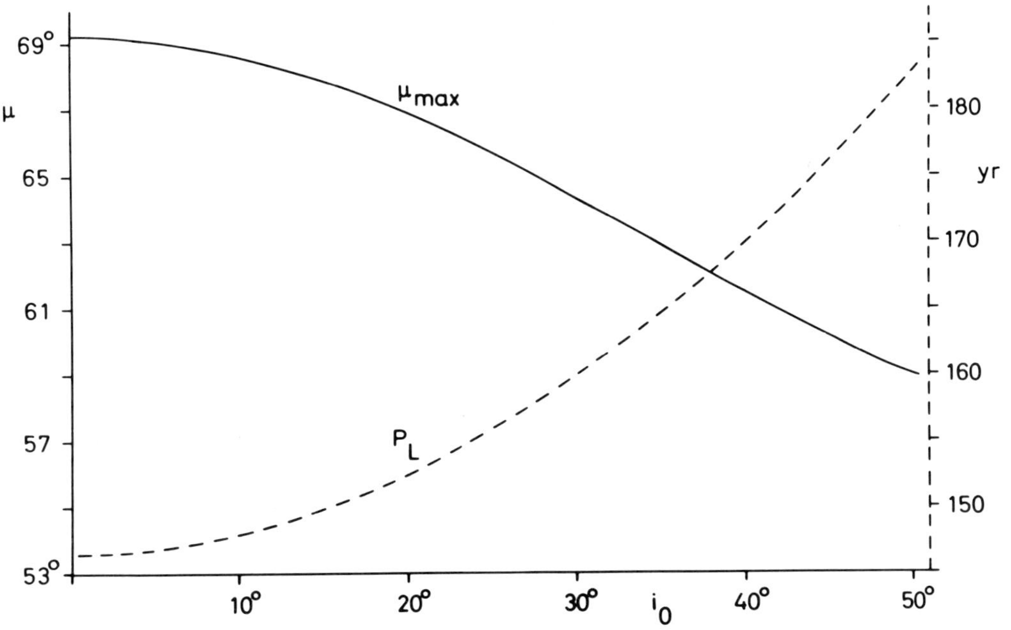

Fig. 1. μ_{max} (left scale) and P_L (dashed line, right scale) are plotted against the starting value of inclination. $\mu_{min} \approx 53°$ coincides approximately with the horizontal axis. Results from "av.ell.3b.".

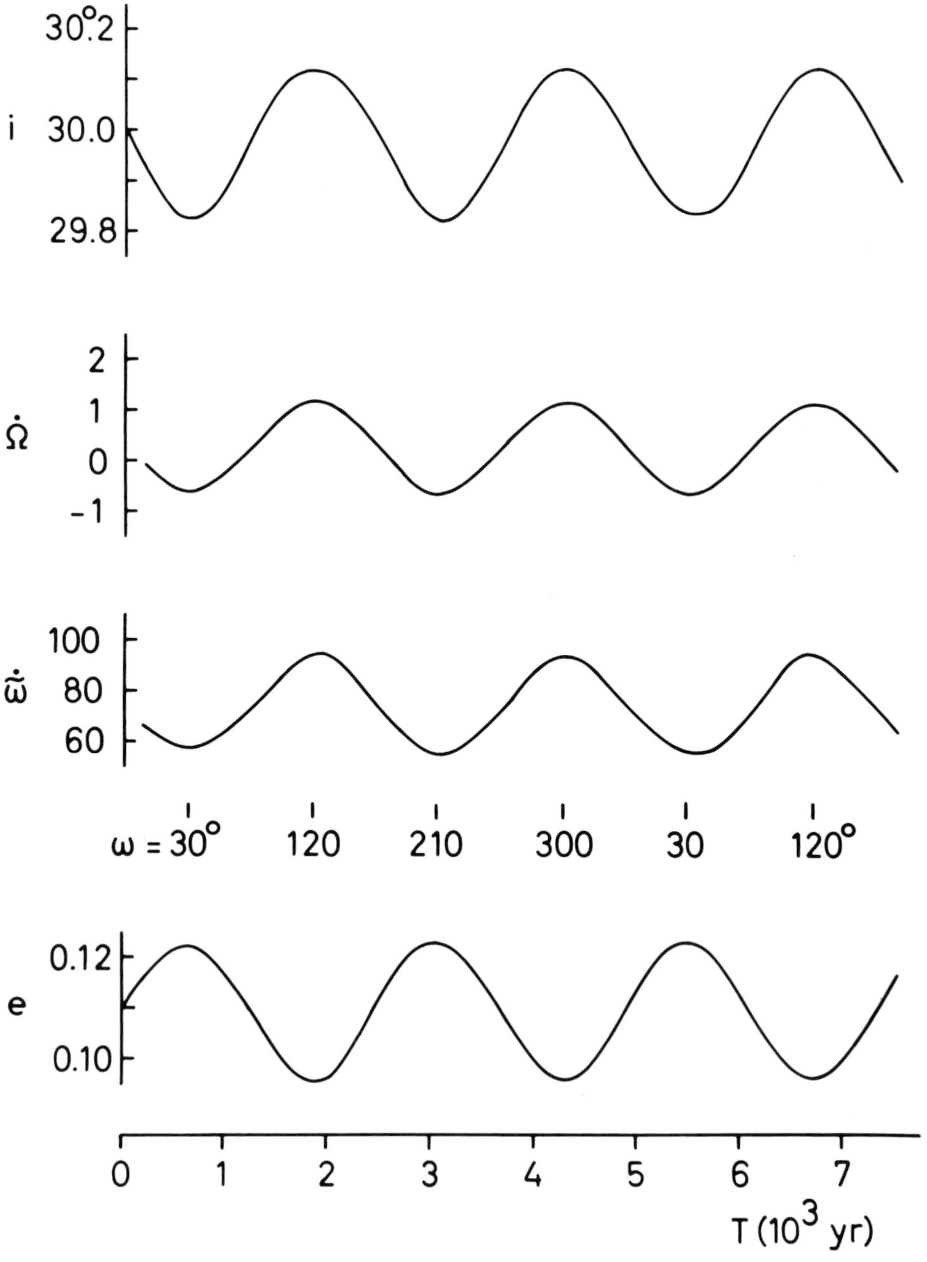

Fig. 2. i, $\dot{\Omega}$, $\dot{\tilde{\omega}}$, e versus T, smoothed curves from "circ.3b.". The unit of $\dot{\Omega}$ and $\dot{\tilde{\omega}}$ is degree/10^3 yr. Special values of ω are marked additionally. The curves correspond to the standard example with $i_0 = 30°$.

described qualitatively by functions $\pm\cos(2\omega - \gamma)$ with $\gamma \stackrel{\sim}{\sim} 60°$. For other examples one finds $\gamma \stackrel{\sim}{\sim} \pm 60°$, where the sign corresponds to that of μ. The mean value of $\dot{\Omega}$ is so close to zero, that the sign of $\dot{\Omega}$ can change. An extremely slow motion of Ω is typical for Trojan orbits.

Fig. 3 is obtained from "av.ell.3b." and shows two stages of the evolution of the ψ_1, ψ_2 curve of the highly inclined following Trojan (1208) Troilus. In order to describe the quality of such a curve we

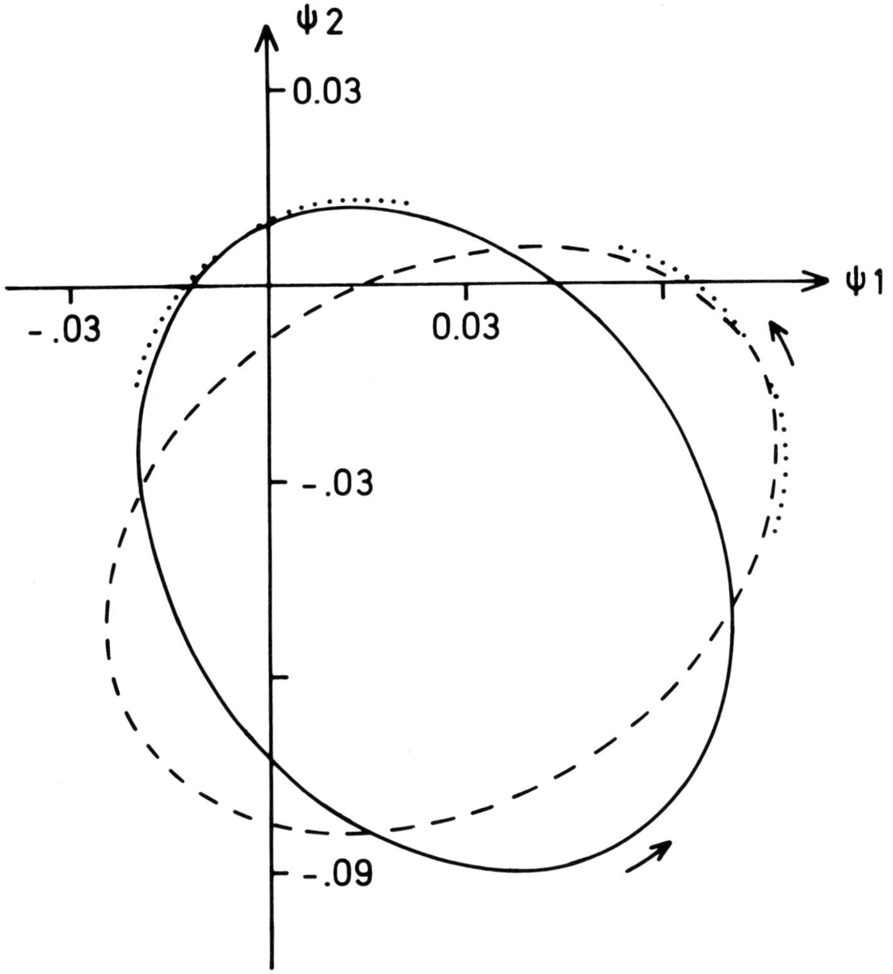

Fig. 3. Two parts of the ψ_1, ψ_2 curve of (1208) Troilus, smoothed results from "av.ell.3b.". The dotted beginning and end of each part indicates that the curve corresponds to a slowly rotating ellipse. The arrows demonstrate the direction of motion. The dashed part is later by 90 000 yr. $P_{\omega} \stackrel{\sim}{\sim} 5\,000$ yr corresponds to the period of circulation around the center of the rotating ellipse.

introduce the proper eccentricity e_p and the proper perihelion $\tilde{\omega}_p$ by the relations

$$\psi_1 - \psi_1^* = e_p \cos(\tilde{\omega}_p - \tilde{\omega}_J) , \quad \psi_2 - \psi_2^* = e_p \sin(\tilde{\omega}_p - \tilde{\omega}_J) ,$$

where ψ_1^*, ψ_2^* is the position of the center of the rotating curve (note: $\tilde{\omega}_J = 0°$ in 3-body cases). This is a generalization of relations used by Bien (1978, 1980a). e_p represents the distance from the center of the curve to the point ψ_1, ψ_2. In analogy to the variations of e shown in Fig. 2, e_p varies qualitatively like the function $\cos(2\omega_p - \gamma)$, where ω_p is introduced by $\omega_p = \tilde{\omega}_p - \Omega$, and $\gamma \approx -60°$ refers to the case of Fig. 3. $P_{\tilde{\omega}}$ is now given by the mean period of revolution of $\tilde{\omega}_p - \tilde{\omega}_J$, and a general definition of P_ω is possible: P_ω equals the mean period of revolution of ω_p. For the three standard examples treated by "ell.3b." we find $P_{\tilde{\omega}} \approx$ 3 700 yr, 4 800 yr, 6 000 yr for $i_0 = 4°$, 30°, 40°, respectively. According to the slow motion of Ω, $P_\omega \approx P_{\tilde{\omega}}$ holds for all Trojans. Since $\frac{1}{2} P_\omega$ rules the oscillation of e_p, two maxima of e_p correspond to one revolution on the ψ_1, ψ_2 curve. In case of Fig. 3 the mean of $\dot{\Omega}$ equals 1° per 10^3 yr. Therefore the values of $\tilde{\omega}_p$ that correspond to maxima of e_p will increase according to this rate. Thus the rotating ellipse of Fig. 3 is qualitatively explained by a combination of effects of $P_{\tilde{\omega}}$ and $\frac{1}{2} P_\omega$.

A study of the variations of i shows an influence of $\frac{1}{2} P_\omega$ and $P_{\tilde{\omega}}$, and also of the period corresponding to the argument $\tilde{\omega}_p + \tilde{\omega}_J - 2\Omega$. An influence of the very long period corresponding to $2\Omega - 2\tilde{\omega}_J$ is indicated. Fig. 4 shows mean values of $\dot{\Omega}$. The curve refers to the examples considered in Fig. 1 and shows an increase of $\dot{\Omega}$ with i_0, but the crosses indicate a dependence on μ_0. Especially a decrease in μ_0, i.e. an increase in the amplitude of libration of μ, causes a considerable decrease in $\dot{\Omega}$. This decrease is smaller for large i_0. The two dots in Fig. 4 refer to following Trojans, but also to fictitious preceding objects according to a symmetry of "av.ell.3b.".

4. SPECIAL RESULTS FROM THE FOUR-BODY MODEL

The addition of Saturn to the system of bodies causes important effects which appear in our studies by "simp.4b.". A secular period of about 54 000 yr causes an oscillation of e_J between 0.03 and 0.06 and a temporary retrograde shift of $\tilde{\omega}_J$, although the mean of $\dot{\tilde{\omega}}_J$ is positive. The approximate 5/2 resonance of Jupiter and Saturn refers also to Trojan and Saturn. In case of Jupiter it causes the well-known perturbation in longitude and smaller effects with periods of about 900 yr in the other elements. The graphs by Cohen et al. (1973) demonstrate all variations in the orbit of Jupiter.

The extended integrations of our standard examples with $i_0 = 4°$, 30°, and 40° by "simp.4b." reveal effects of periods near 900 yr in a, μ, i, and in the ψ_1, ψ_2 curves. Such an effect has temporarily an amplitude of $\approx 0\overset{\circ}{.}02$ in i of the case with $i_0 = 30°$. The earlier results

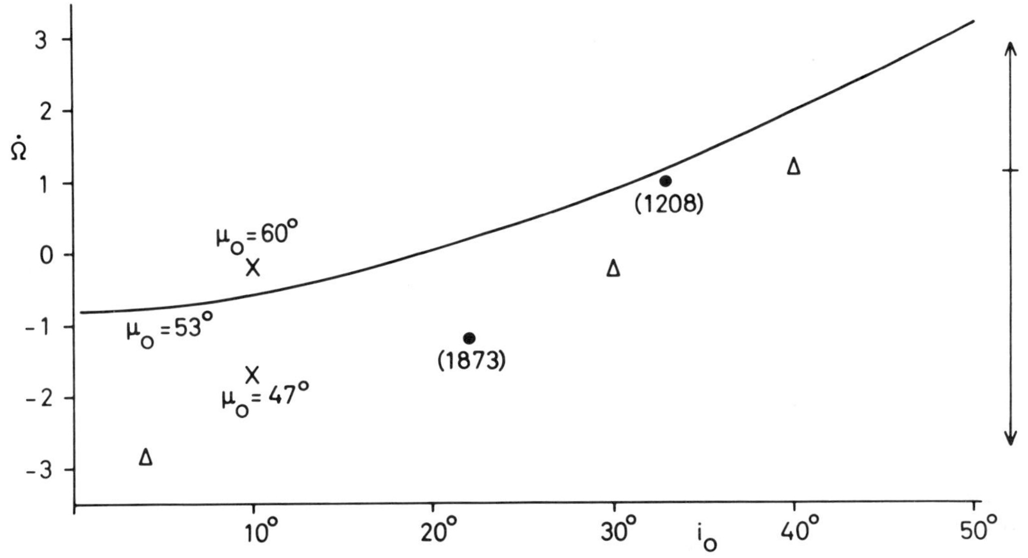

Fig. 4. $\dot{\Omega}$ in degree/10^3 yr is plotted against i_0. The curve corresponds to those of Fig. 1. The crosses demonstrate effects by a variation of μ_0, the dots refer to real objects. Results from "av.ell.3b.", but the triangles and the vertical line demonstrate effects by Saturn (see text).

indicate a dependence of the center of a ψ_1, ψ_2 curve on e_J. Therefore, it is no surprise that the center (ψ_1^*, ψ_2^*) changes with e_J. Now we have to replace Ω by $\Omega - \tilde{\omega}_J$, and this angle causes the forward or backward rotation of the elliptic curves, as in Fig. 3 the angle Ω. An alternation of forward and backward rotation of such an ellipse can occur in special examples, but the shape of the ellipse does not vary. Saturn changes the mean of $\dot{\Omega}$, as demonstrated by triangles for the three standard examples in Fig. 4. Therefore a positive $\dot{\Omega}$ in "ell.3b." can become negative in "simp.4b." (compare Bien, 1980b). The range of the variation of $\dot{\tilde{\omega}}_J$ which is caused by the secular period of 54 000 yr is demonstrated in Fig. 4 by the vertical straight line that refers to the scale of $\dot{\Omega}$. A bar on this line represents the mean of $\dot{\tilde{\omega}}_J$. It turns out that for many high-inclination orbits an alternation between a forward and a backward shift of $\Omega - \tilde{\omega}_J$ can occur. Our example with $i_0 = 40°$ gives a value $\dot{\Omega}$ which agrees nearly with the mean of $\dot{\tilde{\omega}}_J$. If effects due to periods near 900 yr are neglected, the variations in i resemble those found in "ell.3b.". However, the amplitudes of effects depending on e_J show an oscillation that follows the 54 000 yr period of e_J, and this period can also cause variations of the mean of i in case of Trojans. A special feature occurs in the example with $i_0 = 40°$: The slow mean motion of the argument $\Omega - \tilde{\omega}_J$ corresponds to an increase of $\approx 5°$ in 10^5 yr during the computation, and the mean of i with respect to the visible periodicities decreases monotonously by 0°2 in 10^5 yr. We cannot predict the variations of $\Omega - \tilde{\omega}_J$ and i of this example during more extended periods, and we do not know, whether there is a relation between the variations of these two quantities.

5. DISCUSSION AND OUTLOOK

In the motion of Trojan asteroids the following long periods appear: the libration period P_L, the period of perihelion $P_{\tilde{\omega}}$, one half of the period of the argument of perihelion $\frac{1}{2}$ P_ω and periods that correspond to the angular arguments $\tilde{\omega}_p + \tilde{\omega}_J - 2\Omega$ and $2\Omega - 2\tilde{\omega}_J$. The influence of four of these periods is also suggested by terms with long-period arguments listed by Brown and Shook (1933, p.259). Further, periods due to the 5/2 resonance of Jupiter and Saturn, or asteroid and Saturn play an important role, and the period of 54 000 yr in the oscillation of e_J is even more important. We point out that in case of special Trojan orbits the very slow motion of Ω allows a secular resonance with the mean motion of $\tilde{\omega}_J$ which is very small as well. Obviously, our standard example with $i_0 = 40°$ is very close to such a resonance. According to a former computation, the mean of $\dot{\Omega}$ of the real object (2146) Stentor ($i \approx 38°$) is close to +1°.0 per 10^3 yr, which is just 0°.1 smaller than the value corresponding to the secular resonance. However, the attraction of additional major planets can change this difference.

Our results demonstrate that a study of Trojan motion over very long intervals of time must include the effects of Saturn. Fundamentally, the slow nodal motion of a Trojan can make possible a secular resonance with respect to a suitable secular frequency of the major planets, and especially to the mean of $\dot{\tilde{\omega}}_J$ in case of high-inclination orbits. This can be important for the evolution of Trojan orbits (compare Yoder, 1979).

The authors plan a more detailed numerical analysis of the periods and effects in the motion of Trojans. It is planned to extend the number of examples, and to take into account the mutual orbital inclination of Jupiter and Saturn. We thank Mrs.E.Miltenberger and Mrs.I.Seckel for aid in typing and drawing. We used the IBM 370-168 computer at the University of Heidelberg's Rechenzentrum.

REFERENCES

Bien,R. : 1978, Astron.Astrophys. 68, pp.295-301.
Bien,R. : 1980a, Astron.Astrophys. 81, pp.255-259.
Bien,R. : 1980b, Moon and Planets 22, pp.163-166.
Brown,E.W. and Shook,C.A. : 1933, Planetary Theory, The University Press, Cambridge, pp.250-288.
Chebotarev,G.A., Belyaev,N.A., and Eremenko,R.P. : 1974, The Stability of the Solar System and of Small Stellar Systems, Ed.: Y. Kozai, IAU Symp. No.62, pp.63-69.
Cohen,C.J., Hubbard,E.C., and Oesterwinter,C. : 1973, Astron. Papers

American Ephemeris Washington, Vol.22, part 1.
Érdi,B. : 1981, Celes.Mech. 24, pp.377-390.
Schubart,J. : 1978, Dynamics of Planets and Satellites and Theories of their Motion, Ed.: V. Szebehely, IAU Coll. No.41, pp.137-143.
Schubart,J. : 1979, Dynamics of the Solar System, Ed.: R.L. Duncombe, IAU Symp. No.81, pp.207-215.
Schubart,J. and Stumpff,P. : 1966, Veroeffentl.Astron. Rechen-Inst. Heidelberg No.18.
Yoder,C.F. : 1979, Icarus 40, pp.341-344.

RESONANT ASTEROIDAL MOTION IN THE KIRKWOOD GAPS : A THREE-DIMENSIONAL STUDY

H. SCHOLL (*) and C. FROESCHLE (**)

(*) Astronomisches Rechen Institut
Heidelberg, F.R. Germany

(**) Observatoire de Nice, France

ABSTRACT

Resonant asteroidal motion is investigated over 17 000 years in a three-dimensional elliptical model Sun-Jupiter-Asteroid averaged by Schubart's method à la Poincarè. Orbits remain trapped in the resonance over this period. The various stability mechanisms are discussed. With respect to their behaviour of ω and $\tilde{\omega}$, our resulting orbits reveal 5 distinct classes. These 5 classes can be described in Schubart's topology for the planar problem. As compared to the planar case, eccentricities of orbits in the three-dimensional model vary more strongly. This is an important result for the problem of the delivery of meteorites.

V. V. Markellos and Y. Kozai (eds.), Dynamical Trapping and Evolution in the Solar System, 163.

ORBITAL EVOLUTION OF TROJAN ASTEROIDS

Bálint Érdi
Department of Astronomy
Eötvös University
Budapest, Hungary

ABSTRACT

The motion of the orbits of Trojan asteroids are investigated. Four asteroids around L_5 are shown to have librating perihelion. A criterion for the libration of the perihelion is derived and expressed by the initial conditions. This criterion is solved for the initial conditions in two special cases. Thus regions in the configuration space of the initial conditions yielding libration of the orbits of Trojan asteroids are established.

1. INTRODUCTION

According to the Lagrangian solutions of the three-body problem a small body of negligible mass resting at the relative equilibrium points L_4 or L_5 of the Sun-Jupiter system moves around the Sun on an orbit which is similar to Jupiter's orbit but the perihelions of the two orbits are at 60° from each other. If the body suffers a perturbation it leaves L_4 or L_5 and its orbit around the Sun also changes. The stability investigations of the Lagrangian points usually deal with the motion of the small body around L_4 or L_5. This paper studies the behaviour of the orbit of the small body around the Sun.

The well-known examples for the Lagrangian solutions of the three-body problem are the Trojan asteroids. The main perturbations of the eccentricity e and the longitude $\tilde{\omega}$ of the perihelion of the orbits of these asteroids can be described by the equations (Érdi, 1979)

$$\begin{aligned} e \sin\tilde{\omega} &= a - c \sin\chi , \\ e \cos\tilde{\omega} &= b - c \cos\chi \end{aligned} \qquad (1)$$

V. V. Markellos and Y. Kozai (eds.), Dynamical Trapping and Evolution in the Solar System, 165–176.

where a, b and $c \geq 0$ are constants and γ is a slowly changing function of the time. It can be shown that if

$$c \leq \sqrt{a^2 + b^2}$$

then the perihelion of the asteroids librates around a direction which is about at 60° from Jupiter's perihelion. If

$$c > \sqrt{a^2 + b^2}$$

then the perihelion of the asteroids circulates.

For an asteroid which is exactly at L_4, $c = 0$ and $e = e_J$ where e_J is the orbital eccentricity of Jupiter. Moreover, $\tilde{\omega} - \tilde{\omega}_J = 60°$ where $\tilde{\omega}_J$ is the longitude of the perihelion of Jupiter. If the asteroid undergo a small perturbation it leaves L_4 and its orbit begins libration around the equilibrium orbit. The parameter c can be considered as the measure of the perturbation. As c increases so does the amplitude of the libration of the perihelion. When the difference between the two extremum values of $\tilde{\omega}$ exceeds the value 180° the orbit of the asteroid begins to circulate.

The purpose of this paper is to find initial conditions which result in the libration of the perihelion. Knowing the initial position and velocity of an asteroid near L_4 the question is whether the orbit of the asteroid will librate or circulate. For the solution of this problem the results of the author's previous papers (Érdi, 1978, 1979, 1981) are applied.

2. THE MOTION OF THE PERIHELION

Equations (1) were derived from an asymptotic solution for the motion of Trojan asteroids. First a short summary of that solution is given here. Under the assumptions that the motion of a Trojan asteroid around the Sun is perturbed only by Jupiter and Jupiter's orbit around the Sun is an ellipse the equations of motion of the asteroid are (Érdi, 1978)

$$\frac{d^2 r}{dv^2} - r\left(\frac{d\alpha}{dv}\right)^2 - 2r\frac{d\alpha}{dv} = \frac{1}{1+e_J\cos v}\left[r - \frac{1-\mu}{R_1^3}r + \mu\left(\frac{\cos\alpha - r}{R_2^3} - \cos\alpha\right)\right],$$

$$\frac{d}{dv}\left(r^2\frac{d\alpha}{dv} + r^2\right) = \frac{\mu r\sin\alpha}{1+e_J\cos v}\left[1 - \frac{1}{R_2^3}\right], \qquad (2)$$

$$\frac{d^2 z}{dv^2} + z = \frac{z}{1+e_J\cos v}\left[1 - \frac{1-\mu}{R_1^3} - \frac{\mu}{R_2^3}\right]$$

where r, α, z are the cylindrical coordinates of the asteroid (see Figure 1), v is the true anomaly of Jupiter, μ is the mass of Jupiter divided by the total mass of the Sun-

Jupiter system and

$$R_1 = \sqrt{r^2 + z^2}\,, \qquad R_2 = \sqrt{1 + r^2 - 2r\cos\alpha + z^2}\,.$$

The coordinates r and z are dimensionless, the Sun-Jupiter distance is supposed to be unity.

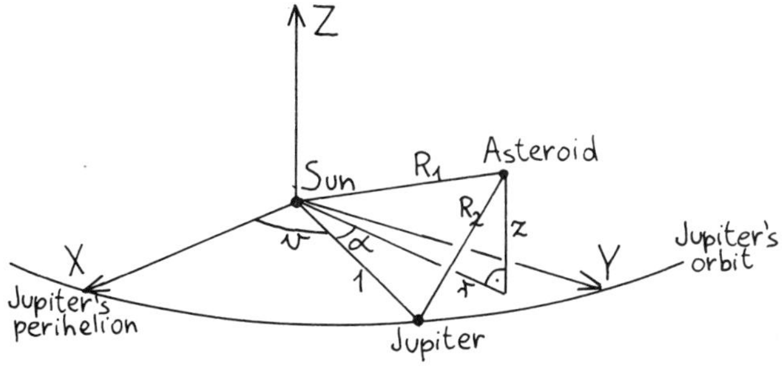

Figure 1. The coordinates r, α, z in the Cartesian coordinate system SXYZ.

The solution of Equations (2) were derived in the form of a three-variable asymptotic expansion (Érdi, 1978, 1981)

$$\begin{aligned} r &= 1 + \sum_{n=1}^{N} \varepsilon^n r_n(v, u, \tau) + O(\varepsilon^{N+1}), \\ \alpha &= \alpha_0(u, \tau) + \sum_{n=1}^{N} \varepsilon^n \alpha_n(v, u, \tau) + O(\varepsilon^{N+1}), \\ z &= \varepsilon^{1/2}\left[\sum_{n=0}^{N} \varepsilon^n z_n(v, u, \tau) + O(\varepsilon^{N+1})\right] \end{aligned} \tag{3}$$

where $\quad \varepsilon = \sqrt{\mu}\,, \quad u = \varepsilon(v - v_0)\,, \quad \tau = \varepsilon^2(v - v_0)$

and v_0 is the initial value of v at the epoch. This solution is a generalization of Kevorkian's two-variable solution for the planar problem (Kevorkian, 1970).

The functions r_n, α_n, z_n depend on the variables v, u, τ representing three different time-scales of the motion of the Trojan asteroids. The variable v corresponds to the orbital revolution of the asteroids around the Sun. The time-scale of the long-periodic librational motion around L_4 or L_5 is represented by the variable u. The variable τ is connected with the motion of the perihelion of the asteroids. The approximate periods are 12, 150, 3600 years.

The determination of the functions r_n, α_n, z_n is deduced to a system of partial differential equations. One of the equations to be solved is

$$\frac{\partial^2 \alpha_0}{\partial u^2} + 3\left[1 - 2^{-3/2}(1 - \cos\alpha_0)^{-3/2}\right]\sin\alpha_0 = 0. \tag{4}$$

The function $\alpha_0(u)$ describing the main part of the librational motion around L_4 or L_5 can be determined from Equation (4). An integral of Equation (4) is

$$\frac{1}{2}\left(\frac{\partial\alpha_0}{\partial u}\right)^2 - 3\left[\cos\alpha_0 - 2^{-1/2}(1-\cos\alpha_0)^{-1/2}\right] = h \tag{5}$$

where h might depend on τ, but it can be shown that h is a constant.

For moderate librational amplitudes which occur among the known Trojan asteroids the solution of Equation (4) is (Érdi, 1978, 1981)

$$\begin{aligned}\alpha_0 = {} & \frac{\pi}{3} + \frac{3\sqrt{3}}{2^3}\ell^2 + \frac{13\sqrt{3}}{2^8}\ell^4 + \\ & + \ell\cos\phi - \left(\frac{\sqrt{3}}{2^3}\ell^2 + \frac{\sqrt{3}}{2^8 3^2}\ell^4\right)\cos 2\phi + \\ & + \left(\frac{5}{2^6}\ell^3 - \frac{65}{2^{12}}\ell^5\right)\cos 3\phi - \frac{25\sqrt{3}}{2^7 3^2}\ell^4\cos 4\phi + \\ & + \frac{1283}{2^{12}\cdot 3\cdot 5}\ell^5\cos 5\phi + O(\ell^6)\end{aligned} \tag{6}$$

where

$$\phi = \sqrt{\frac{27}{2^2}\left(1 - \frac{3}{2^3}\ell^2 - \frac{97}{2^9}\ell^4\right)}\, u + \delta$$

and ℓ is a constant and δ depends on τ. The parameter ℓ means the approximate librational amplitude around L_4. For most of the known Trojan asteroids $\ell < 0.5$.

Substituting the solution (6) into Equation (5) it follows

$$h = \frac{3}{2} + \frac{27}{2^3}\ell^2 - \frac{81}{2^8}\ell^4 + O(\ell^6). \tag{7}$$

Equations (1) were derived from the solution (3) using the formulas of the two-body problem. In Equations (1) the parameters a, b, c and χ mean

$$a = -e_J\frac{A_2}{A_0}, \quad b = -e_J\frac{A_1}{A_0}, \tag{8a}$$

$$c = \varepsilon\varrho_{11}, \tag{8b}$$

$$\chi = A_0\tau + \psi_{11} \tag{8c}$$

where

$$A_0 = \frac{27}{2^3} + \frac{129}{2^6}\ell^2 - \frac{87}{2^7}\ell^4 + O(\ell^6), \tag{8d}$$

$$\frac{A_1}{A_0} = -\frac{1}{2} - \frac{17}{2^4 3}\ell^2 - \frac{329}{2^8 3^3}\ell^4 + O(\ell^6), \tag{8e}$$

$$\frac{A_2}{A_0} = -\frac{\sqrt{3}}{2} + \frac{73\sqrt{3}}{2^4 3^2}\ell^2 - \frac{6233\sqrt{3}}{2^8 3^4}\ell^4 + O(\ell^6) \qquad (8f)$$

and ϱ_{11} and Ψ_{11} are constants. Note, that Equations (1) are valid for asteroids around L_4. In case of asteroids around L_5 the parameter α must be changed by $-\alpha$.

In the paper (Érdi, 1979) Equations (1) were used to study the motion of the perihelion of 30 known Trojan asteroids. The parameter ℓ and the constants ϱ_{11}, Ψ_{11} can be determined from the osculating orbital elements of the asteroids. The parameter ℓ can also be calculated from the endpoints of the libration around L_4 or L_5. Thus it can be determined which Trojan asteroids have librating perihelion. From the 30 investigated asteroids 10 asteroids, all around L_4, proved to show the libration of the perihelion. A numerical integration of the planar elliptic restricted three-body problem (Érdi and Presler, 1980) confirmed the perihelion-libration of the same ten asteroids. In the case of the asteroid PL 4572 the libration of the perihelion was also shown by Bien (1980).

Equations (1) are applied here for 5 Trojan asteroids, all around L_5, which were not included in the earlier investigation. Table I shows the limits of the variations of e and $\tilde{\omega}$ of these asteroids and also the approximate periods of the variations of e and $\tilde{\omega}$, obtained from Equations (1) using the osculating orbital elements at the epoch December 27.0, 1980 ET.

Table 1. Variations of e and $\tilde{\omega}$ of 5 Trojan asteroids around L_5.

	e	$\tilde{\omega}$	$T_{e,\tilde{\omega}}$
1871 Astyanax	0.029-0.059	291°3-331°3	3230 years
1872 Helenos	0.029-0.060	287.7-329.6	3320
1870 Glaukos	0.030-0.066	279.4-323.4	3600
1867 Deiphobus	0.012-0.077	262.9-356.1	3290
1873 Agenor	0.077-0.172	0 -360	3590

Thus in the case of the asteroid 1873 Agenor the perihelion circulates and in the other four cases the perihelion librates. That means that from 35 known Trojan asteroids 14 asteroids have librating perihelion. (It must be mentioned that in Equations (1) and also in Table 1 $\tilde{\omega}$ is counted from the perihelion of Jupiter.)

Figure 2 shows the $e,\tilde{\omega}$ trajectories for 10 known Trojan asteroids around L_5. For those asteroids which are not included in Table 1 the trajectories were taken from the paper (Érdi, 1979).

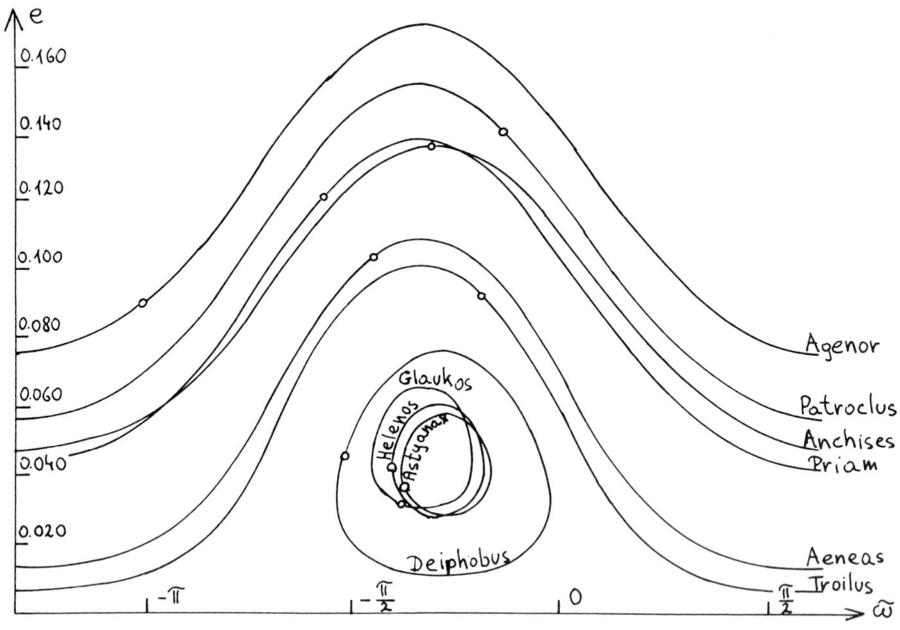

Figure 2. Variations of e and $\tilde{\omega}$ of 10 Trojan asteroids around L_5. The circles indicate the values at the epoch Dec. 27, 1980.

3. A CRITERION FOR THE LIBRATION OF THE PERIHELION

Considering Equations (1), (8a) and (8b) the condition for the libration of the perihelion is

$$0 \leqq \varrho_{11} \leqq e_1 \frac{\sqrt{A_1^2 + A_2^2}}{A_0} \tag{9}$$

where e_1 is defined by the equation

$$e_J = \varepsilon e_1 . \tag{10}$$

In order to determine such initial conditions which result in the libration of the perihelion the condition (9) should be expressed by the initial conditions of the motion. For the sake of simplicity only the planar motion of Trojan asteroids will be considered here.

Let the initial conditions at $v = v_0$ be

$$\begin{aligned} r &= 1 + \varepsilon r_0 , & \alpha &= \alpha_{00} , \\ \frac{dr}{dv} &= \varepsilon \dot{r}_0 , & \frac{d\alpha}{dv} &= \varepsilon \dot{\alpha}_{00} . \end{aligned} \tag{11}$$

Then using the approximate solution (Érdi, 1981)

$$r = 1 + \varepsilon\left[\rho_1 \cos(v + \psi_1) - \frac{2}{3}\frac{\partial \alpha_0}{\partial u}\right] + O(\varepsilon^2),$$

$$\alpha = \alpha_0 + \varepsilon\left[-2\rho_1 \sin(v + \psi_1) + q_1\right] + O(\varepsilon^2)$$

where q_1 is a known function of u and τ , and

$$\frac{dr}{dv} = -\varepsilon \rho_1 \sin(v + \psi_1) + O(\varepsilon^2),$$

$$\frac{d\alpha}{dv} = \varepsilon\left[-2\rho_1 \cos(v + \psi_1) + \frac{\partial \alpha_0}{\partial u}\right] + O(\varepsilon^2),$$

it follows that at $v = v_0$

$$\alpha_0 = \alpha_{00}, \quad \frac{\partial \alpha_0}{\partial u} = -3(2r_0 + \dot{\alpha}_{00}), \tag{12a}$$

$$\begin{aligned} \rho_1 \cos\psi_1 &= -(3r_0 + 2\dot{\alpha}_{00})\cos v_0 - \dot{r}_0 \sin v_0, \\ \rho_1 \sin\psi_1 &= -\dot{r}_0 \cos v_0 + (3r_0 + 2\dot{\alpha}_{00})\sin v_0, \end{aligned} \tag{12b}$$

$$q_1 = -2\dot{r}_0. \tag{12c}$$

Using the equations (Érdi, 1981)

$$\rho_1 \cos\psi_1 = \rho_{10}\cos(\alpha_0 + \psi_{10}) + e_1,$$

$$\rho_1 \sin\psi_1 = \rho_{10}\sin(\alpha_0 + \psi_{10}),$$

$$\rho_{10}\cos\psi_{10} = \rho_{11}\cos(A_0\tau + \psi_{11}) + e_1\frac{A_1}{A_0},$$

$$\rho_{10}\sin\psi_{10} = -\rho_{11}\sin(A_0\tau + \psi_{11}) - e_1\frac{A_2}{A_0}$$

and Equations (12b) , the constant ρ_{11} can be expressed by the initial conditions. Thus the criterion (9) can be substituted by the following criterion for the libration of the perihelion

$$\begin{aligned} -e_1^2\frac{A_1^2 + A_2^2}{A_0^2} \leqq\ & e_1^2 + (3r_0 + 2\dot{\alpha}_{00})^2 + \dot{r}_0^2 + \\ & + 2e_1^2\left(\frac{A_1}{A_0}\cos\alpha_{00} + \frac{A_2}{A_0}\sin\alpha_{00}\right) + \\ & + 2e_1\left[\frac{A_1}{A_0}(3r_0 + 2\dot{\alpha}_{00}) - \frac{A_2}{A_0}\dot{r}_0\right]\cos(v_0 + \alpha_{00}) + \\ & + 2e_1\left[\frac{A_2}{A_0}(3r_0 + 2\dot{\alpha}_{00}) + \frac{A_1}{A_0}\dot{r}_0\right]\sin(v_0 + \alpha_{00}) + \\ & + 2e_1(3r_0 + 2\dot{\alpha}_{00})\cos v_0 + 2e_1\dot{r}_0 \sin v_0 \leqq 0 . \end{aligned} \tag{13}$$

Note, that in this inequality A_1/A_0 and A_2/A_0 depend on the initial conditions through Equations (12a), (5), (7), (8e), (8f). Given the initial conditions r_0, α_{00} , $\dot{r}_0$, $\dot{\alpha}_{00}$ at $v = v_0$ it can be determined from (13) whether the perihelion makes libration (the inequality is satisfied) or circulation (the inequality is not satisfied). Two special cases will be considered next.

4. INITIAL CONDITIONS FOR PERIHELION-LIBRATION

4.1 Limit-velocity curves at L_4

The following problem is considered. Putting a small body into the point L_4 and starting it in a given direction with different initial velocities increasing from zero, determine that critical velocity where the libration of the perihelion of the orbit of the body changes to circulation. The critical velocities in different directions form a limit-velocity curve inside which all velocities, given to the body as initial velocity, will result in the libration of the perihelion. At different values of v_0, that is at different initial configurations of the three bodies, the limit-velocity curves are different.

Substituting $r_0=0$, $\alpha_{00}=\pi/3$ into (13) the condition for the libration of the perihelion in this case is

$$-e_1^2\frac{A_1^2+A_2^2}{A_0^2}\leqq K\leqq 0 \tag{14}$$

where
$$K=\dot{r}_0^2+B\dot{r}_0+C \tag{15}$$

and
$$\left.\begin{aligned}
B&=2e_1(F\sin v_0+G\cos v_0),\\
C&=4\dot{\alpha}_{00}^2+e_1^2(2F-1)+4e_1\dot{\alpha}_{00}(F\cos v_0-G\sin v_0),\\
F&=1+\frac{1}{2}\frac{A_1}{A_0}+\frac{\sqrt{3}}{2}\frac{A_2}{A_0},\\
G&=\frac{\sqrt{3}}{2}\frac{A_1}{A_0}-\frac{1}{2}\frac{A_2}{A_0}.
\end{aligned}\right\} \tag{16}$$

According to Equations (12a) and (5) in the case $r_0=0$, $\alpha_{00}=\pi/3$

$$h=\frac{3}{2}+\frac{9}{2}\dot{\alpha}_{00}^2$$

and from Equation (7) approximately

$$\ell^2=\frac{2^4}{3}\left(1-\sqrt{1-\frac{1}{2}\dot{\alpha}_{00}^2}\right). \tag{17}$$

Considering Equations (8e), (8f) and (16) now it can be seen that in (15) B and C depends only on $\dot{\alpha}_{00}$ and v_0. It can be shown that the minimum of K according to v_0 is

$$K_{min}=4\dot{\alpha}_{00}^2+e_1^2(2F-1)-4e_1\dot{\alpha}_{00}\sqrt{F^2+G^2}$$

and $$-e_1^2 \frac{A_1^2 + A_2^2}{A_0^2} \leqq K_{min} \tag{18}$$

for every $\dot{\alpha}_{00}$ in the interval (-0.6, 0.6) where Equation (17) gives those values of ℓ for which the solution (6) is sufficiently accurate. The equality holds at $\dot{\alpha}_{00}=0$, when $K_{min}=-e_1^2(A_1^2+A_2^2)/A_0^2 = -e_1^2$.

It follows from (15) and (18) that for every values of v_0 and at a given value of $\dot{\alpha}_{00}$ from the interval (-0.6, 0.6)

$$K \leqq 0$$

if $$\dot{r}_{02} \leqq \dot{r}_0 \leqq \dot{r}_{01}$$

where $$\dot{r}_{01} = -\frac{B}{2} + \sqrt{\left(\frac{B}{2}\right)^2 - C}\,, \quad \dot{r}_{02} = -\frac{B}{2} - \sqrt{\left(\frac{B}{2}\right)^2 - C}\,.$$

At $\dot{\alpha}_{00}=0$: $\dot{r}_{01} = e_1$, $\dot{r}_{02} = -e_1$ for every v_0.

Figure 3 shows the limit-velocity curves for $v_0=0$, $\pi/2$, π, $3\pi/2$. The curves for any v_0 and $v_0+\pi$ are mirror-images of each other for the centre of the coordinate system. The curves are open in the investigated region of $\dot{\alpha}_{00}$. For larger values of $|\dot{\alpha}_{00}|$ a numerical integration is necessary to determine the limit-velocity curves.

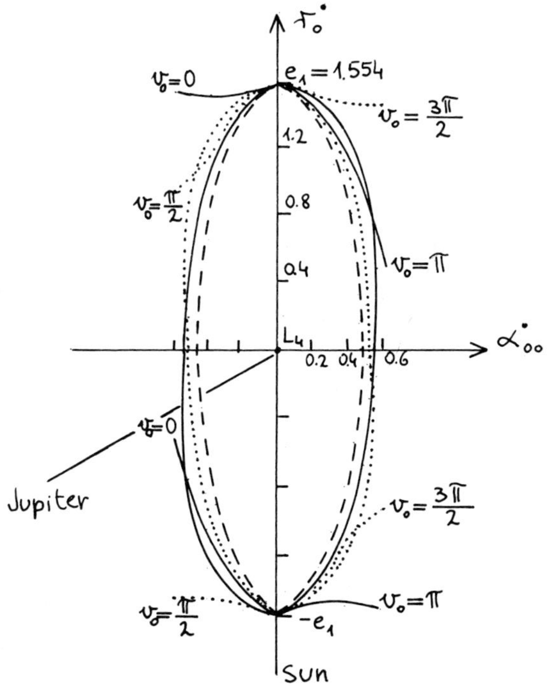

Figure 3. Limit-velocity curves at L_4.

Considering the extremum values of $\dot{r}_{01}$ and $\dot{r}_{02}$ according to v_0, an estimation can be derived for the allowed region of $\dot{r}_0$ at a given value of $\dot{\alpha}_{00}$ which is valid for every v_0. According to this estimation

$$|\dot{r}_0| \leqq -e_1\sqrt{F^2+G^2} + \sqrt{-4\dot{\alpha}_{00}^2 + e_1^2(1-2F) - 4e_1\dot{\alpha}_{00}\sqrt{F^2+G^2}} \quad \text{if } \dot{\alpha}_{00} \geqq 0$$

$$|\dot{r}_0| \leqq -e_1\sqrt{F^2+G^2} + \sqrt{-4\dot{\alpha}_{00}^2 + e_1^2(1-2F) + 4e_1\dot{\alpha}_{00}\sqrt{F^2+G^2}} \quad \text{if } \dot{\alpha}_{00} < 0$$

for every v_0. The corresponding limit-velocity curve is also shown on Figure 3 (dashed line). Inside that curve every velocity as initial velocity at L_4 will result in the libration of the perihelion of the small body for every initial configurations of the three bodies.

4.2 Libration with zero initial velocity

Let us determine that region around L_4 in which a small body starting from any point with zero initial velocity (in the rotating coordinate system in which Equations (2) are valid) will have an orbit with librating perihelion. A similar problem was studied by McKenzie and Szebehely (1981) but they determined initial positions around L_4 and L_5 with zero initial velocities for librational motions around L_4 and L_5 in the Earth-Moon system.

Now $\dot{r}_0 = 0$, $\dot{\alpha}_{00} = 0$, and (13) takes the form

$$-e_1^2\frac{A_1^2 + A_2^2}{A_0^2} \leqq K \leqq 0 \tag{19}$$

where $$K = 9r_0^2 + Br_0 + C, \tag{20}$$

and
$$\left.\begin{aligned} B &= 6e_1(F\cos v_0 - G\sin v_0), \\ C &= e_1^2(2F-1), \\ F &= 1 + \frac{A_1}{A_0}\cos\alpha_{00} + \frac{A_2}{A_0}\sin\alpha_{00}, \\ G &= \frac{A_1}{A_0}\sin\alpha_{00} - \frac{A_2}{A_0}\cos\alpha_{00}. \end{aligned}\right\} \tag{21}$$

Those values of r_0, α_{00} are to be determined for which the inequality (19) is satisfied.

It follows from Equations (12a), (5) and (7) that

$$\ell^2 = \frac{2^4}{3}\left(1 - \sqrt{\frac{7}{6} - 2r_0^2 + \frac{1}{3}\left[\cos\alpha_{00} - 2^{-1/2}(1-\cos\alpha_{00})^{-1/2}\right]}\right) \tag{22}$$

and according to Equations (8e), (8f) and (21) the parameters B and C in (20) depend on both r_0 and α_{00}. Thus the solution of the inequality (19) can be obtained only numerically.

Figure 4. shows a solution which has been obtained by the following method. For a given value of α_{00} the values of K are calculated for different values of τ_0 increasing and decreasing from zero. Then it is decided whether (19) is satisfied or not. The procedure is repeated for different values of α_{00} around $\alpha_{00} = \pi/3$.

According to Equation (22) as $|\tau_0|$ increases ℓ also increases and after a value of ℓ about 0.6 the solution (6) will not be accurate enough. Figure 4. has been obtained by fixing the upper value of ℓ as 0.5 which is valid for most of the known Trojan asteroids. Thus the curve on Figure 4 corresponds to those values of τ_0 and α_{00} for which $\ell = 0.5$. Along and inside this curve the inequality (19) is satisfied for every value of v_0.

The region in which the small body at any point with zero initial velocity will have an orbit with librating perihelion is certanly more extended than shown on Figure 4. However, its accurate size can be determined only by numerical integration.

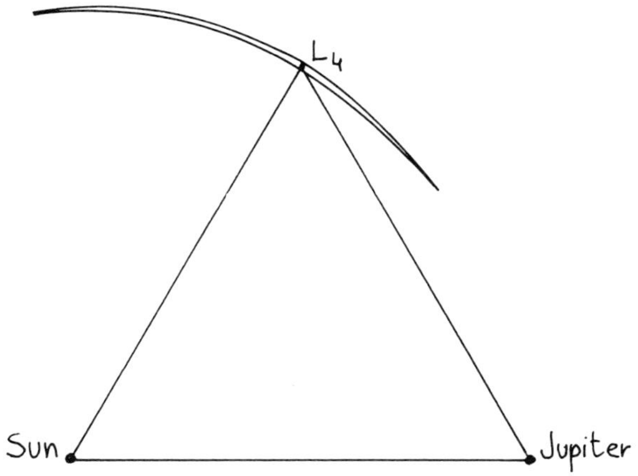

Figure 4. A region around L_4 for the libration of the perihelion with zero initial velocity.

REFERENCES

Bien, R.: 1980, Astron. and Astrophysics, 81, 255.
Érdi, B.: 1978, Celes. Mech., 18, 141.
Érdi, B.: 1979, Celes. Mech., 20, 59.
Érdi, B.: 1981, Celes. Mech., 24, 377.
Érdi, B., and Presler, W. H.: 1980, Astron. Journ., 85, 1670.
Kevorkian, J.: 1970, in G. E. O. Giacaglia (ed.) 'Periodic

Orbits, Stability and Resonances', D. Reidel Publ. Co., Dordrecht, Holland, p. 286.
McKenzie, R., and Szebehely, V.: 1981, Celes. Mech., 23, 223.

COLLISIONAL ORIGIN OF ASTEROID FAMILIES : EFFECTS OF THE TARGET'S GRAVITY

Vincenzo Zappalà[+], Paolo Farinella[++] and Paolo Paolicchi[+++]
[+] Osservatorio Astronomico di Torino, Pino Torinese, Italy
[++] Scuola Normale Superiore and Dipartimento di Matematica dell'Università, Pisa, Italy
[+++] Osservatorio Astronomico di Brera, Merate, Italy

ABSTRACT . The outcomes of asteroidal catastrophic collisions are strongly affected by the target asteroid's gravity, since only the fragments escaping with initial velocities higher than the target's escape velocity are not reaccumulated into "rubble pile" remnants. This idea can be compared with the observational evidence on the properties of family asteroids in several ways : (1) the shape and spin period of the "reaccumulated" family asteroids will roughly fit the relationships valid for self-gravitating fluid bodies; (2) the relative velocities of the few escaping fragments arising from a breakup event marginally overcoming self-gravity will often have an anisotropic distribution, affecting the final distribution of orbital elements; (3) the amount of mass which in a given family escaped to "infinity" will be correlated with the target's size, since only for objects larger than ∿ 100 km self-gravity plays an important role. These predictions are discussed and compared with the available data.

In some recent papers (Farinella et al., 1981; Paolicchi et al., 1982; Farinella et al., 1982) we have suggested that the outcomes of asteroidal catastrophic collisions, at least for targets exceeding some critical size, should be strongly affected by the existence of a significant self-gravitational binding of the target's material. After the breakup event, only the fragments ejected with initial velocities higher than the target's escape velocity will eventually reach independent orbits; all the other fragments will be quickly reaccumulated into "rubble pile" remnant bodies, i.e., loose and internally fractured aggregates of material held together mainly by its own self-gravity and having very low mechanical strength (see also Davis et al., 1979, and Weidenschilling, 1981). These objects will tend to relax toward the equilibrium figures consistent with their spin rate and density; in order of increasing angular momentum of rotation, we will get oblate Maclaurin spheroids, triaxial Jacobi ellipsoids and

V. V. Markellos and Y. Kozai (eds.), Dynamical Trapping and Evolution in the Solar System, 177–188.

Darwin binary systems (Chandrasekhar, 1969). The critical size for which self-gravitational effects become important in determining the collisional response can be estimated by comparing the target's escape velocity V_e with the typical velocity of the fragments ejected after a catastrophic impact (V_{fr}); this latter at present can only be estimated by laboratory impact experiments such as those performed by Fujiwara and Tsukamoto (1980). Following the assumptions of Farinella et al. (1982), we have

$$V_e \simeq 1.2 \times 10^{-3} (\rho/2.5)^{1/2} R \text{ cm/s} , \qquad (1)$$

$$V_{fr} \simeq 2.1 \times 10^{-3} (E/M)^{0.76} \text{ cm/s} , \qquad (2)$$

where R and ρ are the target's radius and density, while E/M is the ratio between the projectile's kinetic energy and the target's mass (all these quantities must be expressed in cgs units). Typically the largest impacts endured by asteroids larger than $\sim$ 10 km occurred at velocities of the order of 5 km/s and involved projectile objects of mass 10^{-3} to 10^{-2} times the mass of the target (Farinella et al., 1982). This implies that for the majority of asteroids the most destructive collision occurred with an impact specific energy in the range 10^8 to 10^9 erg/g. According to the evidence provided by laboratory experiments, this results always into an extensive fragmentation of the target, since most rocky materials are already ruptured at E/M values of the order of 10^7 erg/g. Moreover, the critical size over which $V_e > V_{fr}$ (implying effective gravitational reaccumulation of fragments) comes out to be in the range 50 to 200 km, depending on the energy and geometry of the largest collision. This is a size range where many observational data on the asteroid physical properties are now available; since the previous conclusions are partially based on questionable extrapolations of small-scale laboratory experiments to bodies of asteroidal dimensions, it seems very interesting to try a check of the theoretical predictions with the data.

The general considerations summarized above find a natural and intriguing application in the particular case of the asteroid dynamical families. These sets of objects, having independent heliocentric orbits with very close proper elements, are widely believed to be of collisional origin, i.e., to represent the remnants of the recent collisional disruption of a common parent asteroid. The physical and orbital properties of asteroids belonging to various families have been analysed by several investigators (Wiesel, 1978; Gradie et al., 1979; Ip, 1979; Fujiwara, 1982; and others) in order to reconstruct the properties of the parent body or the mechanism of collisional rupture. We now want to point out the remarkable fact that, by using the orbital data, it is possible to estimate the original relative velocities of the members of a given family : the typical velocity dispersion results to be in the range 0.1 to 0.3 km/s. This is within a factor two of the escape velocity of an object of radius $\sim$ 100 km, as many parent bodies of the asteroid families presumably were. In the case of families, therefore, we are probably observing just the outcome of the limiting case in which the target's self-gravitational barrier is marginally overcome by a part of the broken fragments, which can escape but retain only small relative velocities "at infinity", resulting eventually into a clustering of heliocentric orbital elements. In such conditions, it seems likely that the largest object resulting from the breakup is the core of the parent body covered by a "megaregolith", i.e., a "rubble pile" asteroid of the type described before. This outcome is produced because the low-velocity portion of the ejecta mass distribution conceivably has not sufficient kinetic energy to reach the escape velocity, and therefore falls back onto the principal remnant producing a partially reaccumulated central object within the family.

How can we compare this theoretical scenario with the observational evidence ? First of all, we have to face the difficulty that the exact population of each family is not univocally known : several authors have obtained significantly different family classifications (Carusi and Valsecchi, 1982), or at least their proposed memberships are often contrasting. For our present purposes, we have chosen to use the list of memberships and proper elements given by Williams (1979), limiting ourselves to 33 families having at least five numbered asteroids as members. This seems to us a reasonable compromise between the most restricting classifications, obtaining only the few numerous families originally discovered by Hirayama, and the broadest classifications for which the vast majority of asteroids belongs to some family. We decided to exclude from the analysis a few Williams' families for which the observed distributions of sizes and proper elements among the members suggest that some large interloper is present in the family itself. For instance, families No. 106, 113 and 138 have two or three largest members of comparable size, and this contrasts with the typical mass distribution arising from a collisional rupture (Fujiwara et al., 1977) and generally observed in the asteroid belt (Kresák, 1977). In a similar way, it seems very unlikely that Ceres really belongs to family No. 67 : no plausible physical mechanism could cause the escape from Ceres of a 150-km sized fragment like 39 Laetitia; therefore , this family is considered without Ceres. Our sample of 33 families can then be analyzed from several points of view.

A first investigation can regard the photometrically determined rotational properties (spin period and lightcurve morphology; the maximum lightcurve amplitude is a rough indicator of the asteroid's triaxial elongation) for family asteroids of different sizes, and its main results have been already reported by Paolicchi et al., 1982, and Farinella et al., 1982. In brief, we have verified that for sizes larger than ∿ 100 km the rotational properties of family asteroids show a clear correlation between short periods and highly elongated shapes, which can be satisfactorily interpreted in terms of equilibrium shapes of self-gravitating "fluid" bodies (since the Jacobi triaxial figures can exist only for short rotational periods, ranging from $\sim$ 4 to $\sim$ 6 hr). The non-family asteroids show the same type of correlation only for diameters larger than $\sim$ 200 km, while in the size range from 100 to 200 km several objects having non-equilibrium shapes (for instance elongated but slow rotators) are certainly present (see also Farinella et al., 1981). This suggests that most intermediate-size family asteroids are indeed covered by deep layers of fragmented material whose global shape has roughly relaxed to the gravitational equilibrium figure; this structure appears at smaller sizes than for non-family objects because the breakup events generating families had to be always energetically close to the threshold for gravitational reaccumulation of fragments, since otherwise no clustering of orbital elements would be actually observed.

A second type of analysis is based on the reconstruction of the mass distribution within each family, and on the assumption that the total observed mass corresponds roughly to the original mass of the parent body (PB). Of course a part of this mass could have been missed either due to physical reasons (if very small and/or high-velocity ejecta were produced by the catastrophic breakup), or due to the magnitude bias of the observations (mainly in the case of the outer-belt families), but for our statistical purposes these effects should not be very important and, in any case, for a typical mass distribution of fragments the contribution of the missing members to the total mass of the family is small. The asteroid masses have been obtained by using the diameters listed by Bowell et al. (1979) and a mean density of 2.5 g/cm^3, or alternatively, when no diameter value was available, by estimating albedos via the known taxonomic types in the family and/or the position in the belt (i.e., by using an S-type 0.16 albedo for semimajor axes smaller than 2.6 AU, a C-type 0.037 albedo beyond 2.7 AU and the intermediate value 0.10 in the interval 2.6 to 2.7 AU). Then the mass of the largest remnant (LR) asteroid in each family (M_{LR}) can be compared with the mass of the parent body (M_{PB}), and we can obtain a ratio δ which is closer to one for families whose mass is more concentrated in the largest object (and for which, presumably, the gravitational reaccumulation mechanism was more effective).

Before discussing the resulting values of δ and their correlations with other parameters characterizing the families, we think that it is useful to introduce a qualitative classification by dividing our sample of families into two main subsamples. Following Ip (1979), we can derive from the semimajor axis difference Δa of each body with respect to the LR of its family the along-track component ΔV of the relative velocity when the two Keplerian orbits intersect :

$$\Delta V \simeq n\,\Delta a\,/\,2 \quad , \tag{3}$$

where n is the orbital mean motion of the LR. ΔV can be identified with the ejection velocity after breakup, since the laboratory experiments have shown that in most cases the largest fragment moves very slowly with respect to the target. In this way we can plot the mass distribution of the fragments of each family versus the ejection velocity, in order to get some insight into the kinematics of the fragment ejection. In Figure 1 various examples of such plots are shown, where the vertical axis refers to the percentage of fragmental mass M_{fr} ($\equiv M_{PB} - M_{LR}$) assigned to each ΔV bin of width 0.02 km/s (the family numbers of Williams' classification are indicated in each case in the brackets). From the Figure it is clear that two broad categories can be readily identified : the "asymmetric" families, whose fragments are asymmetrically distributed on one side of the LR, and the "dispersed" families, which have their LR roughly at the center of the fragmental mass distribution. This classification could be made more quantitative by defining an "asymmetry parameter"

$$C \equiv \langle \Delta V^2 \rangle / \langle \Delta V \rangle^2 \quad , \tag{4}$$

where the mean values are weighted over the mass of the various fragments forming the family : for asymmetric families we have values of C of the order of one, whilst for dispersed families $C \gg 1$. As remarked by Ip (1979), a similar conclusion follows when one uses more complex two- or three-dimensional mass vs. ejection velocity distributions, obtained by analysing the differences in proper eccentricity and inclination among the family members. In this case, however, a real reconstruction of the three-dimensional ejection velocity vector would need the knowledge of the PB's angular orbital elements (including true anomaly) at the time of breakup, and therefore some additional assumption must be introduced.

Ip suggested that the asymmetric families arose because, after the catastrophic breakup, fragments were not scattered isotropically, but were ejected with some preferential direction. We agree that the asymmetry effect is connected with the mechanism of fragmentation, but in our opinion it is strongly amplified by the influence of the PB's self-gravity : when most fragments fall back and form a reaccumulated asteroid, it is highly probable that the small high-velocity fraction of escaping bodies is distributed anisotropically, and only these fragments are observed today as minor family members. If we use a reasonable initial velocity distribution of the ejected mass, it is easy to verify that an initially high C value (corresponding to a nearly-isotropic explosion) is reduced by the reaccumulation mechanism by a factor of the same order as the ratio between the initially ejected mass and the mass of the non-reaccumulated fragments. Therefore, we expect that asymmetric families have LRs with a "rubble pile" structure, since the gravitational reaccumulation was

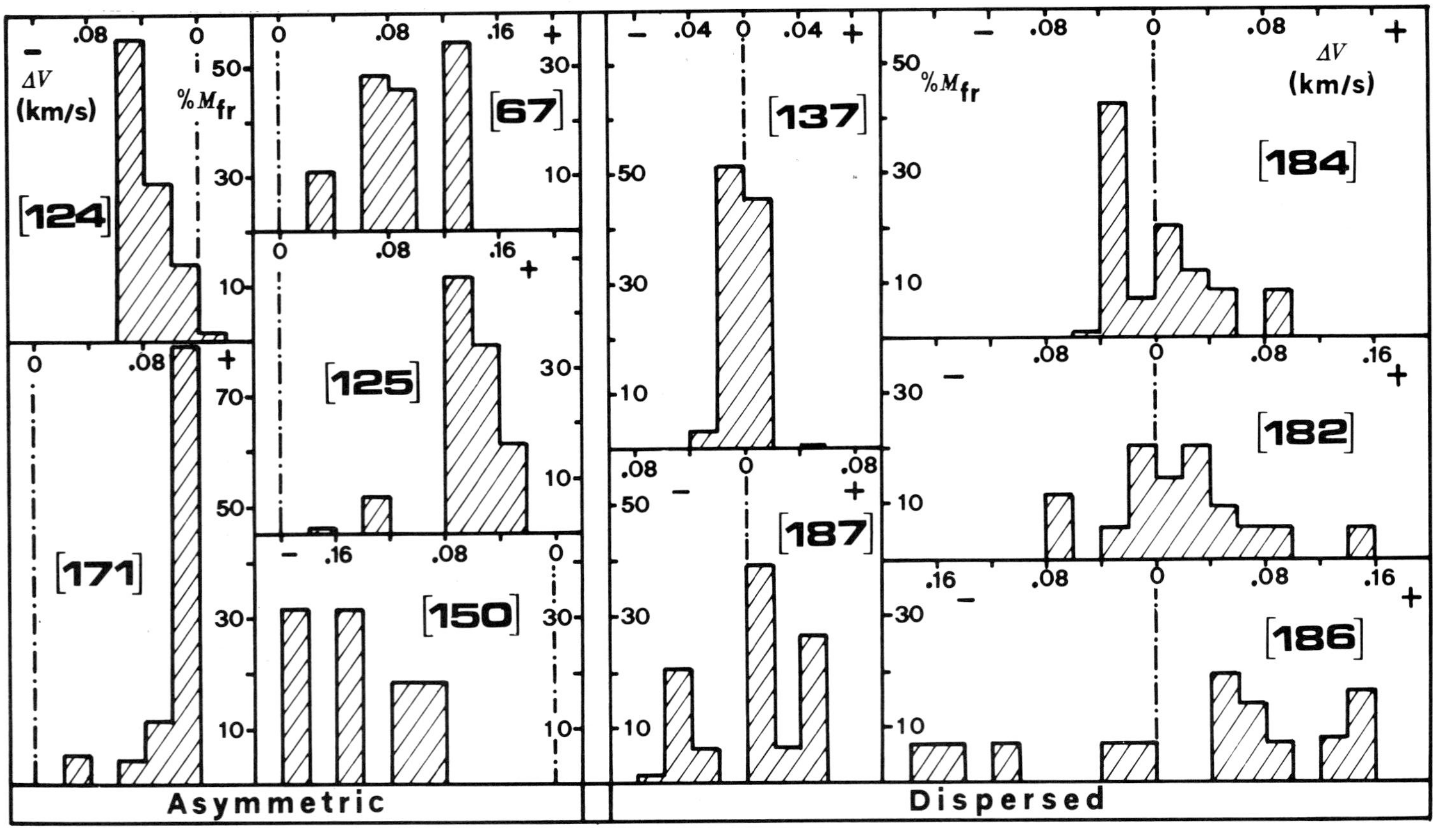

Figure 1. The Figure shows the fragmental mass vs. ΔV (ejection velocity) distributions for 10 Williams' families, 5 of the "asymmetric" and 5 of the "dispersed" type (see text). The dash-dotted vertical lines, corresponding to the zero of the ΔV axes, refer to the location of the largest member of each family (whose mass is not included in the distributions). The numbers in brackets identify the families as listed by Williams (1979).

very likely considerable. As a matter of fact, we find very frequently that the rotational properties of the LRs of asymmetric families are diagnostic of nearly-equilibrium figures, in the sense described earlier.

It is worthwhile to note that the mass distributions of asymmetric families often present a remarkable ΔV gap between the LR and the main secondary concentrations of mass (see Figure 1). This fact can be explained in the following way. Let us assume that the initial mass vs. velocity distribution had the form

$$dM = C\ V_o^{-\alpha}\ dV_o \qquad (\text{for } V_o > V_{min}) \quad , \qquad (5)$$

where V_o is the ejection velocity after breakup, dM is the fragmental mass ejected in the interval (V_o, $V_o + dV_o$), C and α are constants and V_{min} is a cutoff at small speeds (i.e., dM = 0 for $V_o < V_{min}$). For crater ejecta, we have $\alpha \simeq 3$. At "infinity", that is after escaping the target's gravity, the velocity of a fragment will be

$$V = (\ V_o^2 - V_e^2\ )^{1/2} \qquad (6)$$

and the mass distribution, as a function of V (the observable quantity) becomes :

$$dM = C\ V\ (\ V^2 + V_e^2\ )^{-(\alpha+1)/2}\ dV \equiv f(V)\ dV \qquad (7)$$

(provided $V_e > V_{min}$). Note that the function f(V) approaches to zero for small values of V, and has a maximum for $V = V_e/\sqrt{\alpha}$. Due to the non-linear relation between V and V_o, therefore, the probability of observing at "infinity" velocities much lower than V_e is very small (this conclusion follows for every reasonable form of the mass vs. velocity distribution, and does not depend strictly on the use of Eq.(5)). For instance, for a target's escape speed V_e= 0.1 km/s, the only fragments with final velocities smaller than 0.02 km/s will be those ejected initially with V_o in the narrow range 0.1 ÷ 0.102 km/s. This effect could easily explain the observed ΔV gaps; moreover, it favours a concentration of mass for final velocities of the order of the escape speed, and this is also observed in most cases (as noted earlier).

For all the above-mentioned reasons, it is plausible to expect that asymmetric families will show preferentially the marks of an effective gravitational reaccumulation, i.e., besides equilibrium-shaped LRs, also values close to one of the ratio $\delta \equiv M_{LR}/M_{PB}$. In Figure 2 the various families of our sample are represented in a diagram showing δ versus M_{LR}. Full circles correspond to asymmetric families, while open circles indicate dispersed families; some intermediate cases are also shown as half-colored circles. We can note that asymmetric families lie always close to the line $\delta = 1$, and this correlation is particularly strong for objects with nearly-equilibrium shapes. On the contrary, dispersed families are

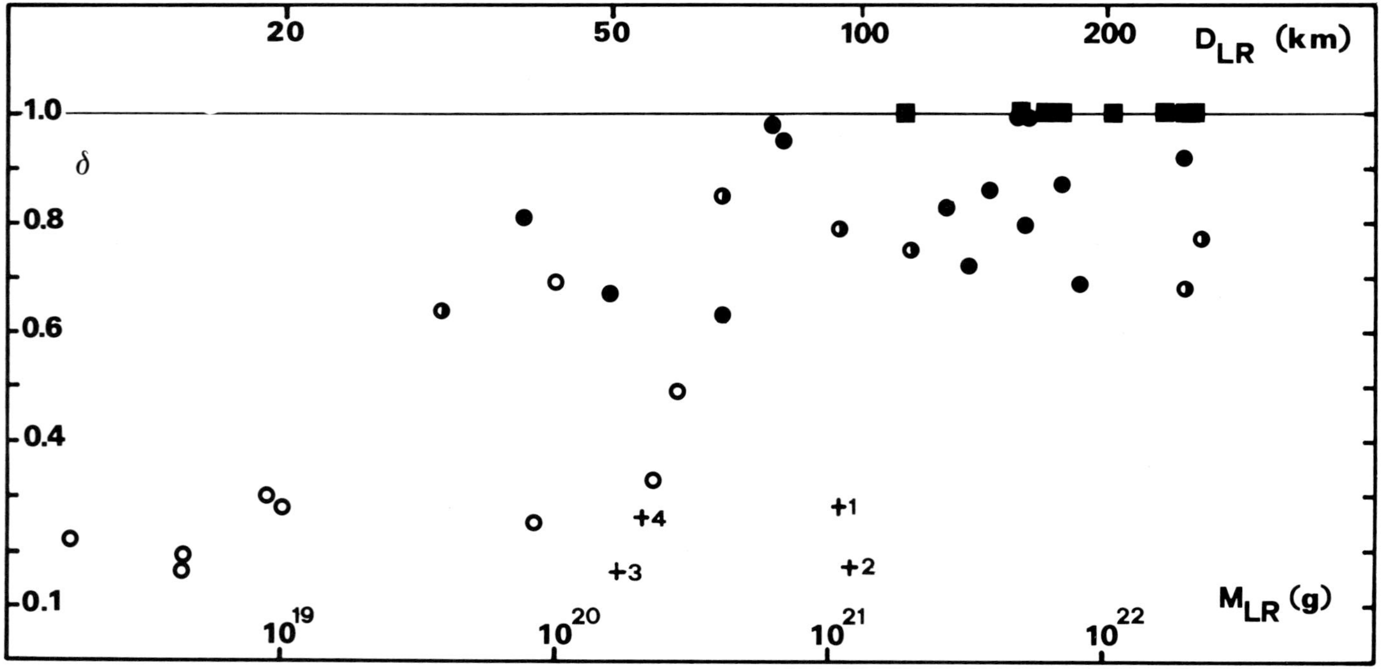

Figure 2. For each family the mass ratio between the largest member and the parent body is shown versus the mass (lower horizontal scale) and the diameter (upper scale) of the largest member. Open circles represent dispersed families, full circles asymmetric families and half-colored circles indicate intermediate cases. The numbered crosses correspond to the first four Hirayama families (as classified by Williams, 1979), while the full squares represent non-family asteroids having presumably triaxial equilibrium shapes (Farinella et al., 1981).

present throughout the range $0.1 \lesssim \delta \lesssim 0.8$. This trend clearly supports our interpretation that for asymmetric families the catastrophic breakup was just energetic enough to allow a very limited fragment escape, while most of the mass was gravitationally reaccumulated. In the case of dispersed families, on the other hand, the mass loss from the PB was much more pronounced, indicating a comparatively weaker effect of gravity.

The same conclusion is supported by another trend clearly evident in Figure 2 : δ is always $\gtrsim 0.7$ for LR diameters $D_{LR} \gtrsim 100$ km, and always $\lesssim 0.3$ for $D_{LR} \lesssim 30$ km. In the intermediate diameter range, a large dispersion of δ values is apparent. Correspondingly, asymmetric families are preferentially located at large diameters, while the opposite is true for dispersed families. These correlations can be easily interpreted by recalling the transition that we previously described between the response to catastrophic impacts of small asteroids, for which the target's gravity is not so important, and that of large asteroids, which are frequently subjected to gravitational reaccumulation of fragments. In this latter case (corresponding to $V_e \gtrsim V_{fr}$) a family, whenever formed, must have $\delta \simeq 1$ and, quite probably, an asymmetric appearance. In Figure 2 we have plotted as full squares also the non-family asteroids whose shapes have been interpreted as Jacobi equilibrium figures by Farinella et al. (1981): they are located obviously along the line $\delta = 1$, but the fact that these objects are found preferentially at large diameters appears now as a natural extrapolation of the trend shown by family asteroids. The non-family equilibrium-shaped objects could be considered as "non-born" families, in the sense that they are probably outcomes of impact events causing fragmentation of the target but for which self-gravitation prevented any dispersion of mass, yielding finally a completely reaccumulated "rubble pile" body (Farinella et al., 1982). In general, the evidence provided by Figure 2 represents a strong indication that the transition towards gravity-dominated collisional outcomes does really occur : it appears to begin for diameters of about 50 km and to be almost completed at diameters $\sim$ 200 km. This result compares satisfactorily with the theoretical estimates based on the results of laboratory impact experiments.

Finally, in Figure 2 we have marked by numbered crosses four of the families originally discovered by Hirayama (the numbers refer to Williams' classification). They are all located in a peculiar region, at the boundary of the zones more populated by the other families, i.e., they have lower values of δ and/or larger values of D_{LR} with respect to the general trend. This means that these families have small LRs though their PB mass was fairly large. Several hypotheses could explain this peculiar feature : (a) the rupture event possibly was more energetic than the average, due to a larger projectile mass or velocity (note also that Fujiwara, 1982, quotes the possibility suggested by experimental results that for the same energy a larger projectile mass results into a more effective energy transfer to fragments); (b) possibly the catastrophic event was not a single one : after a first collision converting the PB into a "rubble pile" remnant, a second impact could have dispersed most of the mass (this could explain why the Themis family consists of a "core" of large objects surrounded by a "cloud" of smaller bodies, as noted by Gradie et al., 1979); (c) possibly these are not single families, but are composed by genetically different groups overlapping each other in the space of proper elements (as it seems the case for the Flora family; see Tedesco, 1979). At any rate, a more detailed analysis of the breakup event which generated the Eos, Themis and Koronis families has led Fujiwara (1982) to the conclusion that also in the case of these families the gravitational reaccumulation mechanism had to be of crucial importance.

Before concluding, we have to point out some problems and difficulties of the interpretation proposed in this paper, which in our opinion do not question its validity but certainly deserve further scrutiny. First of all, we have to be cautious in extrapolating all the small-scale experimental results on catastrophic collisions to impact events between bodies of asteroidal size. As we have seen, the experimental relationship between V_{fr} and E/M is consistent with the data on asteroid properties; on the other hand, Fujiwara and Tsukamoto (1980) found another empirical relationship between E/M and the M_{LR}/M_{PB} mass ratio ($M_{LR}/M_{PB} \simeq 7 \times 10^8$ $(E/M)^{-1.34}$, with E/M expressed in erg/g), which yields clearly inconsistent results. According to this relation, in fact, we should expect almost always a very high degree of fragmentation of the target, with typical values of M_{LR}/M_{PB} of the order of 10^{-3}. But even in the case of dispersed families (for which we have indications that self-gravity has not been very effective), we observe always $\delta \gtrsim 0.1$; the problem is not removed by the fact that obvious selection effects favour the discovery of families with δ not much lower than one. This discrepancy suggests that due to some unknown reason for the same impact energy the asteroids are more resistant than laboratory targets to collisional comminution, even if the broken fragments seem to be ejected with similar speeds in the two cases. The same problem appears for the Saturnian satellite Hyperion, if we assume that its irregular shape is due to a collisional breakup (Farinella et al., 1983), and this fact could indicate that some fundamental physical reason prevents the extrapolation of laboratory results. An alternative possibility could be that the discrepancy is due to a large structural difference between the solid laboratory targets and the asteroids, if the majority of them is converted into "rubble piles" before enduring a really catastrophic impact. Perhaps some comparative experiment on the collisional response of solid vs. fractured targets could test this hypothesis.

Different problems arise when we try to derive the three-dimensional ejection velocities by using the differences in proper eccentricity and inclination among the asteroids of a given family. From a preliminary analysis, it seems that very often the contribution of eccentricity is dominant and while this could be understood in a single case (due to the possible anisotropy of the velocity field and to the unknown angular orbital elements of the PB at the time of breakup), when we "average" over our entire sample of families, the components of the relative velocities in different directions with respect to the original orbit should give comparable contributions. Possibly this problem is due to some bias implicit in the clustering method used to define family memberships. We note also that in some cases of asymmetric families such a bias could explain why the peak of the fragmental mass distribution is shifted at velocities higher than the PB's escape velocity, contrasting with the trend (described before) connected with the process of gravitational reaccumulation.

In conclusion, we can state that more extensive and detailed studies of the asteroid families, coupling physical and dynamical information, have the potential of providing new and significant insights into the physics and history of the asteroid collisional process. The main prerequisites at present appear to be : (1) a more univocal and reliable definition of the family memberships; (2) a better understanding of the rupture mechanisms for bodies of different structure and size; (3) an enlargement and improvement of the data base on the physical and rotational properties of the family asteroids.

ACKNOWLEDGMENTS

We thank for valuable discussions on the subject of this paper J.A. Burns, Z. Knežević and G.B. Valsecchi. This work and the participation of one of us (P.F.) in the I.A.U. Coll. No. 74 were partially supported by the National Research Council of Italy (C.N.R.).

REFERENCES

Bowell, E., Gehrels, T., and Zellner, B.: 1979, in *Asteroids* (T. Gehrels, Ed.), pp. 1108 - 1129, University of Arizona Press, Tucson.

Carusi, A., and Valsecchi, G.B.: 1982, Astron. Astrophys., in press.

Chandrasekhar, S.: 1969, *Ellipsoidal Figures of Equilibrium*, Yale University Press, New Haven and London.

Davis, D.R., Chapman, C.R., Greenberg, R., Weidenschilling, S.J., and Harris, A.W.: 1979, in *Asteroids* (T. Gehrels, Ed.), pp. 528 - 557, University of Arizona Press, Tucson.

Farinella, P., Paolicchi, P., Tedesco, E.F., and Zappalà, V.: 1981, Icarus 46, pp. 114 - 123.

Farinella, P., Paolicchi, P., and Zappalà, V.: 1982, Icarus, in press.

Farinella, P., Milani, A., Nobili, A.M., Paolicchi, P., and Zappalà, V.: 1983, Icarus, in press.

Fujiwara, A., Kamimoto, G., and Tsukamoto, A.: 1977, Icarus 31, pp. 277 - 288.

Fujiwara, A., and Tsukamoto, A.: 1980, Icarus 44, pp. 142 - 153.

Fujiwara, A.: 1982, submitted to Icarus.

Gradie, J.C., Chapman, C.R., and Williams, J.G.: 1979, in *Asteroids* (T. Gehrels, Ed.), pp. 359 - 390, University of Arizona Press, Tucson.

Ip, W.-H.: 1979, Icarus 40, pp. 418 - 422.

Kresák, Ľ.: 1977, Bull. Astron. Inst. Czech. 28, pp. 65 - 82.

Paolicchi, P., Farinella, P., and Zappalà, V.: 1982, in *Sun and Planetary System* (W. Fricke and G. Teleki, Eds.), pp. 295 - 298, D. Reidel, Dordrecht, Holland.

Tedesco, E.F.: 1979, Icarus 40, pp. 375 - 382.

Weidenschilling, S.J.: 1981, Icarus 46, pp. 124 - 126.

Wiesel, W.: 1978, Icarus 34, pp. 99 - 116.

Williams, J.G.: 1979, in *Asteroids* (T. Gehrels, Ed.), pp. 1040 - 1063, University of Arizona Press, Tucson.

ANALYSIS OF A SIMPLE MECHANISM TO DEPLETE THE KIRKWOOD GAPS

Anne LEMAITRE
Department of Mathematics
Facultés Universitaires Notre-Dame de la Paix à Namur, Belgium
Aspirante F.N.R.S.

§1. INTRODUCTION

In most applications resonance problems are implicitely or explicitely modelized by the *Fundamental Model of Resonance* i.e. the pendulum, characterized by its Hamiltonian function :

$$H = \frac{a}{2} I^2 - b \cos \psi \tag{1}$$

This reduction is performed in two steps : first, the action-angles variables are introduced to get a "one degree of freedom" Hamiltonian system, given by (2) :

$$K = K_o(S) + \varepsilon K_1(S, s) \tag{2}$$

where K_1 is 2π - periodic in s (the resonant angle).
Secondly, K_o and K_1 are expanded in Taylor's series with respect to S and then truncated.
This method is easy to apply when $K_1(0, s)$ is a non-constant function of s (its simplest form is $K_1(0, s) = \cos s$, which leads to the pendulum by a simple translation). However in many instances $K_1(S, s)$ possesses the "d'Alembert Characteristic" in $(\sqrt{2S}, s)$. This is the case in many orbit-orbit resonances where $\sqrt{2S}$ is proportional to the eccentricity and in some spin-orbit resonances where $\sqrt{2S}$ is there proportional to the obliquity.
In such cases, the simplest form for the truncated Hamiltonian is :

$$K' = \alpha S + \beta S^2 + \varepsilon \sqrt{2S} \cos s \tag{3}$$

where the reduction to the pendulum (1) is no longer that simple and is only valid on some parts of the phase space and for some values of the three parameters α, β and ε .

The functions (1) and (3) are both *one degree of freedom* hamiltonians; the only advantage of (1) over (3) is that most of the computations can be carried out by means of elliptic functions which are well tabulated.

V. V. Markellos and Y. Kozai (eds.), Dynamical Trapping and Evolution in the Solar System, 189–201.

We think that this advantage is not important enough to justify the approximation and the intricacies involved in the last step.

So we would like to introduce (3) as a *Second Fundamental Model of Resonance* as simple and as well documented as the pendulum (1) but closer to the main resonance problems.

A complete analysis of this model is described elsewhere (Henrard-Lemaître 1983). We plan to summarize this analysis here and to apply it to the problem of depletion of some of the *Kirkwood gaps* in the asteroïd belt.

§2. SCALING THE MODEL

The Hamiltonian (3) depends on three parameters α, β and ε which are not independent. By scaling the time and the actions, we can consider that our model has only one truly independent parameter called δ. Let us define :

$$\begin{aligned} r &= \operatorname{sign}\beta \,.\, s \qquad &\text{if} \quad \beta\varepsilon < 0 \\ r &= \operatorname{sign}\beta \,.\, s + \pi \qquad &\text{if} \quad \beta\varepsilon > 0 \end{aligned} \tag{4}$$

and the scaled time and momentum :

$$T = \left[\frac{|\beta|\ \varepsilon^2}{4}\right]^{1/3} . \ t \tag{5}$$

$$R = \left[\frac{2\ \beta}{\varepsilon}\right]^{2/3} . \ S \tag{6}$$

The Hamiltonian (3) is replaced by :

$$H(r,\ R) = -3(\delta + 1)\ R + R^2 - 2\sqrt{2R}\cos r \tag{7}$$

and δ is given by :

$$\delta = -\left[\frac{4}{27}\ \frac{\alpha^3}{\beta\varepsilon^2}\right]^{1/3} - 1 \tag{8}$$

The Hamiltonian (7) is not differentiable at $R = 0$; this is why we introduce canonical cartesian variables x and y :

$$x = \sqrt{2\ R}\ \cos r \qquad \text{and} \qquad y = \sqrt{2\ R}\ \sin r \tag{9}$$

and the Hamiltonian (7) becomes :

$$H_2(y, x) = -\frac{3}{2}(\delta + 1)(x^2 + y^2) + \frac{1}{4}(x^2 + y^2)^2 - 2x \qquad (10)$$

§3. LIBRATION AND CIRCULATION

The equilibria of the Hamiltonian dynamical system are given by the solutions of :

$$\frac{\delta H}{\delta x} = \frac{\delta H}{\delta y} = 0 \quad . \qquad (11)$$

They are located at points $(x, y) = (x^\star, 0)$, where $x^\star$ is a root of :

$$x^3 - 3(\delta + 1)x - 2 = 0 \quad . \qquad (12)$$

The location of the roots versus δ is given in figure 1. For $\delta > 0$ one of the roots (the leftmost one) is unstable and the other two are stable. Two homoclinic orbits come out of the unstable equilibrium. They enclose an area that we call the *topological libration zone* (see figure 2).

Besides the libration zone we distinguish two circulation zones (an internal one, inside the smallest homoclinic orbit and an external one, outside the largest homoclinic orbit).

To describe the capture into resonance we need a new concept. We shall call the *area index* of a trajectory the area enclosed by it (plus the area enclosed by the smallest homoclinic orbit in the case of libration trajectories). In this way there is a one-to-one correspondence between trajectories and area indexes (see figure 3).

§4. SOLUTIONS FOR MODERATE δ

In figures 4 and 5 some of the trajectories of the problem are plotted for some values of δ. In (Henrard - Lemaître 1983) we give more information about the orbits (periods, area indexes, energy constants).
For large values of δ asymptotic formulae can be calculated, obtained by expansions with respect to Δ, where $\Delta = -3(\delta + 1)$. (13)

§5. EVOLUTION THROUGH RESONANCE

Let us assume that the parameter δ slowly varies with the time, i.e.

$$|\dot{\delta}| \leqslant \eta \qquad |\ddot{\delta}| \leqslant \eta^2 \qquad (14)$$

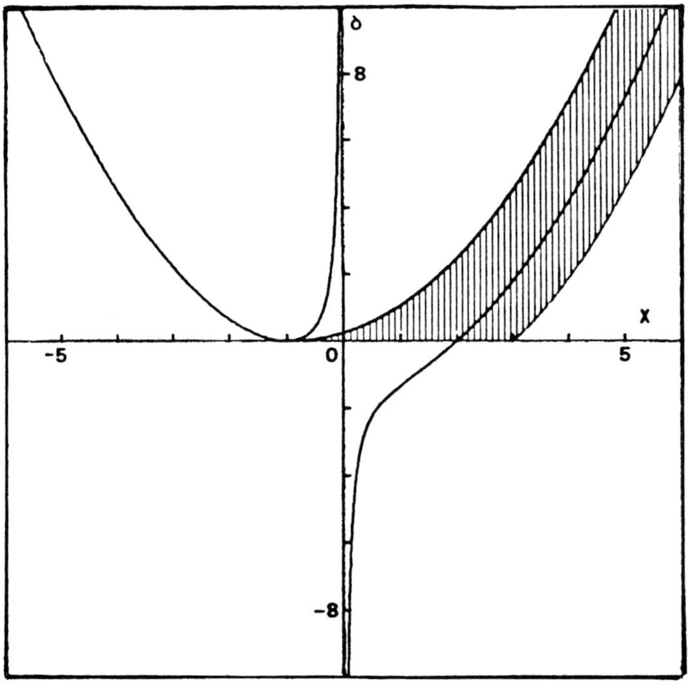

figure 1 : location of the equilibria .

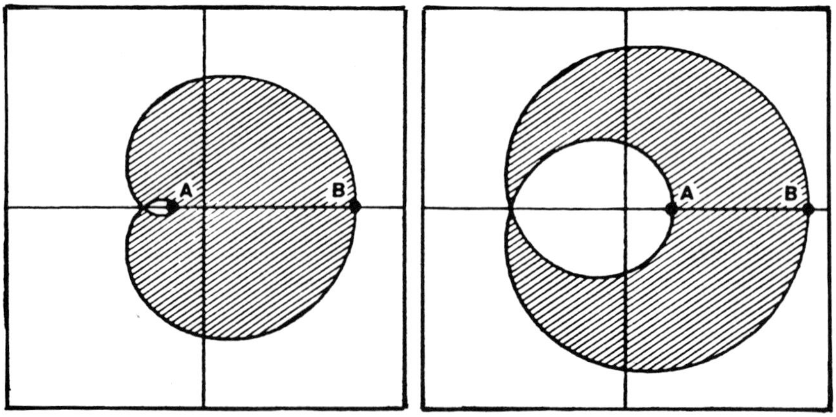

figure 2 ; topological libration zone .

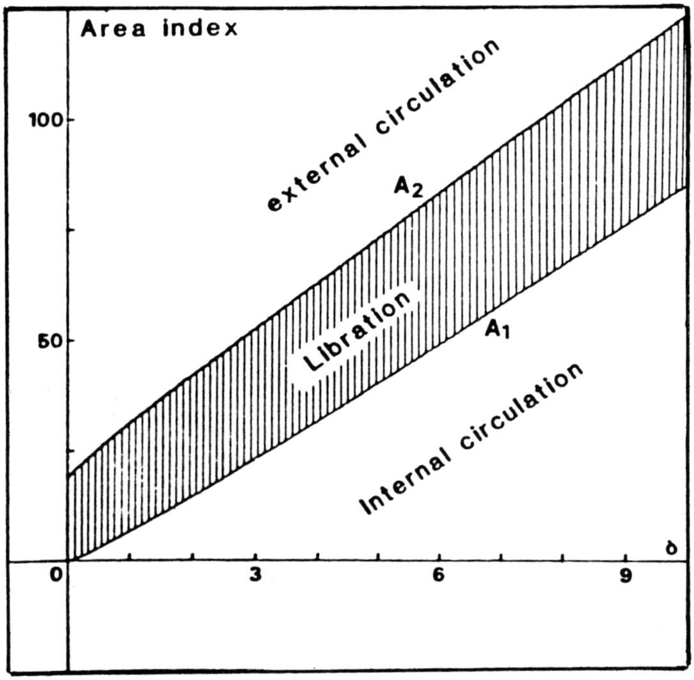

figure 3 : the critical areas and the three zones in the area index diagram .

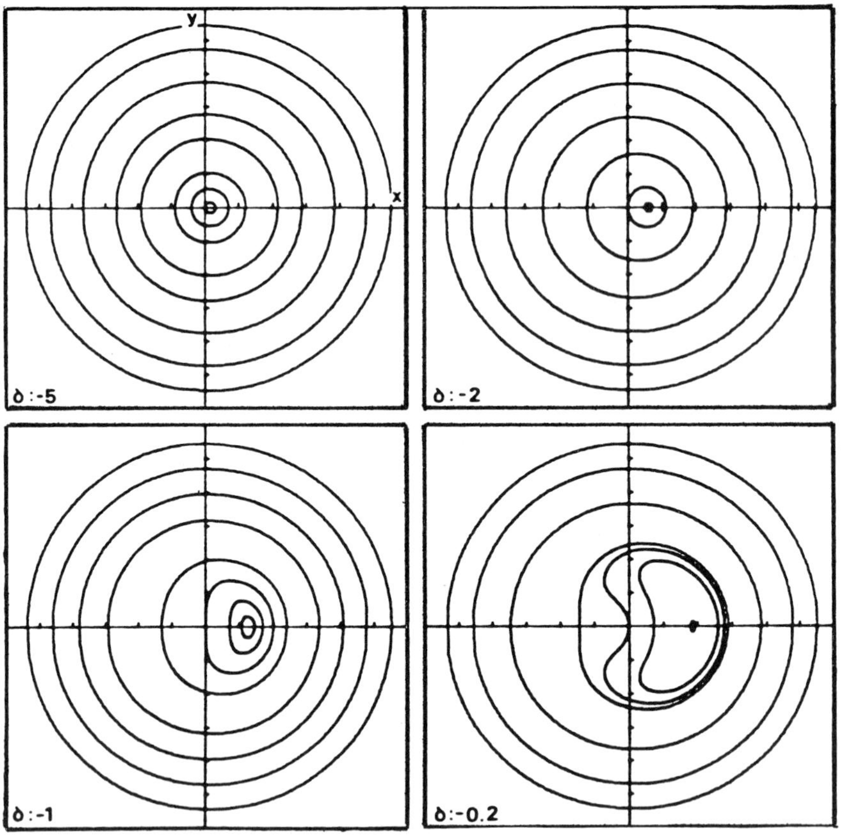

figure 4 : trajectories in the plane (x, y) for negative δ .

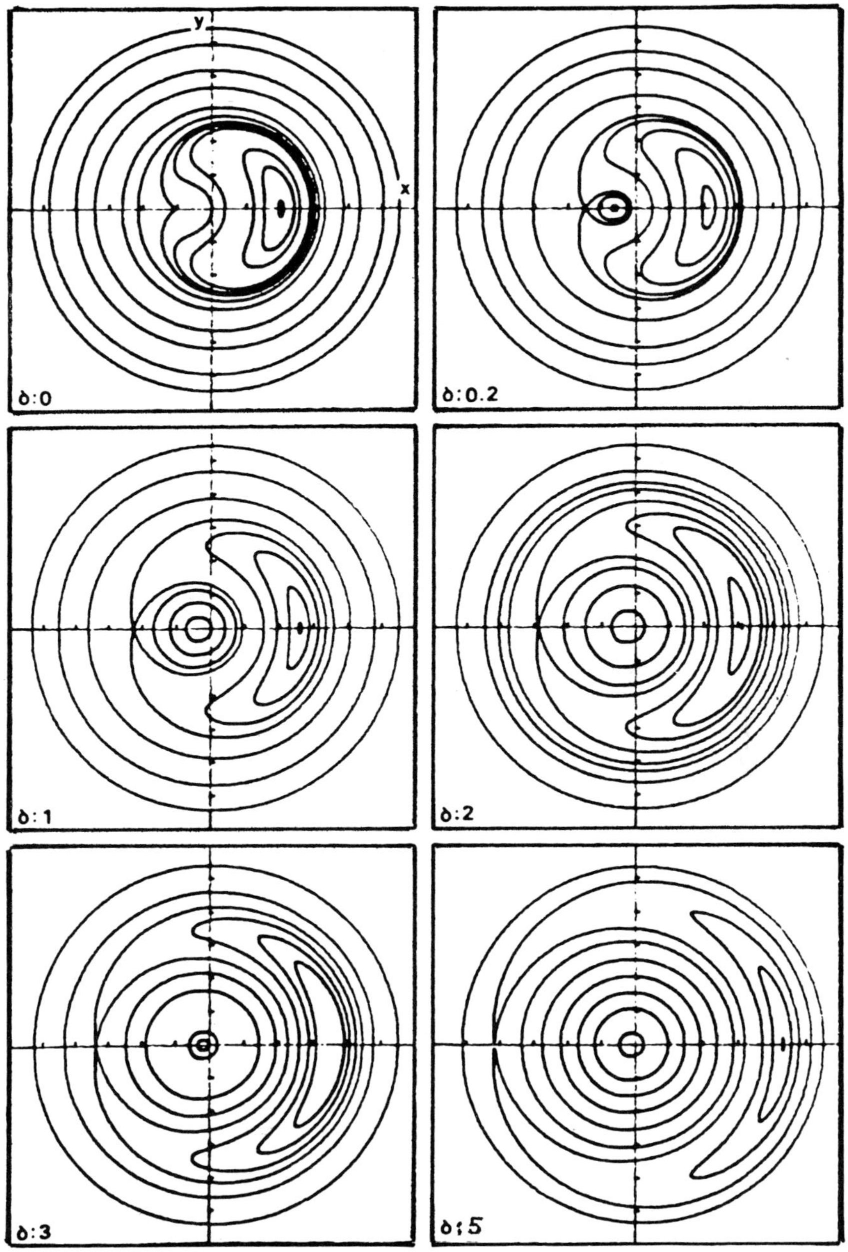

figure 5 : trajectories in the plane (x, y) for positive δ .

where η is some small quantity.

The solutions of system (7) are not anymore the closed curves described above but they are close to them.
They stay for a moderate period of time on a solution of system (7) which we call the *guiding trajectory*.

The guiding trajectory evolves and to identify it, we use the *adiabatic invariant* theory (Henrard 1982a) which states that the area enclosed by the guiding trajectory does not change by more than η for times smaller than η^{-1}

A problem appears when the guiding trajectory becomes an homoclinic orbit at some point in the evolution. In that case the classical theory of the adiabatic invariant breaks down. Under some assumptions (verified for the system (7)) the evolution of the system can be described in the following terms : (Henrard - 1982 b).

Let us assume that the guiding trajectory is first a circulation trajectory. After going through an homoclinic orbit, it may become either a circulation trajectory of the other type or a libration trajectory. This choice depends on the phase at the time of transition. It is expressed in terms of probability, assuming that all phases are equally probable. In both cases the area enclosed by the guiding trajectory undergoes a discontinuity but then stays constant again. A discontinuity in the area index appears, which we represent in terms of the *free vibration amplitude* defined by :

$$F = \left[\frac{\text{Area}}{\pi} \right]^{1/2} \tag{15}$$

This concept is based on the following idea : when δ is large in absolute value (a long time before or after transition) the trajectory is almost a circle, with F as radius and the origin as center. F can be interpreted as normalised eccentricity.

We must distinguish two types of behaviour : δ increasing or δ decreasing with the time. In our paper (Henrard - Lemaître 1983) we give a description of both cases and a detailed example for the first one. This is why we develop here only the second case, with an application of it in section 6 .
First we assume that our guiding trajectory is an internal circulation trajectory. When δ decreases to zero the area of the largest possible internal circulation goes to zero. Hence the guiding trajectory is forced to undergo a transition and to become an external circulation orbit.
No capture into libration is possible, because the area of the libration zone is also decreasing with δ .
Secondly, we assume that our guiding trajectory is an external circulation trajectory. When δ decreases to zero, it must stay an external

circulation orbit, because the area enclosed by the largest homoclinic orbit decreases.
Thirdly, we assume that our guiding trajectory is a libration one. When δ decreases with the time, it may become an external circulation orbit by undergoing a transition or if no transition has occured when $\delta = 0$, it becomes automatically a circulation.

If we note the free vibration amplitude after transition F_f and before transition F_i, we can plot the jump $(F_f - F_i)$ with respect to the initial value F_i (figure 6). This jump is always positive and can be interpreted as a jump in the normalized eccentricity.

§6. APPLICATION

The evolution in resonance we have just described could be of some interest to explain the well-known *Kirkwood Gaps* in the asteroïd belt.

We consider a circular restricted three-body problem Sun-Jupiter-Asteroïd, to describe the motion of a minor planet in a first-order resonance zone with Jupiter's mean motion n_J $(\frac{nA}{nJ} \cong \frac{p+1}{p})$.

In terms of Schubart's variables N, S, U and s (see Schubart - 1966), the Hamiltonian function is given by :

$$\mathbb{K} = - n_J\,(p+1)\,(N-S) - \frac{\mu^2}{2(N-S)^2 p^2} + \mu\, a_J \frac{m_J}{M_A}\, P\left(\frac{a}{a_J}, e, s\right) \qquad (16)$$

where $\mu = G.M_A$, M_A being the *mass acting on the asteroïd*, m_J is Jupiter's mass, a is the semi-major axis, e is the eccentricity and s the resonant angle.

After expansion with respect to S and truncation, we obtain the Hamiltonian (3) :

$$K' = \alpha S + \beta S^2 + \varepsilon \sqrt{2S} \cos s$$

with

$$\alpha = n_J \,.\, (p+1) - \frac{\mu^2}{p^2 N^3}$$

$$\beta = -\frac{3}{2} \frac{\mu^2}{p^2 N^4} \qquad (17)$$

$$\varepsilon = \frac{\mu}{a_J} \frac{m_J}{M_A} \frac{b}{\sqrt{N}}$$

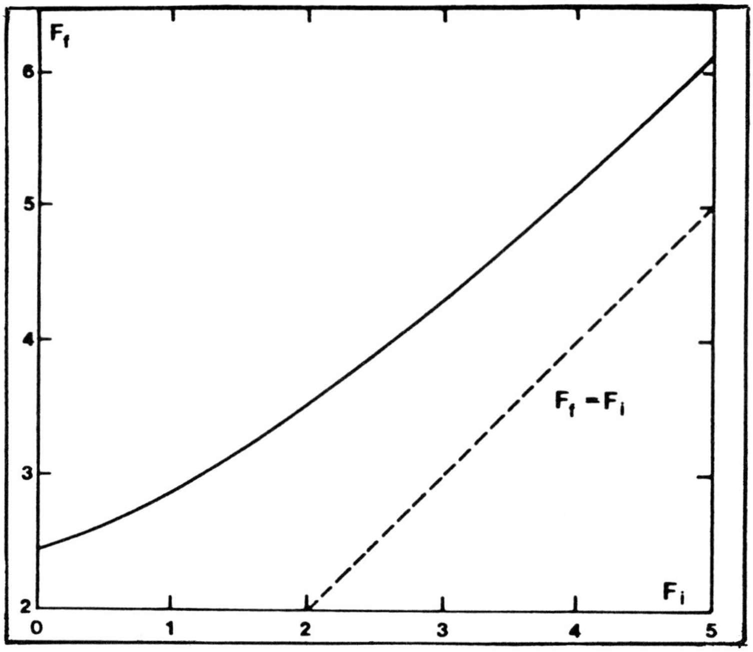

figure 6 : the free vibration amplitude after transition (F_f) versus its value (F_i) before transition for a passage through resonance with decreasing δ (for internal circulation orbits).

(b is the coefficient of $e \cos s$ in the expansion of the perturbation P) and by scaling, we get our "Second Fundamental Model of Resonance" (7), where δ is calculated as :

$$\delta = \frac{2}{3}\left(\frac{n_J}{n_A}\frac{p+1}{p} - 1\right)\left(\frac{1}{3}\frac{a_J^2}{a^2}\frac{M_A^2}{m_J^2}\frac{1}{b^2}\right)^{1/3} - 1 \; . \qquad (18)$$

If δ slowly decreases with the time (a few units is enough) we can reproduce exactly the same behaviour as in section 5 , which can be interpreted as a depletion of the libration zone. We can note an increase in the eccentricity of the asteroïds which undergo a transition from internal to external circulation.

However, a question still remains : how to explain that variation in δ ?

An idea would be to situate this phenomena during the solar nebulae dispersion. In a first approximation at least the effect of the nebula upon Jupiter and upon the asteroïd can be modelized by a change in the mass of the Central body. This change depends upon the amount of material included in the orbit of Jupiter or of the asteroïd and we have to distinguish between M_J : the mass of the pseudo-Sun as seen by Jupiter and M_A : the mass of the pseudo-Sun as seen by the Asteroïd.

Following Torbett and Smoluchowski (1980) we adopt Cameron's model for the description of the early solar system and more precisely we use the numerical data of example C of Torbett and Smoluchowski's paper.

Concurrently with the formation of the protosolar object, an accretion disk of the Lynden-Bell and Pringle(1974) variety forms and ultimately reaches a maximum mass M_D of $\sim 1.7\ M_S$ with a radius R_D of $\sim$ 600 AU.
At this point it reverses its growth and begins to disperse with a fraction of its mass going into the protosolar object and the rest being ejected out of the system.

Before dispersion the masses of the protosolar object (M_C) and of the disk (M_D) are :

$$M_C = 0.7\ M_S$$
$$M_D = 1.7\ M_S \qquad (19)$$

where M_S is the present solar mass.

During dispersion, the variation of M_J and M_A can be simply modelized by (Torbett-Smoluchowski 1980) :

$$M_J = M_C + \frac{\pi}{2} M_D \frac{a_J}{R_D}$$
$$M_A = M_C + \frac{\pi}{2} M_D \frac{a}{R_D} \qquad (20)$$

which gives an initial value to the ratio $\frac{M_J}{M_A}$:

$$\frac{M_J}{M_A} = \frac{0.1044}{0.1 + 0.0044 \frac{a_A}{A_J}} \quad \text{at time } t = 0 \qquad (21)$$

(Today after disparition of the disk the ratio equals 1).

So for an asteroïd initially situated in the 2/1 resonance zone ($p = 1$, $\frac{a_A}{a_J} = 0.629$) this ratio varies from 1.0158 to 1 , which gives a variation in δ of 1.808 .
(We can notice that by angular momentum conservation $\frac{n_A}{n_J}$ is proportional to $\frac{M_A^2}{M_J^2}$).

Doing the same for an asteroïd in the 3/2 resonance zone ($p = 2$, $\frac{a_A}{a_J} = 0.763$) the ratio $\frac{M_J}{M_A}$ varies from 1.010 to 1 and δ decreases of 0.898 .

In conclusion, the mechanism of depletion of the Kirkwood Gaps we have just outlined seems promising.
It apparently predicts the right size for the gaps, the increase in eccentricity and a more marked depletion for the resonance 2/1 than for 3/2 . Of course it has to be investigated much more carefully both in its dynamical and cosmogonical aspects.

REFERENCES

HENRARD, J.; 1982a. "Capture into Resonance : An extension of the use of Adiabatic invariants". Celest. Mech. 27, 3-22.

HENRARD, J.; 1982b. "The Adiabatic Invariant : Its use in Celestial Mechanics" in Application of Modern Dynamics to Celestial Mechanics and Astrodynamics (V. Szebehely editor). D. Reidel Publ. Co.,Holland.

HENRARD, J.; LEMAITRE, A.; 1983. "A second fundamental model for resonance" submitted for publication in Celestial Mechanics, 30, 197.

SCHUBART, J.; 1966 . "Special cases of the restricted problem of three bodies", Proc. I.A.U., Symposium n° 25, 187-193.

TORBETT, M.; SMOLUCHOWSKI, R.; 1980 . "Sweeping of the Jovian Resonances and the Evolution of the Asteroïds", Icarus n°44, 722-729.

ON THE AGES OF ASTEROID FAMILIES

Manabu YUASA
Department of Mathematics & Physics, Kinki University,
Higashi-Osaka, Osaka 577, Japan

Abstract

The ages of asteroid families are estimated by using a second order secular perturbation theory. Their ages are rather young. To explain these results we have some possible interpretations. The first interpretation is that in the theory of secular perturbation we do not include the effects of mutual perturbations of asteroids. The second is that we can regard these values as the minimum ages.

1. Solutions of the second-order secular perturbation theory.

We adopt Poincare's canonical variables:

$$p^* = \sqrt{2(L-G)}\cos(g+h), \qquad q^* = -\sqrt{2(L-G)}\sin(g+h),$$
$$r^* = \sqrt{2(G-H)}\cos h, \qquad s^* = -\sqrt{2(G-H)}\sin h, \tag{1}$$

where L, G, H, g and h are Delauney's variables.
The equations of motion are as follows:

$$dp^*/dt = \partial F/\partial q^*, \qquad dp^*/dt = -\partial F/\partial p^*,$$
$$dr^*/dt = \partial F/\partial s^*, \qquad ds^*/dt = -\partial F/\partial r^*, \tag{2}$$

where F is the Hamiltonian in which the short periodic terms have been eliminated.
The solutions of this system to the second order are given as follows (Yuasa 1973):

$$p^* = \sqrt{L}[\nu \cos((g_0 + 2/L\partial F_1^{**}/\partial\nu^2)t + \beta) + \sum_{j\ell}\sum \nu j\ell/g_0 - g_{j\ell}\cos(g_{j\ell}t + \beta_{j\ell})], \tag{3}$$

V. V. Markellos and Y. Kozai (eds.), Dynamical Trapping and Evolution in the Solar System, 203–210.

$$q^*=-\sqrt{L}[\nu\sin((g_0+2/L\partial F_1^{**}/\partial\nu^2)t + \beta) + \sum_j\sum_\ell \nu_{j\ell}/g_0-g_{j\ell}\sin(g_{j\ell}t+\beta_{j\ell})], \quad (4)$$

$$r^*=\sqrt{L}[\mu\cos((-g_0+2/L\partial F_1^{**}/\partial\mu^2)t + \gamma) + \sum_j\sum_\ell \mu_{j\ell}/g_0+g_{j\ell}\cos(f_{j\ell}t+\gamma_{j\ell})], \quad (5)$$

$$s^*=-\sqrt{L}[\mu\sin((-g_0+2/L\partial F_1^{**}/\partial\mu^2)t + \gamma) + \sum_j\sum_\ell \mu_{j\ell}/g_0+f_{j\ell}\sin(f_{j\ell}t+\gamma_{j\ell})], \quad (6)$$

where

$$\partial F_1^{**}/\partial\nu^2=m_5'[(B_1-A_1/4)\{2\nu^2+4(\xi_{55}^2+\xi_{56}^2)\} + (B_7+A_3/2)(\mu^2+\eta_{55}^2+\eta_{56}^2)$$
$$+(B_{13}+A_5/4)(\eta_{55}\mu_{55}+\eta_{56}\mu_{56})+2(B_{15}-A_6/8)(\xi_{55}\nu_{55}+\xi_{56}\nu_{56})$$
$$+B_2(\nu_{55}^2+\nu_{56}^2)+B_8(\mu_{55}^2+\mu_{56}^2)]+m_5'^2D_1, \quad (7)$$

$$\partial F_1^{**}/\partial\mu^2=m_5'[(B_7+A_3/2)(\nu^2+\xi_{55}^2+\xi_{56}^2)+(B_4-A_3/4)\{2\mu^2+4(\eta_{55}+\eta_{56})$$
$$+2(B_{11}-A_5/8)(\mu_{55}^2\eta_{55}^2+\mu_{56}^2\eta_{56}^2)+B_6(\mu_{55}^2+\mu_{55}^2)$$
$$+B_9(\nu_{55}^2+\nu_{56}^2)+B_{17}(\xi_{55}^2\nu_{55}^2+\xi_{56}^2\nu_{56}^2)] + m_5'^2D_3. \quad (8)$$

The constants g_0, ξ_{55}, ξ_{56}, η_{55} and η_{56} are determined by the disturbing planets and the semi-major axes of the asteroids. And the constants A_i, B_i, and D_i are determined only by the semi-major axes of asteroids. Furthermore ν, μ, β, γ are the constants of integration and their values depend on the initial conditions of asteroids. Other constants $g_{j\ell}$, $\nu_{j\ell}$, $f_{j\ell}$, $\mu_{j\ell}$, $\beta_{j\ell}$ and $\gamma_{j\ell}$ are determined only by the disturbing planets. And m_5' is the mass of Jupiter.

2. The longitude of the proper perihelion Π_1 and the longitude of the proper node θ_1.

The quatities $(g_0+2/L\partial F_1^{**}/\partial\nu^2)t+\beta$ in the equations (3), (4) and and $(-g_0+2/L\partial F_1^{**}/\partial\mu^2)t+\gamma$ in the equations (5), (6) are called the longitude of the proper perihelion and the longitude of the proper node, respectively. Usually they are indicated by Π_1 and θ_1.

The difference of the second order solution and the first order solution appears in Π_1 and θ_1. In the first order theory, $\Pi_1=g_0t+\beta$ and $\theta_1=-g_0t+\gamma$, therefore

$$\Pi_1+\theta_1=\beta+\gamma. \quad (9)$$

On the other hand in the second order theory

$$\Pi_1+\theta_1=2/L(\partial F_1^{**}/\partial\nu^2+\partial F_1^{**}/\partial\mu^2)t + \beta + \gamma. \quad (10)$$

The asteroids belonging to families have similar initial conditions, so that they have similar β and γ. By this reason in the first order theory the quantity $\Pi_1 + \theta_1$ must concentrate around a value in each asteroid family. But the observational data indicates that $\Pi_1 + \theta_1$ does not concentrate but rather equally distributes from 0 to 2π (Table 1. (Brouwer 1950)).

Table 1. Distribution of $(\Pi_1 + \theta_1)/2\pi$

	0.0-0.1	0.1-0.2	0.2-0.3	0.3-0.4	0.4-0.5	0.5-0.6	0.6-0.7	0.7-0.8	0.8-0.9	0.9-1.0
Themis	4	7	3	9	6	5	3	4	5	7
Eos	8	8	10	6	2	1	5	5	2	11
Coronis	2	5	2	2	3	5	3	7	1	3
Maria	4	1	2	3	1	1	0	2	1	2
Phocaea	0	2	2	5	2	5	0	2	3	2
Flora	11	9	8	13	14	12	11	5	8	11

In the second order theory $\Pi_1 + \theta_1$ has the term depending on t. And we can interpret the non-concentration of $\Pi_1 + \theta_1$ in each asteroid family.

3. Ages of asteroid families.

In the solutions of the second order theory, we can see the time dependence of $\Pi_1 + \theta_1$. In order to do so, we plot $\Pi_1 + \theta_1$ against the values $-(\partial F_1^{**}/\partial\mu^2 + \partial F_1^{**}/\partial\mu^2)$ for Coronis, Flora, Eos and Themis families. The results are shown in Figures 1-4. In these figures a remarkable and interesting feature can be found. If we shift vertically the left-side points between 0 and 2π by 2π upwards and right-side points by 2π downwards in necessary case, all the points seem to lie on a straight line. From the slope of these lines we can estimate the ages of asteroid families (Kozai and Yuasa 1974). Their values are 3.7×10^6 years, 9.98×10^5 years, 6.3×10^5 years and 4.6×10^5 years for Coronis, Flora, Eos and Themis families respectively.

4. Discussion.

The ages of asteroid families obtained in the preceding section are rather young. To explain this result, we have some possible interpretations. The first interpretation is that in the theory of secular perturbation we do not include the effects of mutual perturbations of asteroids. If the mutual perturbations of asteroids had been considerably large, there would be a possibility that old families have been destroyed and only young families have remained. The second interpretation is that we can regard the values obtained in the preceding section as the minimum ages of the asteroid families. In fact we can shift the points in Figures 1-4, not 2π but $2n\pi$. Then the slope of the straight lines can become more steep and we may be able to get more large values of the ages for the asteroid famlies. But to execute this, we need to construct a more

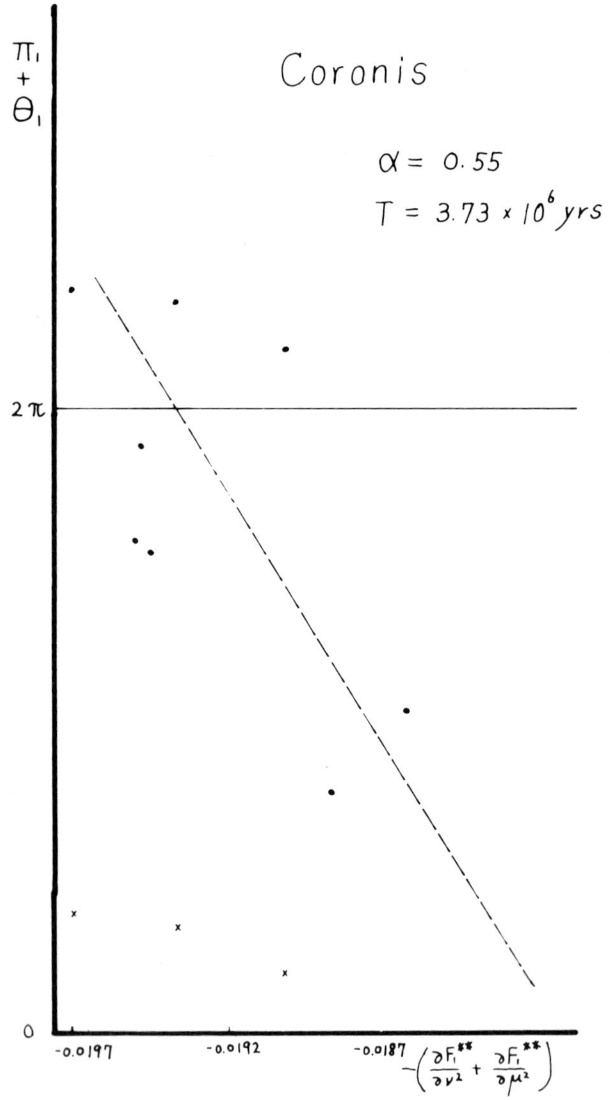

Figure 1. $\Pi_1+\theta_1$ against the coefficient of time for Coronis family. Points x are shifted vertically by 2π upward.

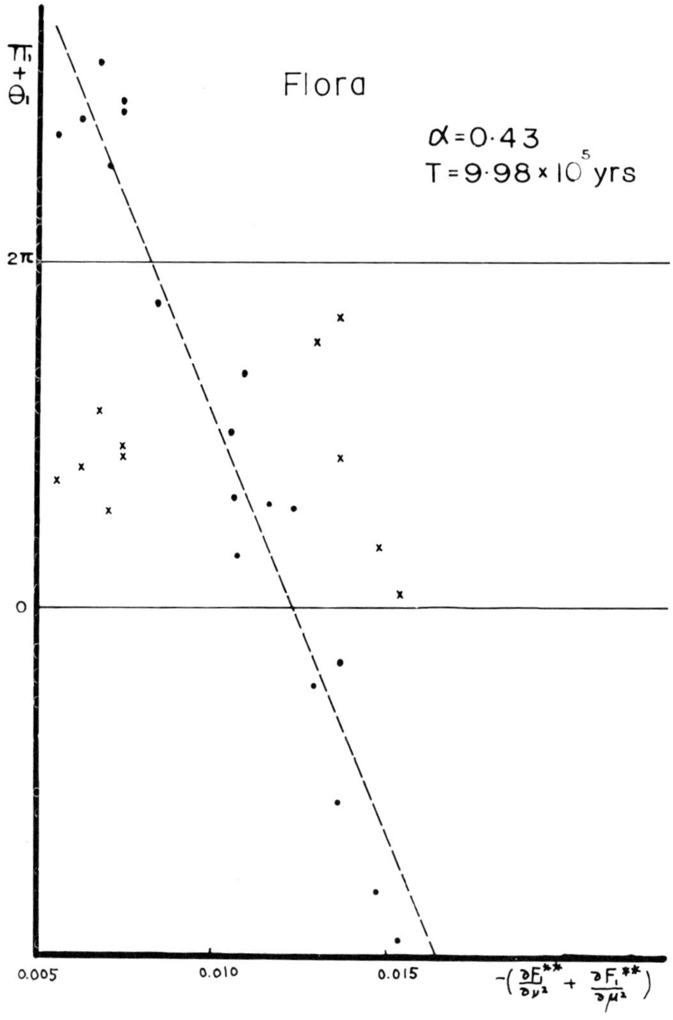

Figure 2. $\Pi_1+\theta_1$ against the coefficient of time for Flora family. Points x are shifted vertiaclly by 2π upwards in the left-side and downwards in the right-side.

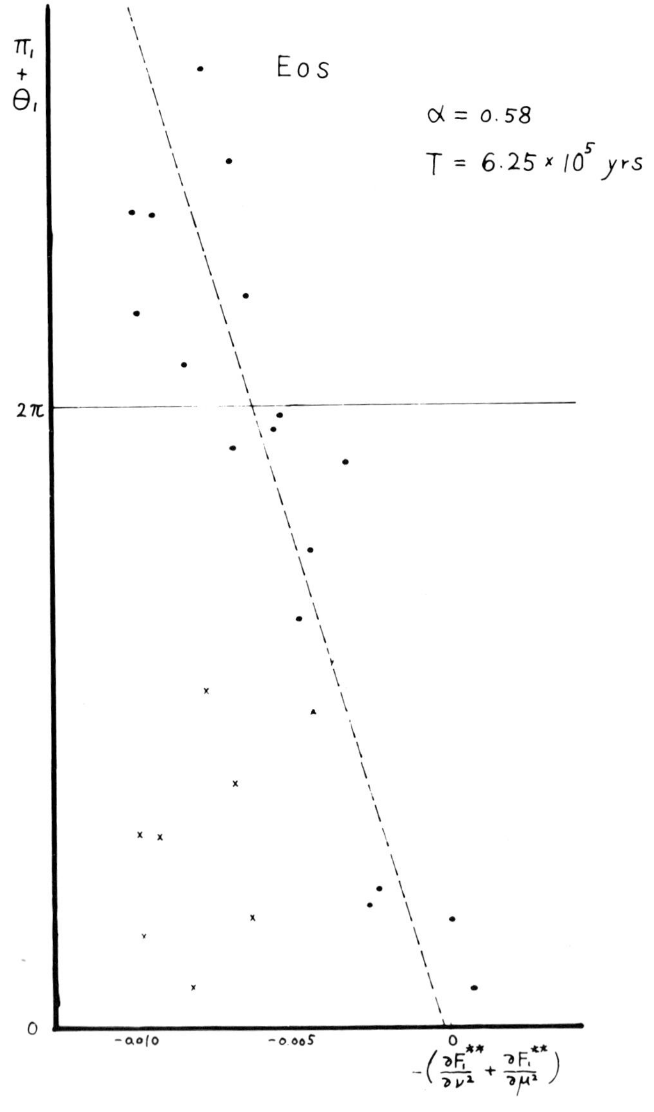

Figure 3. $\Pi_1+\theta_1$ against the coefficient of time for Eos family. Points x are shifted vertically by 2π upwards.

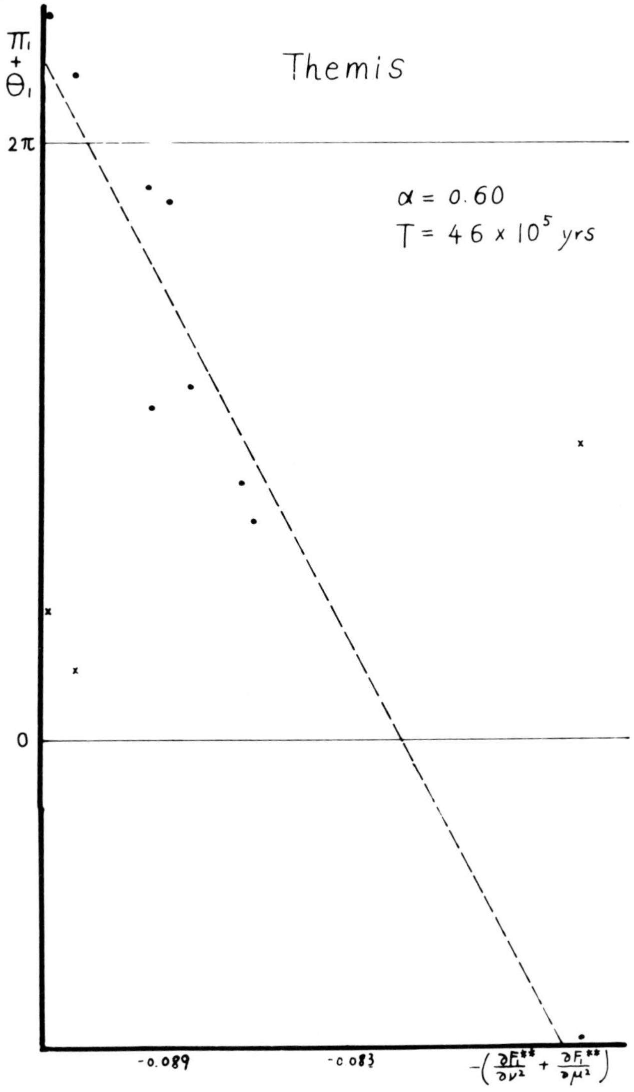

Figure 4. $\Pi_1+\Theta_1$ against the coefficient of time for Themis family. Points x are shifted vertically by 2π upwards in the left-side and downwards in the right-side.

precise theory, namely a more higher order secular perturbation theory. In addition, the data used in Figures 1-4 are not sufficient in number, and now we are preparing the recalculation by adopting more data. When the recalculation is finished whether obtained values are the intrinsic ages or the minimum ages of the asteroid families may become more clear.

References

Brouwer, D. : 1950, Astron. J. 56. 9.
Kozai, Y. and Yuasa, M. : 1974, "Stability of the Solar System and of Small Stellar Systems", ed. Y. Kozai, D. Reidel Publ., p. 81.
Yuasa, M. : 1973, Publ. Astron. Soc. Japn 25, 399.

PART IV

PERIODIC ORBITS

THE MECHANISM OF BRANCHING OF THREE-DIMENSIONAL PERIODIC ORBITS FROM THE PLANE

I.A. ROBIN(*) and V.V. MARKELLOS(**)

(*) University of Glasgow, Glasgow, Scotland
(**) University of Patras, Patras, Greece

ABSTRACT

A linearised treatment is presented of vertical bifurcations of symmetric periodic orbits (bifurcations of plane with three-dimensional orbits) in the circular restricted problem. Recent work on bifurcations from vertical-critical orbits ($a_v = \pm 1$) is extended to deal with the more general situation of bifurcations from vertical self-resonant orbits ($a_v = \cos(2\pi n/m)$ for integer m,n) and it is shown that in this more general case bifurcating families of three-dimensional orbits always occur in pairs, the orbital symmetry properties being governed by the evenness or oddness of the integer m. The applicability of the theory to the elliptic restricted problem is discussed.

1. INTRODUCTION

The occurrence of intersections of planar with three-dimensional periodic orbits of the circular restricted problem ("vertical" bifurcations) at planar orbits for which the vertical stability index $a_v = \pm 1$ ("vertical-critical" orbits) was first proposed by Hénon (1973). Markellos et al (1981) have discussed the mechanism of such bifurcations and calculated entire series of the vertical bifurcations of the basic "Strömgren families" of periodic orbits of the problem. Zagouras and Markellos (1977) and Zagouras and Kalogeropoulou (1978) have presented numerical results on the continuation of vertical-critical orbits into three dimensions, and showed that vertical bifurcations also occur at planar orbits for which

$$a_v = \cos(2\pi n/m), \qquad (1.1)$$

for integer values of m and n. Robin and Markellos (1980) gave examples of families of three-dimensional periodic orbits generated from such "multiple" vertical bifurcations for values of m up to 8, and found that, as had been anticipated by the second author, the bifurcating families always occur in pairs, each pair arising from the same self-

V. V. Markellos and Y. Kozai (eds.), Dynamical Trapping and Evolution in the Solar System, 213–224.

resonant orbit, and that the three-dimensional symmetry properties of these families depend solely on whether the integer m, the "multiplicity" of the bifurcation, is even or odd.

The object of the present paper is to provide an analytical background to the above-mentioned numerical results on "multiple" vertical bifurcations of symmetric periodic orbits, and to show that the observed pattern of occurence of the bifurcating families of three-dimensional orbits is of general validity.

2. BIFURCATION CONDITION

With respect to the usual dimensionless barycentric rotating coordinate system ($x_1 = x$, $x_2 = y$, $x_3 = z$) (see e.g. Robin and Markellos, 1980, hereafter referred to as "Paper I"), the initial conditions for a symmetric planar periodic orbit may be written

$$\underset{\sim}{x}_o = (x_{01},\ 0,0,0,\ x_{05},\ 0). \tag{2.1}$$

This initial state of the massless third body of the system corresponds to a mirror configuration (Roy and Ovenden, 1955) in the horizontal plane, or plane of the primaries. The periodicity conditions for this "unperturbed" orbit can be expressed in the form

$$\begin{aligned} x_2\ (x_{01},\ x_{05},\ T/2) &= 0 \\ x_4\ (x_{01},\ x_{05},\ T/2) &= 0 \end{aligned} \tag{2.2}$$

where T is the orbital period. (Note that the components x_3 and x_6 of the state vector vanish for all values of the epoch t in this unperturbed orbit).

Let us now consider the orbit resulting from small "vertical perturbations" δx_{03} and δx_{06} in the initial conditions (2.1). The initial conditions of this perturbed orbit are

$$\underset{\sim}{x}_0 = (x_{01},\ 0,\ \delta x_{03},\ 0,\ x_{05},\ \delta x_{06}). \tag{2.3}$$

As Hénon (1973) has pointed out, the horizontal components (x_1, x_2, x_4, x_5) of the state vector are, in the linear approximation, unaffected by purely vertical perturbations (see also, Markellos et al, 1981). Denoting the "vertical" components of the state vector in the perturbed orbit by (δx_3, δx_6), we may express these in terms of the initial perturbations (δx_{03}, δx_{06}) as follows:

$$\begin{pmatrix} \delta x_3 \\ \delta x_6 \end{pmatrix} = \begin{pmatrix} v_{33} & v_{36} \\ v_{63} & v_{66} \end{pmatrix} \begin{pmatrix} \delta x_{03} \\ \delta x_{06} \end{pmatrix}, \tag{2.4}$$

where the $v_{k\ell}$'s are the first-order "variations" $\partial x_k / \partial x_{0\ell}$ for the unperturbed orbit.

We now seek to establish periodicity conditions for the perturbed orbit in terms of the Periodicity Theorem of Roy and Ovenden (1955). Referring to Equations (2.1) and (2.2) of Paper I, we see that the initial conditions of the perturbed orbit correspond to a mirror configuration if and only if

$$\delta x_{06} = 0 \tag{2.5}$$

or

$$\delta x_{03} = 0 \tag{2.6}$$

according as the mirror configuration is of type (P) (in-plane) or type (A) (on-axis). (Note that this distinction arises because we are now considering a "three-dimensional" rather than a planar mirror configuration, as was the case for the unperturbed orbit).

By the Periodicity Theorem, the perturbed orbit resulting from initial perturbations (δx_{03}, δx_{06}) satisfying either of the above conditions will be periodic if, at some epoch $t \neq 0$, another mirror configuration occurs. Since, as we have just seen, the horizontal part of the perturbed motion is (to first order) unaffected, it is clear that a mirror configuration can only take place at those epochs corresponding to the occurrence of a mirror configuration in the unperturbed planar periodic orbit: that is, for values of t given by

$$t = N(\frac{T}{2}), \tag{2.7}$$

where N is some positive integer. The condition for a mirror configuration in the perturbed orbit at epoch t satisfying Equation (2.7) is then either

$$\delta x_6 = 0 \tag{2.8}$$

or

$$\delta x_3 = 0, \tag{2.9}$$

again depending on the type of configuration. Combining Equations (2.4)-(2.9), we see that the periodicity conditions for the perturbed orbit can be written

$$\delta x_j = v_{ji}(NT/2) \quad \delta x_{0i} = 0 , \tag{2.10}$$

where $i = 3$ for a type (P) and $i = 6$ for a type (A) mirror configuration at the initial epoch, while $j = 6$ for a type (P) and $j = 3$ for a type (A) mirror configuration at the final epoch (as in Table I of Paper I), and $v_{ji}(NT/2)$ denotes the variation $\partial x_j/\partial x_{0i}$ evaluated at $t = NT/2$ on the unperturbed orbit. Thus, δx_{0i} is always the non-zero initial perturbation; $\delta x_{0i} = 0$ is the trivial solution of Equation (2.10) corresponding to the original unperturbed orbit.

Equation (2.10) expresses the condition that in the linear approximation, there exists a family of three-dimensional symmetric periodic orbits parametrised by the perturbation δs_{0i}, bifurcating from the planar periodic orbit. The condition for the occurrence of a vertical bifurcation from a symmetric planar periodic orbit is therefore that one (at least) of the four elements (v_{33}, v_{36}, v_{63}, v_{66}) of the vertical submatrix V_v of the full variational matrix $V = [v_{k\ell}]_{6x6}$ vanishes at an epoch t equal to an integer number of half-periods of the orbit.

3. PROPERTIES OF BIFURCATING FAMILIES

The symmetry properties of the bifurcating three-dimensional orbits depend on the mirror configuration types occurring at the initial ($t = 0$) and final ($t = NT/2$) epochs, and hence on the values of the subscripts i and j in Equation (2.10). This means that we can predict the symmetry class (plane symmetric, axisymmetric or doubly-symmetric) of the bifurcating family by identifying which of the four elements of the matrix V_v vanishes at the appropriate epoch. For example, a family of plane symmetric three-dimensional orbits would be expected to bifurcate from a planar periodic orbit for which $v_{63}(NT/2) = 0$ for some value of N. The four possible cases are listed in Table I.

The interval between successive mirror configurations for the three-dimensional periodic orbits in the neighbourhood of the bifurcation, as we have seen, is equal to NT/2 for some integer N. This interval is equal to half of the orbital period for a three-dimensional orbit of simple symmetry (plane symmetric or axisymmetric), and equal to a quarter of the period for a doubly-symmetric orbit. Thus, in the linear approximation, the period of the three-dimensional orbits arising from a vertical bifurcation is equal to NT or 2NT according to whether the orbits are of simple or double symmetry, respectively (T being the period of the planar orbit at which the bifurcation takes place). The final column of Table I gives the values of the periods in each case.

Table I

Case	Type of Mirror Congiguration at: Initial Epoch	Final Epoch	Symmetry Class	i	j	Orbital Period
1	P	P	Plane Symmetric	3	6	NT
2	A	A	Axisymmetric	6	3	NT
3	A	P	Doubly-	6	6	2NT
4	P	A	Symmetric	3	3	2NT

The elements of the "vertical variational matrix"

$$V_v = \begin{pmatrix} v_{33} & v_{36} \\ v_{63} & v_{66} \end{pmatrix} \tag{3.1}$$

satisfy the well-known area-preserving property

$$\det V_v = v_{33}v_{66} - v_{36}v_{63} = 1 \tag{3.2}$$

(see, e.g. Hénon, 1973). The terms $v_{33}v_{66}$ and $v_{36}v_{63}$ cannot both vanish, and so zero elements of V_v, indicating a vertical bifurcation, can only occur either singly, or as one of the diagonal pairs (v_{33}, v_{66}) or (v_{36}, v_{63}). Table I shows that in consequence of this fact, a vertical bifurcation orbit can only give rise either to a single family of three-dimensional orbits, corresponding to one and only of Cases 1-4 of the table, or else to two families of three-dimensional orbits, corresponding to Cases 1 and 2 or to Cases 3 and 4 (that is, one family of plane symmetric and one of axisymmetric orbits, or a pair of families of doubly-symmetric orbits). As we shall see presently, the former situation applies in general to vertical-critical orbits, and the latter to vertical self-resonant (non-critical) orbits.

4. BIFURCATION FROM VERTICAL-CRITICAL ORBITS

Let us first of all consider the case of a planar orbit which has a zero element appearing in the matrix $V_v(NT/2)$ for $N = 1$. The full set of vertical stability indices a_v, b_v, c_v, d_v is defined by

$$V_v(T) = \begin{pmatrix} a_v & b_v \\ c_v & d_v \end{pmatrix}, \tag{4.1}$$

and the elements of $V_v(T/2)$ are denoted

$$V_v(T/2) = \begin{pmatrix} A_v & B_v \\ C_v & D_v \end{pmatrix}. \tag{4.2}$$

It can easily be shown that for a symmetric orbit, these two sets of quantities are related by

$$\begin{pmatrix} a_v & b_v \\ c_v & d_v \end{pmatrix} = \begin{pmatrix} A_vD_v + B_vC_v & 2B_vD_v \\ 2A_vC_v & A_vD_v + B_vC_v \end{pmatrix} \quad (4.3)$$

(Hénon, 1973). This incidentally shows the important property $a_v = d_v$ for a symmetric planar periodic orbit.

By Equations (2.10) and (4.2), with $N = 1$, Cases 1-4 of Table I correspond respectively to

$$C_v = 0; \; B_v = 0; \; D_v = 0; \; A_v = 0. \quad (4.4)$$

Now from Equations (3.2) and (4.3), the vertical stability index a_v is given by

$$a_v = 2A_vD_v - 1 = 2B_vC_v + 1. \quad (4.5)$$

In each of the cases (4.4), $|a_v| = 1$: we are therefore dealing with bifurcation from vertical-critical orbits, which has been discussed by Hénon (1973). In the first two cases ($C_v = 0$ and $B_v = 0$), a_v has the value ± 1, corresponding to the values $m = n = 1$ in Equation (1.1). This can be described as a "simple bifurcation", since in the neighbourhood of the bifurcation, the period of the three-dimensional orbits is (to first order) equal to T, the period of the vertical-critical orbit; the orbital multiplicities are of course also equal. In the second two cases of (4.4), $A_v = 0$ and $D_v = 0$, Equation (4.5) shows that $a_v = -1$, corresponding to the values $m = 2$, $n = 1$ in Equation (1.1): this can be described as a "double bifurcation", since it involves a doubling of the period and orbital multiplicity of the vertical-critical orbit (the "multiplicity" of an orbit being defined as half the number of crossings of the (x_1,x_3)-plane occurring in one period).

Since the parameters A_v, B_v, C_v and D_v are all independent, zero elements of the matrix $V_v(T/2)$ will in general occur singly: thus, a vertical-critical orbit will as a rule give rise to only one family of three-dimensional periodic orbits. The summetry properties of the bifurcating family depend on which of the four elements of $V_v(T/2)$ vanishes, as indicated in Table I; this has been clearly illustrated by Hénon (op. cit.).

5. BIFURCATION FROM VERTICAL SELF-RESONANT ORBITS

Let us now consider the case of a planar orbit for which $V_v(NT/2)$ $(N > 1)$ has at least one zero element, such that all of the elements A_v, B_v, C_v and D_v of $V_v(T/2)$ are non-zero. It can be seen from Equation (4.5) that this latter constraint excludes from consideration the special case of vertical-critical orbits ($a_v = \pm 1$), which were dealt with separately in the previous section. As we shall see later, we are now dealing with vertical self-resonant orbits, for which a_v is given by Equation (1.1) with values of the integer m greater than 2.

In order to relate the occurrence of a zero element of $V_v(NT/2)$ for some $N > 1$ to the value of the vertical stability index a_v, we make use of the well-known property of the variational matrix

$$V(t + T) = V(t)V(T) \tag{5.1}$$

(e.g. Wintner, 1946); this general property can be applied in particular to the "vertical submatrix" V_v of V. It is convenient to consider separately the cases of even and of odd values of the integer N. Using Equation (5.1), we may express $V_v(NT/2)$ for odd values of $N = 2r+1$ as

$$V_v(NT/2) = V_v(T/2 + rT) = V_v(T/2)\,[V_v(T)]^r \tag{5.2}$$

$$(r = 0,1,2,\ldots).$$

Similarly, for even values of $N = 2r$ we have

$$V_v(NT/2) = V_v(rT) = [V_v(T)]^r \tag{5.3}$$

$$(r = 0,1,2,\ldots).$$

Note that although we are restricting our attention to values of $N > 1$, the above formulae are valid for all non-negative values of N.

It is easily shown by induction that the vertical variational matrix V_v computed at $t = NT/2$ $(N = 0,1,2,\ldots)$ satisfies the following two equations:

$$V_v(T/2 + rT) = \begin{pmatrix} \alpha_r A_v & \beta_r B_v \\ \beta_r C_v & \alpha_r D_v \end{pmatrix} \quad (r \geqq 0) \tag{5.4}$$

where α_r and β_r are functions of A_v, B_v, C_v and D_v only;

$$V_v(rT) = \begin{pmatrix} \gamma_r & 2\delta_r B_v D_v \\ 2\delta_r A_v C_v & \gamma_r \end{pmatrix} \quad (r \geqq 0), \tag{5.5}$$

where γ_r and δ_r are functions of A_v, B_v, C_v and D_v only. We may therefore state the following:

If the elements A_v, B_v, C_v and D_v of $V_v(T/2)$ are all non-zero, then for all values of $N > 1$, $V_v(NT/2)$ has either no zero elements, or exactly two zero elements on the same diagonal.

This follows from Equation (3.2), together with Equation (5.4) for odd values of N $(= 2r + 1,\ r = 1,2,3,\ldots)$, and Equation (5.5) for even values of N $(= 2r,\ r = 1,2,3,\ldots)$, the important point being the appearance of the common factors (α_r, β_r), (γ_r, δ_r) in the diagonal pairs of elements: since A_v, B_v, C_v and D_v are assumed to be all non-zero, an element of $V_v(NT/2)$ can vanish only if one of the functions α_r, β_r, γ_r or δ_r is zero.

We therefore have the important result that families of three-dimensional periodic orbits bifurcating from a vertical self-resonant periodic orbit for which $N > 1$ (that is, excluding vertical-critical orbits) always occur in pairs, and as we have already seen, both families must consist either of simply-symmetric or of doubly-symmetric orbits.

Using Equations (5.1) (as applied to V_v), (4.1) and (4.3), together with Equations (5.4) and (5.5), pairs of simultaneous recurrence relations can be established for the functions α_r, β_r, γ_r and δ_r:

$$\alpha_r = a_v\alpha_{r-1} + (a_v - 1)\beta_{r-1}\,,$$
$$\beta_r = a_v\beta_{r-1} + (a_v + 1)\alpha_{r-1}\,, \tag{5.6}$$

$$\gamma_r = a_v\gamma_{r-1} + (a_v^2 - 1)\delta_{r-1}\,,$$
$$\delta_r = a_v\delta_{r-1} + \gamma_{r-1}. \tag{5.7}$$

Since $|a_v| \neq 1$, the β's can be eliminated from Equations (5.6), and the δ's from Equations (5.7), giving

$$\alpha_{r+1} - 2a_v\alpha_r + \alpha_{r-1} = 0,$$
$$\gamma_{r+1} - 2a_v\gamma_r + \gamma_{r-1} = 0. \tag{5.8}$$

The general solutions of these two identical second-order recurrence relations are

$$\alpha_r = Ae^{ir\varphi} + Be^{-ir\varphi}\,,$$
$$\gamma_r = Ce^{ir\varphi} + De^{-ir\varphi}\,, \tag{5.9}$$

where

$$\cos\varphi = a_v, \tag{5.10}$$

and A, B, C, D are constants to be determined from the initial conditions

$$\alpha_0 = 1, \qquad \alpha_1 = 2a_v - 1;$$
$$\gamma_0 = 1, \qquad \gamma_1 = a_v . \tag{5.11}$$

Calculation of the four constants yields

$$\alpha_r = \frac{\sin (r+1)\varphi - \sin r\varphi}{\sin \varphi} , \tag{5.12}$$

$$\gamma_r = \cos r\varphi , \tag{5.13}$$

and the associated solutions for β_r, δ_r are found to be

$$\beta_r = \frac{\sin (r+1)\varphi + \sin r\varphi}{\sin \varphi} , \tag{5.14}$$

$$\delta_r = \frac{\sin r\varphi}{\sin \varphi} . \tag{5.15}$$

Note that since $|\cos \varphi| = |a_v| \neq 1$ for vertical self-resonant orbits, $\sin \varphi \neq 0$.

Let us consider the conditions for the occurrence of a pair of zero elements of the matrix $V_v(NT/2)$, for odd values of $N = 2r+1$ ($r = 1,2,3, \ldots$). It is clear from Equation (5.4) that this requires either α_r or β_r to vanish, for some $r > 0$. By Equation (5.12), the function α_r is equal to zero if and only if

$$\sin (r+1)\varphi = \sin r\varphi , \tag{5.16}$$

with solutions

$$\varphi = \left(\frac{2k+1}{2r+1}\right)\pi , \tag{5.17}$$

where k is an arbitrary integer, such that $(2k+1)/(2r+1)$ is not an integer. Substitution of the solutions (5.17) into Equation (5.10) gives

$$a_v = \cos \left(\frac{2k+1}{2r+1}\right)\pi , \tag{5.18}$$

and with $0 \leq k < r$, the complete set of r roots of α_r, a polynomial in a_v of degree r, is obtained. (A duplicate set of r solutions is obtained for $r < k \leq 2r$).

In a similar way, the roots of β_r, also an rth degree polynomial in a_v, are found to be given by

$$a_v = \cos\left(\frac{2k}{2r+1}\right)\pi , \tag{5.19}$$

which with $0 < k \leq r$ gives the complete set of r roots of β_r. Note that the two sets of solutions (5.18) and (5.19) are mutually exclusive, as would be expected, since α_r and β_r cannot vanish simultaneously.

We now consider the conditions for the occurrence of a bifurcation for even values of $N = 2r$ $(r = 1,2,3,\ldots)$, that is, for the appearance of a pair of zero elements of $V_v(rT)$. This requires that either γ_r or δ_r vanishes, for some $r > 0$. From Equation (5.13), the roots of γ_r are found to be

$$a_v = \cos\left(\frac{2k+1}{2r}\right)\pi, \tag{5.20}$$

the complete set of r solutions being given by $0 \leq k < r$. Similarly, from Equation (5.15), δ_r has roots

$$a_v = \cos\left(\frac{k}{r}\right)\pi , \tag{5.21}$$

the complete set of roots of the polynomial δ_r, of degree r-1 in a_v, being given by $0 < k < r$.

Let us now relate these results to the vertical self-resonance condition (1.1),

$$a_v = \cos\left(\frac{2\pi n}{m}\right), \tag{5.22}$$

m and n being mutually prime integers with $0 < n \leq m$. The vertical-critical cases $m = n = 1 (a_v = \pm 1)$ and $m = 2$, $n = 1 (a_v = -1)$ are excluded, having been dealt with already. From Equations (5.18)-(5.21), the condition for the occurrence of a bifurcation associated with the vanishing of one of the functions α_r, β_r, γ_r, δ_r (and therefore of one of the diagonal pairs of elements (v_{33}, v_{66}), (v_{36}, v_{63}) of the matrix $V_v(NT/2)$) can be expressed in the form (5.22), with the values of the integers m and n in each case as given in Table II, with r any positive integer.

The final entry of Table II, corresponding to $\delta_r = 0$, is essentially redundant, since all the possible combinations of values of m and of n can be constructed from the entries corresponding to the cases $\alpha_r = 0$ and $\gamma_r = 0$. This redundancy of solutions reflects the fact that a doubly-symmetric periodic orbit can be regarded as simply-symmetric if one of its symmetries is ignored; the bifurcation of doubly-symmetric orbits corresponding to $v_{33}(N_1T/2) = v_{66}(N_1T/2) = 0$, for some $N_1 > 1$ (α_r or γ_r equal to zero), automatically gives $v_{36}(N_2T/2) = v_{63}(N_2T/2) = 0$, where $N_2 = 2N_1$ is even (that is, $\delta_{N_1} = 0$). The occurrence of a bifurca-

Table II

Function	m	n	
α_r	$2(2r+1)$	$2k+1$	: $k = 0,1,2,\ldots,r-1$
β_r	$2r+1$	k	: $k = 1,2,\ldots,r$
γ_r	$4r$	$2k+1$	: $k = 0,1,2,\ldots,r-1$
δ_r	$2r$	k	: $k = 1,2,\ldots,r-1$

tion of genuinely simply-symmetric orbits is associated with the vanishing of the function β_r, for some $r > 1$.

The possible values of the integer m in Equation (5.22) in each case of Table II (with $r = 1,2,3,\ldots$) are

$$\alpha_r = 0 : \quad m = 6,10,14,\ldots$$

$$\beta_r = 0 : \quad m = 3,5,7,\ldots$$

$$\gamma_r = 0 : \quad m = 4,8,12,\ldots$$

which together account for all integer values greater than 2 ; the values $m = 1$ and $m = 2$ applying to the special case of vertical-critical orbits. It is evident that an even value of m corresponds to the case of doubly-symmetric three-dimensional bifurcating orbits ($v_{33} = v_{66} = 0$), while an odd value of m corresponds to a bifurcation with a family of simply-symmetric orbits ($v_{36} = v_{63} = 0$). The following conclusion may therefore be stated:

> A vertical self-resonant orbit, with vertical stability index given by Equation (5.22), gives rise to one family of axisymmetric and one of plane symmetric three-dimensional orbits if m is odd, or to two families of doubly-symmetric orbits if m is even.

6. REMARK

The foregoing discussion, which is of general validity in the circular restricted problem, can easily be extended to the elliptic restricted problem, the only difference being that the orbital period of the three-dimensional orbits arising from a vertical bifurcation in the elliptic case have fixed period (an integer multiple of the period of the primaries), the eccentricity of the orbit of the primaries varying along the bifurcating family instead of the period (Robin,1981). The pattern of vertical bifurcations in the two versions of the problem would therefore appear to be identical, in terms of the occurrence of pairs of vertical branches whose symmetry properties are governed by the evenness or oddness of the "multiplicity" m of the bifurcation and the special nature of bifurcation from vertical-critical orbits.

7. REFERENCES

Hénon, M.: 1973, *Astron. Astrophys.* 28, 415.

Markellos, V.V., Goudas, C.L., and Katsiaris, G.A.: 1981, in F.D. Kahn (ed.), *Investigating the Universe*, D. Reidel Publ. Co., Dordrecht-Holland, p. 321.

Robin, I.A. and Markellos, V.V.: 1980, *Celes. Mech.* 21, 395.

Robin, I.A.: 1981, *Celes. Mech.* 23, 97.

Roy, A.E. and Ovenden, M.W.: 1955, *Monthly Notices Roy. Astron. Soc.* 115, 296.

Wintner, A.: 1946, *Analytical Foundations of Celestial Mechanics*, Princeton University Press.

Zagouras, C. and Markellos V.V.: 1977, *Astron. Astrophys.* 59, 79.

Zagouras. C.G. and Kalogeropoulou, M.: 1978, *Astron. Astrophys.* (Suppl.)

STABILITY AND BIFURCATIONS OF SYMMETRIC PERIODIC ORBITS IN THE RESTRICTED 3-BODY PROBLEM

A. Milani
Department of Mathematics, University of Pisa, Italy

ABSTRACT. The continuation of symmetric periodic orbits can be described in terms of "symmetry functions"; the branching of the zero-level lines in a neighbourhood of a critical point gives rise to the transition from "first kind" to "second kind" periodic orbits. When the families are parametrized with the Jacobi integral, the bifurcations can be described as "catastrophes" of the generating functions. However bifurcations of higher order are more complex than the generic catastrophes with one parameter: both symmetric and asymmetric bifurcations occur.

In this way the symmetric periodic orbits that do not have close approaches to the secondary body can be described in an analytic way and their stability can be deduced from simple bifurcation rules. However numerical experiments are required to determine the "natural termination" of the families.

1. CONTINUATION OF SYMMETRIC PERIODIC ORBITS

The "synodic" two-body problem can be described by the hamiltonian:

$$H_o = -m_1^2/2\Lambda^2 - \Lambda + (\eta^2+\xi^2)/2 \tag{1}$$

in the "synodic Poincaré variables", which are defined in terms of the usual keplerian elements a, e, ω, M of the orbit around m_1, by:

$$\begin{aligned}
\lambda &= M+\omega-t \\
\eta &= [(4m_1 a)^{\frac{1}{2}}(1-(1-e^2)^{\frac{1}{2}}]^{\frac{1}{2}}\cos(\omega-t) \\
\Lambda &= (m_1 a)^{\frac{1}{2}} \\
\xi &= [(4m_1 a)^{\frac{1}{2}}(1-(1-e^2)^{\frac{1}{2}}]^{\frac{1}{2}}\sin(\omega-t)
\end{aligned} \tag{2}$$

For $e=0$, λ is defined as the angle formed by the two bodies and the rotating x axis; in this way $(\lambda, \eta, \Lambda, \xi)$ are defined for negative energy and positive angular momentum of the osculating orbit.

We will deal with "symmetric perturbations" to H_o, i.e. hamiltonian problems of the form

V. V. Markellos and Y. Kozai (eds.), Dynamical Trapping and Evolution in the Solar System, 225–233.

$$H=H_o+R(\lambda,\eta,\Lambda,\xi,\mu) \tag{3}$$

such that R=0 for $\mu=0$, and the involutive transformation

$$\sigma:(\lambda,\eta,\Lambda,\xi)\longrightarrow(-\lambda,\eta,\Lambda,-\xi) \tag{4}$$

leaves R invariant. Every perturbing function R depending only on the "synodic" coordinates x,y and even in y has this property; e.g. the perturbing function of the restricted 3-body problem in synodic eliocentric coordinates x,y:

$$R=\mu[x-1/((x-1)^2+y^2)^{\frac{1}{2}}] \qquad \mu=1-m_1 \tag{5}$$

We will anyway suppose that T is real-analytic apart from a finite number of singularities for each value of μ.

As a corollary of the "mirror theorem" (Roy and Oveden,1955), if on an orbit of H there are two symmetric configurations (i.e. $\lambda=0$ or π, $\xi=0$) at two different times, e.g. t=0 and t=T, the orbit is periodic of period 2T and symmetric with respect to σ.

This "symmetry condition" can be computed using the "surface of section" method: chosen an hypersurface $S_\ell:\lambda=\ell\pi$ as a cross section ($\ell=0,1$), we call the first crossing of $S_{\ell+s}$ the "Poincaré map of order s" (s an integer $\neq 0$):

$$\theta_{\ell,s}:(\eta,\Lambda,\xi)\longrightarrow(\eta_1,\Lambda_1,\xi_1) \tag{6}$$

then we impose $\xi_1=0$ for $\xi=0$, and we have a "symmetry function" $\Phi_{\ell,s}(\eta,\Lambda,\mu)=0$ defining for every μ the "characteristic lines" in the η, Λ plane, corresponding to symmetric periodic orbits of order s (s synodic revolutions in a period).

For $\mu=0$ the symmetric periodic orbits are easily computed because the general integral of (1) can be described as $\Lambda=\Lambda_o$, $\lambda=n(\Lambda_o)t+\lambda_o$, and (η,ξ) rotating clockwise with angular velocity 1; where $n(\Lambda)=m_1^2/\Lambda^3$ is the usual mean motion; then:

$$\Phi_{\ell,s}(\eta,\Lambda,0)=-\eta\,\sin[s\pi/(n(\Lambda)-1)]=0 \tag{7}$$

has the solutions $\eta=0$, $\xi=0$ (circular orbits) and $\xi=0$, $n(\Lambda)=(s+k)/k$, k an ineteger $\neq 0$ (resonant orbits of order s).

In this setting we can state a "continuation theorem" which gives more informations than the classical ones (Poincaré,1892; Arenstorf, 1963; Barrar,1965). We say that a subset W of the η,Λ plane in S_ℓ is "s-safe" if every point $(\ell\pi,\eta,\Lambda,0)$ with $(\eta,\Lambda)\in W$ lies on an orbit that for $\mu=0$ does not hit a singular point of R (lying on $\mu=0$) before it crosses $S_{\ell+s}$. Note that the singular points of R, as an analytic function, lying on $\mu=0$, do exist even if $R\equiv 0$ on $\mu=0$; e.g. the restricted 3-body problem for $\mu=0$ is not the two-body problem, but the two-body problem with collisions (Brjuno,1978).

Continuation theorem: let W be an open subset of the η,Λ plane in

S_ℓ, such that its closure $U\cup\partial W$ is compact, s-safe, contained in $n(\Lambda)\neq 1$, $\eta^2<2\Lambda$; suppose that there an no circular resonant orbits of order s (i.e. points with $\mu=0$, $n(\Lambda)=(s+k)/k$, k integer $\neq 0$) on the boundary ∂W.

Then for $|\mu|<\mu_1$, μ_1 depending on W,s,ℓ only, the symmetry function $\Phi_{\ell,s}(\eta,\Lambda,\mu)$ is a Morse function of η,Λ on W (i.e. it has only a finite number of nondegenerate critical points); moreover all these critical points are of saddle type, and are "near" the circular resonant orbits of order s.

Proof.: $n(\Lambda)\neq 1$ and s-safety ensure the smoothness of the symmetry function; (7) shows that for $\mu=0$ it has only critical points of saddle type, and Morse functions are "stable".

However, the qualitative behaviour of the characteristics is determined not only by the location of the critical points, but also from the value of the symmetry function at the critical points:

Shallow resonance lemma: let Λ_o be a resonant value (of order s), i.e. $n(\Lambda_o)=(s+k)/k$, k an integer $\neq 0$, and let the first order perturbation to the symmetry function at the corresponding circular orbit be nonzero:

$$\frac{\partial}{\partial\mu}\Phi_{\ell,s}(0,\Lambda_o,0)\neq 0 \qquad (8)$$

Then the characteristics of order s are smooth curves in a neighbourhood of $(0,\Lambda_o)$ for $0<\mu<\mu_2$, μ_2 depending on s,ℓ,k only.

Proof.: In the Taylor expansion of the critical value of the symmetry function in powers of μ, (8) turns out to be the first order term. Therefore for μ small enough and $\neq 0$ the critical point does not lie on the characteristic.

"Shallow resonance" refers to the hyperbola-like branches that connect the periodic orbits of the "first kind" (continuation of the $\eta=0$ branch) with those of the "second kind" (continuation of the $n(\Lambda)=(s+k)/k$ branch). This "quadratic approximation" was discussed by Guillaume (1969), who also computed (in a different coordinate system) the first-order perturbation to the symmetry function for the restricted 3-body problem and for s=1, obtaining a formula equivalent to:

$$n(\Lambda_o)=\frac{k+1}{k}\Rightarrow\frac{\partial\Phi_{\ell,1}}{\partial\mu}(0,\Lambda_o,0)\cdot(-1)^{k+\ell}\frac{k}{|k|}>0 \qquad (9)$$

and this is enough to understand the structure of shallow resonances;in figure 1 we adapted from Colombo et al.(1969) the plot of the characteristic lines for $\mu\simeq 1/1048$ ("asteroidal problem"); the plot refers to the a, $\tilde{e}=e\cos(\omega-t)$ plane instead of the Λ,η plane, but the qualitative features of the characteristics are the same, and are much like the theory although μ is not "very small".

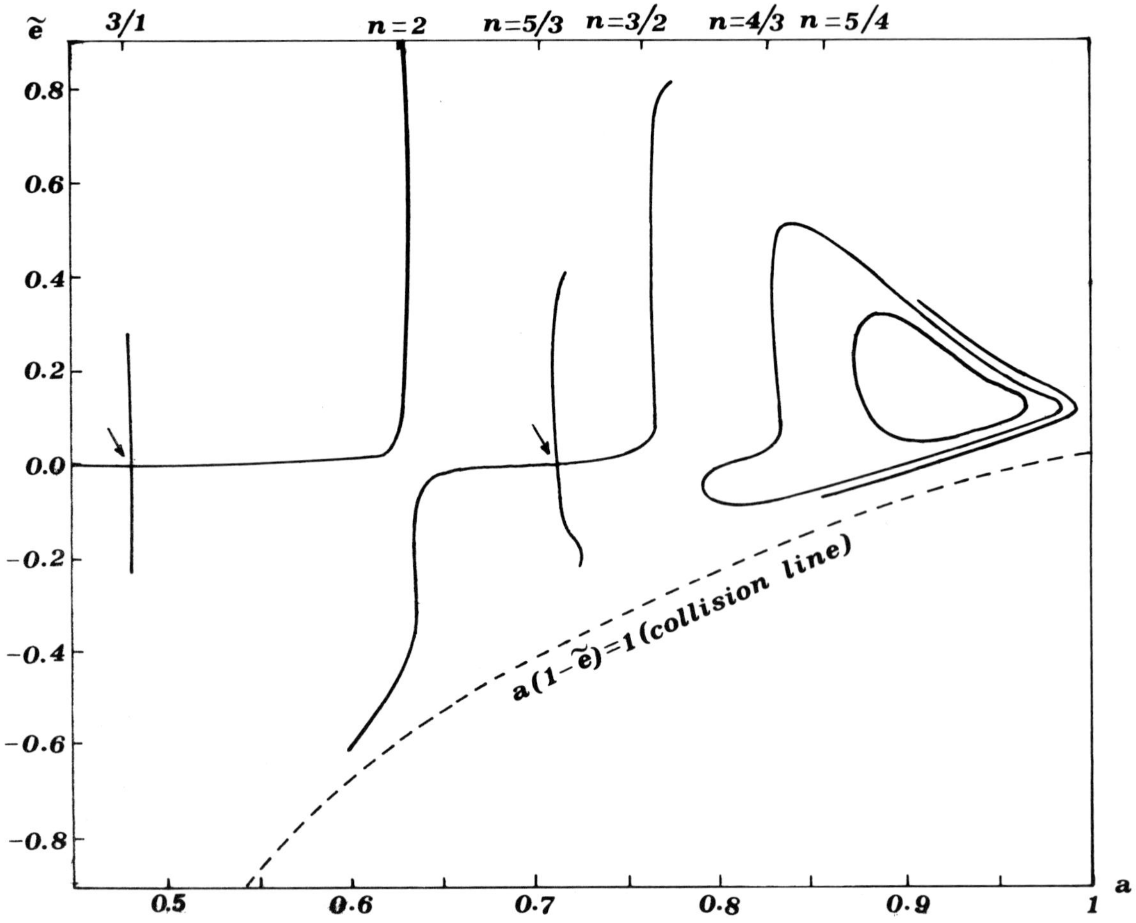

Figure 1

Symmetric periodic orbits on the a, $\tilde{e}$=e cos(ω−t) plane; λ=0. n=2, 3/2, 4/3 and 5/4 families from Colombo et al. (1969), n=3 and 5/3 from numerical experiments performed by the author.

For order s>1, the situation is different:

Deep resonance lemma: let Λ_o be a resonant value of order s strictly > 1, i.e. $n(\Lambda_o)=(s+k)/k$, k integer but $n(\Lambda_o)\neq(1+k)/k$, k any integer. Then there is no shallow resonance, i.e. 0 is a critical value corresponding to a critical point near $(0,\Lambda_o)$ for the symmetry function and the two branches corresponding to first and second kind periodic orbits cross transversally, for $0<\mu<\mu_3$, μ_3 depending only on k,s.

Proof.: If (η,Λ) defines a symmetric periodic orbit of order 1, then it defines also a symmetric periodic orbit of order s, for every s integer $\neq 0$; since the characteristic of order 1 is smooth, far from circular resonances of order 1, we conclude that the symmetry function of order 1 divides the symmetry function of order s, the quotient being a smooth function with zero level corresponding to the second kind branch.

2. STABILITY AND BIFURCATIONS

Sinçe the Hamiltonian $H=H_o+R$ is an integral, the Poincaré map (6) can be "reduced" to a mapping of a two-dimensional manifold in itself; this can be done by solving for Λ in H=h, provided $H_\Lambda=\dot{\lambda}\neq 0$. In this way we can define a map, depending on two parameters h,η:

$$T^{h,\mu}_{\ell,s}:(\eta,\xi)\longrightarrow(q,p)=(Q(\eta,\xi),P(\eta,\xi)) \tag{10}$$

with $q=\eta_1$, $\xi=p_1$; if s is even, T maps S_ℓ in itself, and periodic orbits of order s/2 appear as fixed points. The map T is symplectic, or area-preserving, because λ and Λ are conjugate variables (Siegel and Moser, 1971, §22): its jacobian matrix A satisfies:

$$\det A=Q_\eta P_\xi-Q_\xi P_\eta=1 \tag{11}$$

and it can be (locally) defined with a generating function; if T is not too far from identity, such that $Q_\eta>0$, $P_\xi>0$, we can use a generating function $\mathcal{S}(\xi,q)=-\xi q-S(\xi,q)$ to define (10) by:

$$\eta=q+S_\xi(\xi,q) \qquad p=\xi+S_q(\xi,q)\ ; \tag{12}$$

then the jacobian matrix $A=A(\eta,\xi,h,\mu)$ of (10) can be computed as a function of the derivatives of S:

$$\begin{aligned} Q_\eta&=(1+S_{\xi q})^{-1} & Q_\xi&=-S_{\xi\xi}(1+S_{\xi q})^{-1}\\ P_\eta&=S_{qq}(1+S_{\xi q})^{-1} & P_\xi&=1+S_{q\xi}-S_{qq}S_{\xi\xi}(1+S_{\xi q})^{-1} \end{aligned} \tag{13}$$

To extract stability informations from this canonical formalism we use the Arnold (1976, appendix 9) method, defining an auxiliary function $F(\eta,\xi)=S(\xi,Q(\eta,\xi))$; all the critical points $F_\eta=F_\xi=0$ of the function F are fixed points of the map (10), (12); to investigate the linearized map A at the fixed points, we can compute the hessian matrix B of F at the critical points:

$$F_{\eta\eta}=S_{qq}Q_{\eta}^{2} \qquad F_{\eta\xi}=(S_{q\xi}+S_{qq}Q_{\xi})Q_{\eta} \tag{14}$$
$$F_{\xi\xi}=S_{\xi\xi}+S_{\xi q}Q_{\xi}+(S_{q\xi}+S_{qq}Q_{\xi})Q_{\xi}$$

and by using (13), (14) and (11) we get B as a function of A:

$$B=\begin{bmatrix} Q_{\eta}P_{\eta} & Q_{\eta}(P_{\xi}-1) \\ Q_{\eta}(P_{\xi}-1) & Q_{\xi}(P_{\xi}-2) \end{bmatrix} \tag{15}$$

We will say that a fixed point of T is <u>linearly stable</u> whenever its characteristic multipliers (eigenvalues of A) are complex numbers with absolute value 1, different from ± 1; that it is <u>linearly unstable</u> whenever the multipliers are real numbers ν, $1/\nu$ different from ± 1; linearly critical whenever the multipliers are (twice) +1 or -1. I remember that a linearly unstable periodic orbit is unstable in the ordinary sense, by the Hartmann and Grobman theorem (Hartmann,1964); in the linearly stable case more information is needed to apply KAM theory.

<u>Critical point theorem</u>: let the map (10) (with $Q_{\eta}>0$, $P_{\xi}>0$) be represented by (12); then the auxiliary function F has saddles corresponding to linearly unstable fixed points of T, extrema corresponding to linearly stable ones, and degenerate critical points corresponding to linearly critical fixed points.

<u>Proof</u>.: By (15) and (11):

$$\det B=Q_{\eta}(2-Q_{\eta}-P_{\xi})=Q_{\eta}[2-\mathrm{Tr}\,A]$$

and because the multipliers satisfy the equation:

$$\nu^{2}-\mathrm{Tr}\,A\nu+1 = 0$$

$\det B \gtrless 0$ correspond to complex or real multipliers.

To investigate the linearly critical case, we use also the so-called "<u>Henon stability criterium</u>": if the fixed point of T is on the symmetry line $\xi=0$, then $Q_{\eta}=P_{\xi}$, and linear stability corresponds to $Q_{\eta}=P_{\xi}<1$. Then the linearly critical case for a symmetric periodic orbit is characterized by $P_{\eta}=0$ or $Q_{\xi}=0$; we will say that a linearly critical, symmetric orbit is undergoing a <u>symmetric degeneracy</u> whenever $Q_{\eta}=P_{\xi}=1$, $P_{\eta}=0$; i.e. whenever the eigenspace with eigenvalue 1 contains the symmetry axis $\xi=0$.

Symmetric bifurcation theorem: suppose that the number of symmetric periodic orbits of order s relative to the cross section $S_\ell (\ell=0,1)$, lying on the level surface H=h, changes at the value h_o near the point (η_o,Λ_o) (i.e. this change occurs regardless of the neighbourhood of (η_o,Λ_o) to which we restrict the count). Then (η_o,Λ_o) corresponds to a symmetric periodic orbit of order s, relative to S_ℓ, satisfying each of the following equivalent properties:
(A) the periodic orbit through $(\ell\pi,\eta_o,\Lambda_o,0)$ is undergoing symmetric degeneracy: $Q_\eta=P_\xi=1$, $P_\eta=0$.
(B) the auxiliary function F (can be defined for h near h_o), and has a degenerate critical point at $(\eta_o,0)$ with hessian matrix:

$$B=\begin{bmatrix} 0 & 0 \\ 0 & -Q_\xi \end{bmatrix} \tag{16}$$

(C) either the level lines of H in the η,Λ plane are tangent at (η_o,Λ_o) to the characteristic line $\Phi_{\ell,s}(\eta,\Lambda,\mu)=0$, or the characteristic line itself has a singularity at (η_o,Λ_o).

Proof.: Since the symmetry function is real-analytic, a symmetric periodic orbit cannot disappear, but only bifurcate: this requires det (A-Id)=0; moreover the eigenspace with eigenvalue 1 of A must contain the direction from which the bifurcating fixed points approach, i.e. (A). Then F can be defined, and (15) gives (B). Let us compute P_η as the total derivative of ξ_1 with respect to η on $H=h_o$:

$$P_\eta = \frac{\partial\xi_1}{\partial\eta} - \frac{\partial\xi_1}{\partial\Lambda}\frac{\partial H}{\partial\eta}\left(\frac{\partial H}{\partial\Lambda}\right)^{-1} \tag{17}$$

and $P_\eta=0$ implies (C).

A kind of inverse statement allowing an explicit description of a generic bifurcation is the following:

Fold catastrophe theorem: let for $H=h_o$ there be a symmetric periodic orbit (η_o,Λ_o) of order s, relative to S_ℓ, with Q_η, $P_\xi>0$; suppose that the characteristic $\Phi_{\ell,s}(\eta,\Lambda,\mu)$ is smooth at (η_o,Λ_o) and that the level line $H=h_o$ has a contact of order 2 with it at (η_o,Λ_o), i.e. they have the same tangent but a different curvature, in such a way that they do not cross each other (in a neighbourhood of (η_o,Λ_o)); suppose also that there is no asymmetric degeneracy, i.e. $Q_\xi\neq 0$ in a neighbourhood. Then:
(A) the number of symmetric periodic orbits of order s, relative to S_ℓ, lying on H=h changes at the value h_o near (η_o,Λ_o): two symmetric periodic orbits, one linearly stable and one linearly unstable, collide at (η_o,Λ_o) and disappear.
(B) for $h=h_o$ the auxiliary function F has a degenerate critical point at $(\eta_o,0)$ with hessian matrix B as in (16), with the third derivative

$F_{\eta\eta\eta}(\eta_o,0)\neq 0$; for the parameter h varying across h_o, F undergoes a "fold catastrophe", i.e. an extremum collides with a saddle and disappears.
(C) the periodic orbit through $(\ell\pi,\eta_o,\Lambda_o,0)$ undergoes a symmetric degeneracy, i.e. for $h=h_o$ $P_\eta(\eta_o,0)=0$, with the second derivative

$P_{\eta\eta}(\eta_o,0)\neq 0$.

Proof.:The condition on the curvatures gives $P_{\eta\eta}\neq 0$;i.e.(C);by (15), $F_{\eta\eta\eta}=Q_\eta P_{\eta\eta}+Q_{\eta\eta}P_\eta$ and because of the degeneracy $P_\eta=0$, $P_{\eta\eta}\neq 0$ implies $F_{\eta\eta\eta}\neq 0$; the sign of $F_{\eta\eta}$ changes in the prescribed way, giving (B). Then (A) follows, by the critical point theorem.

In this way, since first kind periodic orbits are known to be linearly stable outside the resonances of order 1,2, if there are no "asymmetric" degeneracies and bifurcations the linear stability character of the symmetric periodic orbits can be determined by only plotting the characteristic lines and the level lines of H in the η,Λ plane.

Higher codimension catastrophes are not to be expected "generically", since the bifurcations depend on only one parameter h;however, this applies strictly only to order s=1, because the auxiliary functions F that are related to symplectic map squared are not an open set in the space of smooth functions. Therefore the bifurcations of higher order resonant orbits from "first kind" periodic orbits are more complex. We will now discuss the case of order s=2.

Even order bifurcation theorem: let s be even and s/2 be odd, and let $\eta=0$, $\Lambda=\Lambda_o$ correspond to a circular orbit resonant of order s (and not of order $<$ s); let $h_o=H(\ell\pi,0,\Lambda_o,0,0)$. Let P, Q be the components of the map $T^{h_o,\mu}_{\ell,2s}$ and let their first order perturbations satisfy:

$$\left.\frac{\partial P_\eta}{\partial\mu}\right|_{\mu=0}+\left.\frac{\partial Q_\xi}{\partial\mu}\right|_{\mu=0}\neq 0 \tag{18}$$

then for μ small enough and $\neq 0$ the family of "first kind" symmetric periodic orbits (continuation of the $\ell=0$ branch) undergoes two separate bifurcations, one symmetric and one asymmetric. However the asymmetric bifurcation for $\ell=0$ is the symmetric one for $\ell=1$ and viceversa.

Proof.: For $\mu=0$ let us plot the lines $P_\eta=0$, $Q_\xi=0$ on the (η,Λ) plane:

$$P_\eta=-\sin t+O(\eta^2)\qquad Q_\xi=\sin t+O(\eta^2)\qquad t=2\pi s/(n(\Lambda)-1) \tag{19}$$

therefore (18) ensures that the two curves $P_\eta=0$ and $Q_\xi=0$ cross transversally the $\eta=0$ line in two distinct point for $\mu\neq 0$ small enough; they also must cross the "first kind" branch in two distinct points,giving one symmetric and one asymmetric degeneracy. By the "deep resonance lemma" the symmetric degenerate periodic orbit must be the crossing point between the first kind and the second kind periodic orbits. On the other hand the map $T^{h_o,0}_{\ell,s/2}$ has a linear part of the form $\begin{bmatrix} 0 & 1 \\ -1 & 0 \end{bmatrix}$ at the resonance, therefore the asymmetric degeneracies on $\lambda=0$ are transformed into symmetric degeneracies on $\lambda=\pi s/2$ and viceversa for $\mu=0$; $h=h_o$:

$$\begin{bmatrix} 0 & 1 \\ -1 & 0 \end{bmatrix} \begin{bmatrix} a & 0 \\ c & a \end{bmatrix} \begin{bmatrix} 0 & -1 \\ 1 & 0 \end{bmatrix} = \begin{bmatrix} a & -c \\ 0 & a \end{bmatrix}$$

Because of the Henon property $Q_\eta=P_\xi$, the eigenspace of eigenvalue -1 for the derivative of $T^{h,\mu}_{1,s}$ must be either the η axis or the ξ axis, therefore it is the η axis for μ near 0; this shows that $Q_\xi=0$ on $\lambda=0$ corresponds to $P_\eta=0$ on $\lambda=\pi$ for μ small enough: this is the point of symmetric bifurcation at the opposition.

The study of the auxiliary function F for $T^{h,\mu}_{\ell,2s}$ shows that the stability of the bifurcating second kind orbits is determined also by the sign of the expression (18): e.g. in the interior case $n(\Lambda)>1$, the bifurcation occurring for smaller Λ gives rise to stable second kind orbits, that occurring for larger Λ gives rise to unstable second kind orbits. Numerical experiments easily show that, in the restricted 3-body problem, for s=2 condition (18) holds and moreover the sign is such that in the interior case the stable resonant orbits of order 2 are symmetric at the oppositions (since pericenter and apocenter at the conjunction gives rise to close approach).

In fig. 1 we have plotted not only the orbits of order 1 found by Colombo et al. (1969), but also the n=3/1 and the n=5/3 families;since fig.1 refers to conjunctions, the asymmetric bifurcations occur "before" the symmetric ones, and they are marked with arrows.

3. REFERENCES

Arnold,V.:1976,*Méthodes mathematiques de la mécanique classique*, MIR, Moscow.

Arenstorf,R.F.:1963, Am.J.Math. 85, p.27.

Barrar,R.B.:1965, Astron.J. 70, p.3.

Brjuno,A.D.:1978, Celestial Mechanics 18, p.9.

Colombo,G. et al.:1968, Astron.J. 73, p.111.

Guillame,P.:1969, Astron.Astrophys. 3, p.57.

Hartmann,P.:1964, *Ordinary differential equations*, J.Wiley & Sons.

Poincaré,H.:1892, *Les méthodes nouvelles de la Mécanique Céleste*,vol.I, Gauthier-Villars.

Roy,A.E.,Ovenden,M.W.:1955, Mon.Nat.R.Astron.Soc. 115, p.297.

Siegel,C.L., Moser,J.K.:1971, *Lectures on Celestial Mechanics*,Springer.

RESONANT THREE-DIMENSIONAL PERIODIC SOLUTIONS ABOUT THE TRIANGULAR EQUILIBRIUM POINTS IN THE RESTRICTED PROBLEM

C.G. ZAGOURAS and V.V. MARKELLOS
University of Patras, Patras, Greece

ABSTRACT

In the three-dimensional restricted three-body problem, the existence of resonant periodic solutions about L_4 is shown and expansions for them are constructed for special values of the mass parameter, by means of a perturbation method. These solutions form a second family of periodic orbits bifurcating from the triangular equilibrium point. This bifurcation is the evolution, as μ varies continuously, of a regular vertical bifurcation point on the corresponding family of planar periodic solutions emanating from L_4.

1. INTRODUCTION

It is known that for values of the mass parameter less than the critical value of Routh, the general solution of the linearized equations of motion around the triangular equilibrium point L_4 has the form

$$\begin{aligned} x(t) &= A_1 \cos \sigma_1 t + A_2 \sin \sigma_1 t + A_3 \cos \sigma_2 t + A_4 \sin \sigma_2 t, \\ y(t) &= B_1 \cos \sigma_1 t + B_2 \sin \sigma_1 t + B_3 \cos \sigma_2 t + B_4 \sin \sigma_2 t, \\ z(t) &= C_1 \cos t + C_2 \sin t, \end{aligned} \tag{1}$$

where

$$\sigma_1 = \left[\frac{1-\sqrt{\Delta}}{2}\right]^{\frac{1}{2}}, \quad \sigma_2 = \left[\frac{1+\sqrt{\Delta}}{2}\right]^{\frac{1}{2}}, \quad \Delta = 1 - 27\mu(1-\mu). \tag{2}$$

Due to the difference in the values of the two frequencies, long and short period terms are recognized, corresponding to small (σ_1) and large (σ_2) values of the frequency .

By a suitable choice of the initial conditions three particular solutions of the linearized equations are obtained in three dimensions.

V. V. Markellos and Y. Kozai (eds.), Dynamical Trapping and Evolution in the Solar System, 235–247.

These are:

$$1. \quad x(t) = 0, \quad y(t) = 0, \quad z(t) = C_1 \cos t + C_2 \sin t, \tag{3}$$

$$2. \quad \begin{aligned} x(t) &= A_1 \cos \sigma_1 t + A_2 \sin \sigma_1 t, \\ y(t) &= B_1 \cos \sigma_1 t + B_2 \sin \sigma_1 t, \\ z(t) &= C_1 \cos t + C_2 \sin t, \end{aligned} \tag{4}$$

$$3. \quad \begin{aligned} x(t) &= A_3 \cos \sigma_2 t + A_4 \sin \sigma_2 t, \\ y(t) &= B_3 \cos \sigma_2 t + B_4 \sin \sigma_2 t, \\ z(t) &= C_1 \cos t + C_2 \sin t. \end{aligned} \tag{5}$$

The first solution is periodic and is continued to periodic orbits of finite size for every value of the mass parameter. A small part of this family has been given by Buck (1920). The second and the third solutions are not periodic unless the period $T_{x,y}$ of the planar motion is commensurate to the period T_z of the motion along the Oz-axis.

We suppose that

$$T_{x,y} = \frac{p}{q} T_z \tag{6}$$

where p and q are mutually prime integers, or equivalently that

$$\sigma_i(\mu) = \frac{p}{q} , \tag{7}$$

with i = 1 for the case of long period and i = 2 for the case of short period planar periodic solutions.

Relation (7) is valid for "special" values of the mass parameter μ. For these values of μ the corresponding linearized equations admit a periodic solution which, as we show in this article, is continued to a family of periodic solutions of the non-linear equations. This family which bifurcates from the triangular equilibrium point is not an isolated dynamical phenomenon occuring for these "special" values of μ but it is the "arrival" at L_4, as μ varies continuously, of a vertical bifurcation point on either the family of planar-short-period, or the planar-long-period solutions.

2. SECOND ORDER EXPANSIONS FOR THE RESONANT THREE DIMENSIONAL PERIODIC SOLUTIONS

The Equations of the three-dimensional motion of the third particle, when expanded to second order terms with respect to x,y and z, take the form

$$x'' - 2(1+\alpha)y' = (1+\alpha)^2\left[\frac{3}{4}x + \frac{3\sqrt{3}}{4}\rho y + \frac{21}{16}\rho x^2 - \frac{3\sqrt{3}}{8}xy - \frac{33}{16}y^2 + \frac{3}{4}\rho z^2\right],$$

$$y'' + 2(1+\alpha)x' = (1+\alpha)^2\left[\frac{3\sqrt{3}}{4}\rho x + \frac{9}{4}y - \frac{3\sqrt{3}}{16}x^2 - \frac{33}{8}\rho xy - \frac{9\sqrt{3}}{16}y^2 + \frac{3\sqrt{3}}{4}z^2\right], \tag{8a}$$

$$z'' = -(1+\alpha)^2\left[z - \frac{3}{2}\rho xz - \frac{3\sqrt{3}}{2}yz\right],$$

where

$$t = (1+\alpha)\tau, \qquad \alpha = \alpha_1\varepsilon + \alpha_2\varepsilon^2, \tag{8b}$$

and $\rho = 1 - 2\mu$.

The solution of Equations (8a) is expressed as

$$x(\tau) = x_1(\tau)\varepsilon + x_2(\tau)\varepsilon^2 + \dots,$$
$$y(\tau) = y_1(\tau)\varepsilon + y_2(\tau)\varepsilon^2 + \dots, \tag{9}$$
$$z(\tau) = z_1(\tau)\varepsilon + z_2(\tau)\varepsilon^2 + \dots,$$

where $x_i(\tau)$, $y_i(\tau)$, $z_i(\tau)$, $i = 1,2,\dots$, are functions of τ to be determined and ε is a small orbital parameter.

Expressions (9) are now substituted into Equations (8a), and the coefficients of the same powers of ε are equated.

The coefficients of the first power of ε are solutions of the "linearized" Equations:

$$x_1'' - 2y_1' = \frac{3}{4} x_1 + \frac{3\sqrt{3}}{4} \rho y_1 ,$$
$$y_1'' + 2x_1' = \frac{3\sqrt{3}}{4} \rho x_1 + \frac{9}{4} y_1 , \tag{10}$$
$$z_1'' = - z_1 .$$

We consider as a particular solution of Equations (10) the solution (4) or (5), i.e.,

$$x_1(\tau) = A_j \cos \sigma_i \tau + A_{j+1} \sin \sigma_i \tau ,$$
$$y_1(\tau) = B_j \cos \sigma_i \tau + B_{j+1} \sin \sigma_i \tau , \tag{11}$$
$$z_1(\tau) = C_1 \cos \tau + C_2 \sin \tau ,$$

which is assumed periodic because of condition (6) which we assume to hold. From condition (6), or (7), "special" values of the mass parameter μ are determined.

The coefficients of the second power of ε are solutions of the Equations

$$x_2'' - 2y_2' - \frac{3}{4} x_2 - \frac{3\sqrt{3}}{4} \rho y_2 = 2\alpha_1 y_1' + \frac{3}{2} \alpha_1 x_1 + \frac{3\sqrt{3}}{2} \alpha_1 \rho y_1$$
$$+ \frac{21}{16} \rho x_1^2 - \frac{3\sqrt{3}}{8} x_1 y_1 - \frac{33}{16} \rho y_1' + \frac{3}{4} \rho z_1^2 ,$$

$$y_2'' + 2x_2' - \frac{3\sqrt{3}}{4} \rho x_2 - \frac{9}{4} y_2 = - 2\alpha_1 x_1' + \frac{3\sqrt{3}}{2} \rho \alpha_1 x_1 + \frac{9}{2} \alpha_1 y_1 \tag{12}$$
$$- \frac{3\sqrt{3}}{16} x_1^2 - \frac{33}{8} \rho x_1 y_1 - \frac{9\sqrt{3}}{16} y_1^2 + \frac{3\sqrt{3}}{4} z_1^2 ,$$

$$z_2'' + z_2 = - 2\alpha_1 z_1 + \frac{3}{2} \rho x_1 z_1 + \frac{3\sqrt{3}}{2} y_1 z_1 .$$

By substitution of expressions (11) into the second members of Equations (12) we obtain the following system of Equations

$$(D^2 - \frac{3}{4}) x_2 - (2D + \frac{3\sqrt{3}}{4} \rho) y_2 = K_1 \sin \sigma_i \tau + K_2 \cos \sigma_i \tau + K_3 \cos^2 \sigma_i \tau$$
$$+ K_4 \sin^2 \sigma_i \tau + K_5 \sin 2\sigma_i \tau + K_6 \sin^2 \tau \triangleq f_2(\tau) ,$$

$$(2D - \frac{3\sqrt{3}}{4} \rho) x_2 + (D^2 - \frac{9}{4}) y_2 = \Lambda_1 \sin \sigma_i \tau + \Lambda_2 \cos \sigma_i \tau + \Lambda_3 \cos^2 \sigma_i \tau$$

$$+ \Lambda_4 \sin^2 \sigma_i \tau + \Lambda_5 \sin 2\sigma_i \tau + \Lambda_6 \sin^2 \tau \triangleq g_2(\tau) , \tag{13}$$

$$(D^2 + 1) z_2 = E_1 \sin (\sigma_i + 1)\tau - E_1 \sin (\sigma_i - 1)\tau +$$

$$+ E_2 \cos(\sigma_i + 1)\tau - E_2 \cos (\sigma_i - 1)\tau \triangleq h_2(\tau) ,$$

where we have abbreviated:

$$\begin{aligned}
K_1 &= \alpha_1 (-2B_j \sigma_i + \frac{3}{2} A_{j+1} + \frac{3\sqrt{3}}{2} \rho B_{j+1}) , \\
K_2 &= \alpha_1 (2B_{j+1} \sigma_i + \frac{3}{2} A_j + \frac{3\sqrt{3}}{2} \rho B_j) , \\
K_3 &= \frac{21}{16} \rho A_j^2 - \frac{3\sqrt{3}}{8} A_j B_j - \frac{33}{16} \rho B_j^2 , \\
K_4 &= \frac{21}{16} \rho A_{j+1}^2 - \frac{3\sqrt{3}}{8} A_{j+1} B_{j+1} - \frac{33}{16} \rho B_{j+1}^2 , \\
K_5 &= \frac{21}{16} \rho A_j A_{j+1} - \frac{3\sqrt{3}}{16} A_{j+1} B_j - \frac{3\sqrt{3}}{16} A_j B_{j+1} - \frac{33}{16} \rho B_j B_{j+1} , \\
K_6 &= \frac{3}{4} \rho C_1^2 , \\
\Lambda_1 &= \alpha_1 (2A_j \sigma_i + \frac{3\sqrt{3}}{2} \rho A_{j+1} + \frac{9}{2} B_{j+1}) , \\
\Lambda_2 &= \alpha_1 (-2A_{j+1} \sigma_i + \frac{3\sqrt{3}}{2} \rho A_j + \frac{9}{2} B_j) , \\
\Lambda_3 &= - (\frac{3\sqrt{3}}{16} A_j^2 + \frac{33}{8} \rho A_j B_j + \frac{9\sqrt{3}}{16} B_j^2) , \\
\Lambda_4 &= - (\frac{3\sqrt{3}}{16} A_{j+1}^2 + \frac{33}{8} \rho A_{j+1} B_{j+1} + \frac{9\sqrt{3}}{16} B_{j+1}^2) , \\
\Lambda_5 &= - (\frac{33}{16} \rho A_j B_{j+1} + \frac{33}{16} \rho A_{j+1} B_j + \frac{3\sqrt{3}}{16} A_j A_{j+1} + \frac{9\sqrt{3}}{16} B_j B_{j+1}) , \\
\Lambda_6 &= \frac{3\sqrt{3}}{4} C_1^2 , \\
E_1 &= \frac{3}{4} C_1 (\rho A_j + \sqrt{3} B_j) , \\
E_2 &= \frac{3}{4} C_1 (\rho A_{j+1} + \sqrt{3} B_{j+1}) .
\end{aligned} \tag{14}$$

A periodic solution of Equations (14) is given by:

$$x_2(\tau) = \frac{\Gamma_1}{\Theta - 12} + \frac{\Gamma_4}{\Phi} \cos 2\sigma_i\tau + \frac{\Gamma_5}{\Phi} \sin 2\sigma_i\tau + \frac{\Gamma_6}{\Theta} \cos 2\tau + \frac{\Gamma_7}{\Theta} \sin 2\tau ,$$

$$y_2(\tau) = \frac{\Delta_1}{\Theta - 12} + \frac{\Delta_4}{\Phi} \cos 2\sigma_i\tau + \frac{\Delta_5}{\Phi} \sin 2\sigma_i\tau + \frac{\Delta_6}{\Theta} \cos 2\tau + \frac{\Delta_7}{\Theta} \sin 2\tau , \quad (15)$$

$$z_2(\tau) = \frac{E_1}{-\sigma_i^2 - 2\sigma_i} \sin(\sigma_i + 1)\tau - \frac{E_1}{-\sigma_i^2 + 2\sigma_i} \sin(\sigma_i - 1)\tau + \frac{E_2}{-\sigma_i^2 - 2\sigma_i} \cos(\sigma_i + 1)\tau - \frac{E_2}{-\sigma_i^2 + 2\sigma_i} \cos(\sigma_i - 1)\tau ,$$

where, supressing the secular terms, we have forced $\alpha_1 = \alpha_2 = 0$ (in (8b)) and, therefore, $t = \tau$. The quantities Θ and Φ are given by

$$\Theta = 12 + \frac{27}{4}\mu(1-\mu), \qquad \Phi = 16\sigma_i^4 - 4\sigma_i^2 + \frac{27}{4}\mu(1-\mu). \quad (16)$$

We have also abbreviated:

$$\Gamma_1 = -\frac{9}{8}(K_3 + K_4 + K_6) + \frac{3\sqrt{3}}{8}\rho(\Lambda_3 + \Lambda_4 + \Lambda_6) ,$$

$$\Gamma_4 = -(2\sigma_i^2 + \frac{9}{8})(K_3 - K_4) + 4\Lambda_5\sigma_2 + \frac{3\sqrt{3}}{8}\rho(\Lambda_3 - \Lambda_4) ,$$

$$\Gamma_5 = -(4\sigma_i^2 + \frac{9}{4})K_5 - 2(\Lambda_3 - \Lambda_4)\sigma_i + \frac{3\sqrt{3}}{4}\rho\Lambda_5 ,$$

$$\Gamma_6 = \frac{25}{8}K_6 - \frac{3\sqrt{3}}{8}\rho\Lambda_6 ,$$

$$\Gamma_7 = \Lambda_6 , \quad (17)$$

$$\Delta_1 = -\frac{3}{8}(\Lambda_3 + \Lambda_4 + \Lambda_6) + \frac{3\sqrt{3}}{8}\rho(K_3 + K_4 + K_6) ,$$

$$\Delta_4 = -(2\sigma_i^2 + \frac{3}{8})(\Lambda_3 - \Lambda_4) - 4K_5\sigma_i + \frac{3\sqrt{3}}{8}\rho(K_3 - K_4) ,$$

$$\Delta_5 = -(4\sigma_i^2 + \frac{3}{4})\Lambda_5 + 2(K_3 - K_4)\sigma_i + \frac{3\sqrt{3}}{4}\rho K_5 ,$$

$$\Delta_6 = \frac{19}{8}\Lambda_6 - \frac{3\sqrt{3}}{8}\rho K_6 ,$$

$$\Delta_7 = 2K_6 .$$

It has been verified numerically that for values of the small parameter ε in the interval $(0, 0.05]$, the periodic functions (15) represent periodic solutions of the problem to an accuracy of at least six significant figures. Furthermore, these solutions can be "corrected" and "continued" by numerical methods. In this way the existence of these resonant periodic orbits which had been questioned by the classical workers (Buck, 1920), has been demonstrated.

3. SOLUTION OF A HILL EQUATION FOR VERTICAL STABILITY ALONG THE FAMILIES OF PLANAR PERIODIC ORBITS

An important question arising here is whether the family of periodic solutions constructed in the previous paragraph exists only for the resonant value of μ.

As we shall see the answer is that it also exists for other values of μ. However, for these other values it does not bifurcate from L_4. Rather, it bifurcates from a vertical-bifurcation point on the planar family of (short - or long - period) periodic orbits. Hereafter we use the term "family of planar periodic solutions" to indicate either the short - period family or the long - period family of planar periodic solutions, the two cases been formally identical.

First we consider the family of planar periodic solutions and we derive second order expansions for them. The derivation of these expansions is similar to the above derivation of the resonant three-dimensional orbits and the resulting expressions differ from expressions (11), (15) only in the absence of the π-periodic terms.

The second order expansions for the planar orbits are:

$$x(t) = (A_j \cos \sigma_i t + A_{j+1} \sin \sigma_i t)\varepsilon + (G_1 + G_2 \cos 2\sigma_i t + G_3 \sin 2\sigma_i t)\varepsilon^2 , \quad (18a)$$

$$y(t) = (B_j \cos \sigma_i t + B_{j+1} \sin \sigma_i t)\varepsilon + (H_1 + H_2 \cos 2\sigma_i t + H_3 \sin 2\sigma_i t)\varepsilon^2 , \quad (18b)$$

where

$$G_1 = \Gamma_1/\Theta - 12, \qquad G_2 = \Gamma_4/\Phi, \qquad G_3 = \Gamma_5/\Phi,$$
$$H_1 = \Delta_1/\Theta - 12, \qquad H_2 = \Delta_4/\Phi, \qquad H_3 = \Delta_5/\Phi, \tag{19}$$

with $i = 1$, $j = 1$ for the long period solutions and $i = 2$, $j = 3$ for the short period ones.

Along each family of planar solutions we can determine the parameter s_v which characterizes every periodic solution as vertically stable or unstable. If the periodic solution is vertically stable,

$$|s_v| < 1,$$

and if there are integers p and q such that

$$s_v = \cos 2\pi \frac{p}{q}, \tag{20}$$

then this planar periodic orbit is vertically self-resonant and a bifurcation point of a three-dimensional family.

In the present case where we know the analytical expression of the family of planar solutions we can in fact calculate s_v analytically as a function of the orbital parameter ε. Indeed, the value of s_v results from two linearly independent solutions of the Hill equation

$$\ddot{v} + Q(t)v = 0,$$

with

$$Q(t) = \frac{1-\mu}{r_1^3} + \frac{\mu}{r_2^3}.$$

(see, e.g. Markellos, 1977). Using the second order expansions (18) for x and y we obtain for the periodic function Q the expression

$$Q(t,\varepsilon) = 1 + (Q_2 \cos\sigma_i t + Q_3 \sin\sigma_i t)\varepsilon$$
$$+ (Q_1 + Q_4 \cos 2\sigma_i t + Q_5 \cos 2\sigma_i t)\varepsilon^2 \tag{21}$$

with:

$$Q_1 = -\frac{3}{2}(1-2\mu)G_1 - \frac{3\sqrt{3}}{2} H_1 + \frac{3}{16}(A_j^2 + A_{j+1}^2) + \frac{33}{16}(B_j^2 + B_{j+1}^2)$$
$$+ \frac{15\sqrt{3}}{8}(1-2\mu)(A_j B_j + A_{j+1}B_{j+1}), \tag{22}$$

$$Q_2 = -\frac{3}{2}(1-2\mu)A_j - \frac{3\sqrt{3}}{2} B_j, \tag{23}$$

$$Q_3 = -\frac{3}{2}(1-2\mu)A_{j+1} - \frac{3\sqrt{3}}{2}B_{j+1}, \tag{24}$$

$$Q_4 = -\frac{3}{2}(1-2\mu)G_2 - \frac{3\sqrt{3}}{2}H_2 + \frac{3}{16}(A_j^2 - A_{j+1}^2) + \frac{33}{16}(B_j^2 - B_{j+1}^2) + \frac{15\sqrt{3}}{8}(A_jB_j - A_{j+1}B_{j+1}), \tag{25}$$

$$Q_5 = -\frac{3}{2}(1-2\mu)G_3 - \frac{3\sqrt{3}}{2}H_3 + \frac{3}{8}A_jA_{j+1} + \frac{33}{8}B_jB_{j+1} + \frac{15\sqrt{3}}{8}(A_jB_{j+1} + A_{j+1}B_j). \tag{26}$$

We seek solutions $v(t)$ of the Equation

$$\ddot{v} + Q(t,\varepsilon)v = 0, \tag{27}$$

in the form

$$v(t) = v_0(t) + v_1(t)\varepsilon + v_2(t)\varepsilon^2. \tag{28}$$

By substitution into Equation (27), neglecting terms of order higher than the second in ε and solving the resulting differential equations, we obtain as the general solution of Equation (27) the expression

$$\begin{aligned} v(t) = {} & (\mu_1 \cos t + \mu_2 \sin t)(1+\varepsilon+\varepsilon^2) + \left[(w_1\cos(1+\sigma_i)t \right. \\ & \left. + w_2\sin(1+\sigma_i)t + w_3\cos(1-\sigma_i)t + w_4\sin(1-\sigma_i)t\right](\varepsilon+\varepsilon^2) \\ & + \left[w_5 t\cos t + w_6 t\sin t + w_7\cos(2\sigma_i+1)t + w_8\sin(2\sigma_i+1)t \right. \\ & \left. + w_9\cos(2\sigma_i-1)t + w_{10}\sin(2\sigma_i-1)t\right]\varepsilon^2, \end{aligned} \tag{29}$$

where we have abbreviated:

$$w_1 = -\frac{u_1}{\sigma_i(\sigma_i+2)}, \qquad w_2 = -\frac{u_2}{\sigma_i(\sigma_i+2)},$$

$$w_3 = -\frac{u_3}{\sigma_i(\sigma_i-2)}, \qquad w_4 = -\frac{u_4}{\sigma_i(\sigma_i-2)},$$

$$w_5 = -\frac{u_5}{2}, \qquad w_6 = -\frac{u_6}{2}, \tag{30}$$

$$w_7 = -\frac{u_7}{4\sigma_i(\sigma_i+1)}, \qquad w_8 = -\frac{u_8}{4\sigma_i(\sigma_i+1)},$$

$$w_9 = -\frac{u_9}{4\sigma_i(\sigma_i-1)}, \qquad w_{10} = -\frac{u_{10}}{4\sigma_i(\sigma_i-1)},$$

and

$$u_1(\mu_1, \mu_2) = -\frac{1}{2} Q_2\mu_1 + \frac{1}{2} Q_3\mu_2 ,$$

$$u_2(\mu_1, \mu_2) = -\frac{1}{2} Q_3\mu_1 - \frac{1}{2} Q_2\mu_2 ,$$

$$u_3(\mu_1, \mu_2) = -\frac{1}{2} Q_2\mu_1 - \frac{1}{2} Q_3\mu_2 ,$$

$$u_4(\mu_1, \mu_2) = \frac{1}{2} Q_3\mu_1 - \frac{1}{2} Q_2\mu_2 ,$$

$$\begin{aligned} u_5(\mu_1, \mu_2) = {} & - Q_1\mu_1 + \frac{Q_2u_1}{2\sigma_i(\sigma_i+2)} + \frac{Q_3u_2}{2\sigma_i(\sigma_i+2)} \\ & + \frac{Q_2u_3}{2\sigma_i(\sigma_i-2)} - \frac{Q_3u_4}{2\sigma_i(\sigma_i-2)} , \end{aligned} \tag{31}$$

$$\begin{aligned} u_6(\mu_1, \mu_2) = {} & - Q_1\mu_2 - \frac{Q_3u_1}{2\sigma_i(\sigma_i+2)} + \frac{Q_2u_2}{2\sigma_i(\sigma_i+2)} \\ & + \frac{Q_3u_3}{2\sigma_i(\sigma_i-2)} + \frac{Q_2u_4}{2\sigma_i(\sigma_i-2)} , \end{aligned}$$

$$u_7(\mu_1, \mu_2) = -\frac{1}{2}\mu_1Q_4 + \frac{1}{2}\mu_2Q_5 + \frac{Q_2u_1}{2\sigma_i(\sigma_i+2)} - \frac{Q_3u_2}{2\sigma_i(\sigma_i+2)} ,$$

$$u_8(\mu_1, \mu_2) = -\frac{1}{2}\mu_2Q_4 - \frac{1}{2}\mu_1Q_5 + \frac{Q_3u_1}{2\sigma_i(\sigma_i+2)} + \frac{Q_2u_2}{2\sigma_i(\sigma_i+2)} ,$$

$$u_9(\mu_1, \mu_2) = -\frac{1}{2}\mu_1Q_4 - \frac{1}{2}\mu_2Q_5 + \frac{Q_2u_3}{2\sigma_i(\sigma_i-2)} + \frac{Q_3u_4}{2\sigma_i(\sigma_i-2)} ,$$

$$u_{10}(\mu_1, \mu_2) = \frac{1}{2}\mu_2Q_4 - \frac{1}{2}\mu_1Q_5 + \frac{Q_3u_3}{2\sigma_i(\sigma_i-2)} - \frac{Q_2u_4}{2\sigma_i(\sigma_i-2)} .$$

4. PARAMETER OF VERTICAL STABILITY AS FUNCTION OF ε

If $v^*(t)$ and $v^{**}(t)$ are two linearly independent solutions of Equation (27), with

$$\begin{pmatrix} v^*(0) & v^{**}(0) \\ \dot{v}^*(0) & \dot{v}^{**}(0) \end{pmatrix} = \begin{pmatrix} 1 & 0 \\ 0 & 1 \end{pmatrix} , \tag{32}$$

then

$$2s_v = v^*(t) + \dot{v}^{**}(t) , \tag{33}$$

where T is the period of the planar periodic orbit.

From Equations (29) and (33) we finally obtain, for the vertical stability parameter, the expression

$$s_v = s_0(\mu) + s_1(\mu)\varepsilon + s_2(\mu)\varepsilon^2 , \tag{34}$$

with

$$s_0(\mu) = \cos \frac{2\pi}{\sigma_i} , \tag{35}$$

$$s_1(\mu) = - \frac{Q_3}{2\sigma_i} \sin \frac{2\pi}{\sigma_i} , \tag{36}$$

$$s_2(\mu) = \frac{1}{2} \left[\Omega_1 \cos \frac{2\pi}{\sigma_i} + \Omega_2 \sin \frac{2\pi}{\sigma_i} \right] , \tag{37}$$

where

$$\frac{2\pi}{\sigma_i} = T ,$$

is the period of the planar periodic orbit.

The quantities Ω_1 and Ω_2 involved in Equation (37) are given by the following expressions:

$$\Omega_1 = 4 + \frac{Q_4}{2(\sigma_i^2 - 4)} + \left[Q_1 + \frac{Q_2^2 + Q_3^2}{2(\sigma_i^2 - 4)} \right] \frac{2\pi}{\sigma_i}$$

$$+ \frac{20\,Q_3^2 + 4Q_2^2 - (5Q_2^2 + 13Q_3^2)\sigma_i^2}{4\sigma_i^2(\sigma_i^2 - 4)^2} \; ,$$

$$\Omega_2 = \frac{-2\sigma_i^2 - \sigma_i + 5}{\sigma_i(\sigma_i^2 - 4)} Q_3 - \frac{\sigma_i}{2(\sigma_i^2 - 4)} Q_5$$

$$+ \frac{-2\sigma_i^3 + 10\sigma_i^2 - 3\sigma_i - 18}{2\sigma_i^2(\sigma_i^2 - 4)^2} Q_2 Q_3 + \frac{Q_2^2 + Q_3^2}{4(\sigma_i^2 - 4)} \; . \qquad (38)$$

Equating the expression for s_v to the bifurcation value (20) we obtain the relation

$$s_0(\mu) + s_1(\mu)\varepsilon + s_2(\mu)\varepsilon^2 = \cos 2\pi \frac{p}{q} \; , \qquad (39)$$

connecting the mass parameter μ with the orbital parameter ε for any given resonance p/q.

Thus, given a value of μ say μ^* near the resonant value $\mu_{p/q}$, we can determine the value ε for which the "vertical bifurcation" occurs. In other words, we find how the resonant family has evolved in going from $\mu_{p/q}$ to μ^*, i.e. from a branching at the equilibrium point to a branching at a vertical self-resonant orbit of the planar family.

We have therefore demonstrated how this "peculiar" resonant family of periodic orbits exists not only at the resonant value of μ but also at the neighboring values. It has a natural evolution as a tree-dimensional branch of the family of planar orbits. This is true for the short period family as well as for the long period family of planar periodic orbits.

Numerical results and further details of the evolution of these families of three-dimensional periodic orbits will be published elsewhere.

5. REFERENCES

Buck, T.: 1920, in F.R. Moulton, *Periodic Orbits*, Carnegie Inst. of Washington, J. Reprint Co., p. 299

Markellos, V.V.: 1977, *Monthly Notices Roy. Astron. Soc.* 180, 103

ASYMMETRIC PERIODIC ORBITS IN THE THREE-BODY PROBLEM AND THEIR STABILITY

A. Tsouroplis and C.G. Zagouras
University of Patras, Patras, Greece

ABSTRACT

An algorithm for the numerical determination of asymmetric periodic solutions of the planar general three body problem is described. The elements of the "variational" matrix which are used in this algorithm are computed by numerical integration of the corresponding "variational" equations. These elements are also used in the study of the linear iso-energetic stability. A number of asymmetric periodic orbits are presented and their stability parameters are given.

1. NUMERICAL DETERMINATION OF ASYMMETRIC PERIODIC SOLUTIONS

We use a rotating system of dimensionless coordinates with origin at the center of mass of the two more massive bodies P_1 and P_2.

The position of the three-body system is fully determined in terms of the coordinates x,y of the third body P_3, the distance x_2 of P_2 from the origin and the angle θ between the rotating and a non-rotating system.

In the rotating coordinate system the Equations of motion of the planar general three body problem are

$$\begin{aligned} \ddot{x} &= Bx + x\dot{\theta}^2 + 2\dot{\theta}\dot{y} + \ddot{\theta}y + \mu A x_2 \\ \ddot{y} &= (B+\dot{\theta}^2)y - x\ddot{\theta} - 2\dot{x}\dot{\theta}, \\ \ddot{x}_2 &= (m_3 B^* + \dot{\theta}^2)x_2 - (1-m_3)(1-\mu)^3/x_2^2 + m_3(1-\mu)Ax, \\ \ddot{\theta} &= -2\dot{\theta}\dot{x}_2/x_2 + m_3(1-\mu)Ay/x_2, \end{aligned} \tag{1}$$

or in first-order form:

$$\frac{dX_1}{dt} = X_4 \triangleq f_1, \quad \frac{dX_2}{dt} = X_5 \triangleq f_2, \quad \frac{dX_3}{dt} = X_6 \triangleq f_3$$

V. V. Markellos and Y. Kozai (eds.), Dynamical Trapping and Evolution in the Solar System, 249–256.

$$\frac{dX_4}{dt} = BX_1 + X_1X_8^2 + 2X_8X_5 + X_8X_2 + \mu AX_3 \triangleq f_4,$$

$$\frac{dX_5}{dt} = (B + X_8^2)X^2 - X_1\dot{X}_8 - 2X_4X_8 \triangleq f_5,$$

$$\frac{dX_6}{dt} = (m_3B^* + X_8^2)X_3 - (1-m_3)(1-\mu)^3/X_3^2 + m_3(1-\mu)AX_1 \triangleq f_6, \tag{2}$$

$$\frac{dX_7}{dt} = X_8 \triangleq f_7,$$

$$\frac{dX_8}{dt} = -2X_8X_6/X_3 + m_3(1-\mu)AX_2/X_3 \triangleq f_8,$$

where

$$(X_1, X_2, X_3, X_4, X_5, X_6, X_7, X_8) = (x, y, x_2, \dot{x}, \dot{y}, \dot{x}_2, \Theta, \dot{\Theta}).$$

A periodic solution $\underline{X}(\underline{X}_o; t)$ of the above Equations will satisfy

$$X_i(\underline{X}_o; t + T) = X_i(\underline{X}_o; t), \quad i \neq 7 \tag{3}$$

where T is the period and $\underline{X}_o = (X_{o1}, \ldots, X_{o8})$ is the initial-conditions vector. Further, without loss of generality, we shall fix initial values of y, Θ and $\dot{\Theta}$ as follows: $y_o = 0$, $\Theta_o = 0$, $\dot{\Theta}_o = 1$.
The periodicity conditions are written in the form:

$$x(x_0, x_{20}, \dot{x}_0, \dot{y}_0, \dot{x}_{20}; T) = x_0, \tag{a}$$

$$y(x_0, x_{20}, \dot{x}_0, \dot{y}_0, \dot{x}_{20}; T) = y_0, \tag{b}$$

$$x_2(x_0, x_{20}, \dot{x}_0, \dot{y}_0, \dot{x}_{20}; T) = x_{20}, \tag{c}$$

$$\dot{x}(x_0, x_{20}, \dot{x}_0, \dot{y}_0, \dot{x}_{20}; T) = \dot{x}_0, \tag{d) (4}$$

$$\dot{y}(x_0, x_{20}, \dot{x}_0, \dot{y}_0, \dot{x}_{20}; T) = \dot{y}_0, \tag{e}$$

$$\dot{x}_2(x_0, x_{20}, \dot{x}_0, \dot{y}_0, \dot{x}_{20}; T) = x_{20}, \tag{f}$$

$$\dot{\Theta}(x_0, x_{20}, \dot{x}_0, \dot{y}_0, \dot{x}_{20}; T) = \dot{\Theta}_0 \tag{g}$$

In practice condition (4b) is satisfied "by force" since we start and terminate the numerical integration when the orbit crosses the Ox axis. Further, due to the integrals of the problem only four of the remaining six periodicity conditions are trully independent. Essentially, therefore, the periodicity conditions are only four and in this work we have used the conditions (4a, c,d,f).

From these periodicity conditions corrector-predictor algorithms can be established for the numerical determination of entire series of asymmetric periodic solutions. In the corrector phase we assume an initial state vector $\underline{X}_0$ which approximately leads to a periodic orbit of (approximate) period T, and seek to adjust this state vector by differential corrections to improve iteratively the accuracy of periodicity.

If we integrate the Equations of motion and stop at the second crossing with the Ox-axis (after one full revolution), we have in general

$$\underline{X}(\underline{X}_0 \;;\; T) \neq \underline{X}_0 \;.$$

We seek corrections $\delta\underline{X}_0 = (\delta x_0,\ 0,\ \delta x_{02},\ \delta\dot{x}_0,\ \delta\dot{y}_0,\ \delta\dot{x}_{20},\ 0,\ 0)$ such that

$$\underline{X}\ (\underline{X}_0 + \delta\underline{X}_0 \;;\; T + \delta T) = \underline{X}_0 + \delta\underline{X}_0 \;. \tag{5}$$

Expanding in Taylor series and neglecting terms of order higher that the first, we shall have

$$x_i + \frac{\partial x_i}{\partial x_{01}}\delta x_{01} + \frac{\partial x_i}{\partial x_{03}}\delta x_{03} + \frac{\partial x_i}{\partial x_{04}}\delta x_{04} + \frac{\partial x_i}{\partial x_{05}}\delta x_{05}$$

$$+ \frac{\partial x_i}{\partial x_{06}}\ x_{06} + \frac{\partial x_i}{\partial T}\delta T = x_{0i} + \delta x_{0i} \;,$$

$$(\,i = 1,2,3,4,6\,) \,. \tag{6}$$

For i=2 we obtain in particular,

$$\frac{\partial x_2}{\partial x_{01}}\delta x_{01} + \frac{\partial x_2}{\partial x_{03}}\delta\, x_{03} + \frac{\partial x_2}{\partial x_{04}}\delta x_{04} + \frac{\partial x_2}{\partial x_{05}}\delta x_{05}$$

$$+ \frac{\partial x_2}{\partial x_{06}}\delta x_{06} + \frac{\partial x_2}{\partial T}\delta T = 0 \;, \tag{7}$$

since, for t=T, $x_2 = y = 0$ while $\delta x_{02} = \delta y_0 = 0$. Solving now Equations (7) for δT and substituting into relations (6) we get

$$x_i + u_{i1}\delta x_{01} + u_{i3}\delta x_{03} + u_{i4}\delta x_{04} + u_{i5}\delta x_{05} + u_{i6}\delta x_{06}$$

$$= x_{0i} + \delta x_{0i}, \qquad i = 1,3,4,6. \tag{8}$$

where

$$u_{ij} = \frac{\partial x_i}{\partial x_{0j}} - \frac{\partial x_2}{\partial x_{0j}} \frac{f_i}{f_2}, \quad i=1,3,4,6, \tag{9}$$

("variations at the crossing"; Markellos, 1977).

Assuming X_{04} constant or equivalently $\delta X_{04} = 0$, Equations (8) become

$$\begin{aligned}
(u_{11}-1)\,\delta X_{01} + u_{13}\delta X_{03} + u_{15}\delta X_{05} + u_{16}\delta X_{06} &= X_{01}-X_1, \\
u_{31}\delta X_{01} + (u_{33}-1)\delta X_{03} + u_{35}\delta X_{05} + u_{36}\delta X_{06} &= X_{03}-X_3, \\
u_{41}\delta X_{01} + u_{43}\delta X_{03} + u_{45}\delta X_{05} + u_{46}\delta X_{06} &= X_{04}-X_4, \\
u_{61}\delta X_{01} + u_{63}\delta X_{03} + u_{65}\delta X_{05} + (u_{66}-1)\delta X_{06} &= X_{06}-X_6 .
\end{aligned} \tag{10}$$

This system is the corrector of the algorithm. It is solved for the corrections δX_{01}, δX_{03}, δX_{05}, δX_{06}, which are then added to the corresponding components of the initial state vector to obtain a better approximation to the periodic orbit with period $T + \delta T$.

After repeated applications of the corrector we find (assuming convergence) the periodic (to the desired accuracy) solution characterized by the value X_{04} which is kept constant during the correction process.

We then proceed to a single application of the predictor:

$$\begin{aligned}
(u_{11}-1)\Delta X_{01} + u_{13}\Delta X_{03} + u_{15}\Delta X_{05} + u_{16}\Delta X_{06} &= -u_{14}\Delta X_{04}, \\
u_{31}\Delta X_{01} + (u_{33}-1)\Delta X_{03} + u_{35}\Delta X_{05} + u_{36}\Delta X_{06} &= -u_{34}\Delta X_{04} , \\
u_{41}\Delta X_{01} + u_{43}\Delta X_{03} + u_{45}\Delta X_{05} + u_{46}\Delta X_{06} &= (1-u_{44})\Delta X_{04}, \\
u_{61}\Delta X_{01} + u_{63}\Delta X_{03} + u_{65}\Delta X_{05} + (u_{66}-1)\Delta X_{06} &= -u_{64}\Delta X_{04}.
\end{aligned} \tag{11}$$

This predictor is designed to obtain the approximate initial state vector $\underline{X}_0 + \Delta\underline{X}_0$ corresponding to another periodic orbit (along the family), characterized by the value $X^*_{04} = X_{04} + \Delta X_{04}$, where the "increment" $\Delta\underline{X}_{04}$ is arbitrary but small so that convergence of the subsequent application of the corrector is secured. The values of the "sensitivities" u_{ij} involved in Equations (10) and (11) are computed from relations (9), where the "variations" $\partial X_i/\partial X_{0j}$ are known through numerical integration of the linear variational Equations:

$$\frac{dV}{dt} = P\,V,$$

where

$$V = (v_{ij}) = (\partial x_i / \partial x_{0j})$$

and

$$P = \left(\frac{\partial f_i}{\partial x_j} \right), \qquad i,j = 1,\ldots,8.$$

2. STABILITY

IF $\underline{X}_0$ is the vector, in phase space, corresponding to a periodic orbit and $\underline{X}_0 + \delta\underline{X}_0$ is the vector of a neighboring orbit corresponding to the same value of the energy and angular momentum integrals, then a transformation T is constructed which transforms the initial state $\underline{X}_0$ to the state $\underline{X}$ when the orbit crosses the surface of section $X_2 = Y = \bar{0}$ for the second time (simple orbits). This transformation is expressed as

$$\underline{X} = \underline{\sigma}(\underline{X}_0), \tag{13}$$

where

$$\underline{\sigma} = (\sigma_1, \sigma_3, \sigma_4, \sigma_6). \tag{14}$$

After linearization, the transformation (13) is written

$$\delta\underline{X} = A\, \delta\underline{X}_0 \tag{15}$$

where

$$\begin{aligned} \delta\underline{X} &= (\delta X_1, \delta X_3, \delta X_4, \delta X_6)^T, \\ \delta\underline{X}_0 &= (\delta X_{01}, \delta X_{03}, \delta X_{04}, \delta X_{06})^T, \end{aligned} \tag{16}$$

and A is the 4x4 matrix with elements the first partial derivatives of the functions $(\sigma_1, \sigma_3, \sigma_4, \sigma_6)$ with respect to the initial conditions, i.e.

$$A = (\alpha_{ij}) = \left(\frac{\partial \sigma_i}{\partial x_{0j}} \right), \quad i,j = 1,3,4,6 \tag{17}$$

The conditions for stability are:

$$\Delta > 0, \quad |p| < 2, \quad |q| < 2, \tag{18}$$

where

$$\Delta = \alpha^2 - 4(\beta - 2), \quad p = \frac{1}{2}(\alpha + \sqrt{\Delta}), \quad q = \frac{1}{2}(\alpha - \sqrt{\Delta}) \tag{19}$$

TABLE I: The series A_{20} of asymmetric periodic orbits of the planar general three body problem for $\mu = 0.25$ and $x_{04} = -0.17292$.

	m_3	x_{01}	x_{03}	x_{05}	x_{06}	E	p	q
1	0.000103	-2.33048	0.749905	1.90139	-0.00037	-0.093803	-1.998	-39.19
2	0.001203	-2.32672	0.749145	1.89876	-0.002430	-0.094241	-2.037	-36.25
3	0.014009	-2.31258	0.743557	1.89264	-0.013446	-0.097700	-2.629	-36.98
4	0.036509	-2.31138	0.736489	1.90284	-0.024301	-0.102142	-3.983	-46.22
5	0.050629	-2.31501	0.732556	1.91394	-0.029892	-0.104733	-5.030	-54.27
6	0.078649	-2.32598	0.725141	1.93860	-0.038778	-0.109138	-7.234	-73.04
7	0.100269	-2.33609	0.719556	1.95911	-0.044647	-0.112165	-9.119	-90.75
8	0.119689	-2.34574	0.714556	1.97801	-0.049404	-0.114642	-10.93	-109.3
9	0.134089	-2.35307	0.710838	1.99217	-0.052676	-0.116338	-12.37	-124.7
10	0.150109	-2.36131	0.706676	2.00797	-0.056093	-0.118088	-14.07	-143.9
11	0.170189	-2.37167	0.701409	2.02777	-0.060084	-0.120083	-16.38	-171.2
12	0.190009	-2.38185	0.696143	2.04723	-0.063737	-0.121838	-18.87	-201.8
13	0.200000	-2.38694	0.693459	2.05699	-0.065480	-0.122642	-20.21	-218.9

The period of the orbits varies from T = 12.4773 (orbit 1) to T = 10.9992 (orbit 13).

and

$$\alpha = -(\alpha_{11} + \alpha_{33} + \alpha_{44} + \alpha_{66}) \tag{20}$$

$$\beta = \begin{vmatrix} \alpha_{11} & \alpha_{13} \\ \alpha_{31} & \alpha_{33} \end{vmatrix} + \begin{vmatrix} \alpha_{11} & \alpha_{14} \\ \alpha_{41} & \alpha_{44} \end{vmatrix} + \begin{vmatrix} \alpha_{11} & \alpha_{16} \\ \alpha_{61} & \alpha_{66} \end{vmatrix}$$

$$+ \begin{vmatrix} \alpha_{33} & \alpha_{34} \\ \alpha_{43} & \alpha_{44} \end{vmatrix} + \begin{vmatrix} \alpha_{33} & \alpha_{36} \\ \alpha_{63} & \alpha_{66} \end{vmatrix} + \begin{vmatrix} \alpha_{44} & \alpha_{46} \\ \alpha_{64} & \alpha_{66} \end{vmatrix}, \tag{21}$$

(Hadjidemetriou, 1975). The elements a_{ij} can be determined as functions of the elements v_{ij} of the "variational" matrix through the expressions:

$$a_{1i} = (v_{1i} - \frac{x_4}{x_5} v_{2i}) + (v_{15} - \frac{x_4}{x_5} v_{25})D_{i5} + (v_{18} - \frac{x_4}{x_5} v_{28})D_{i8},$$

$$a_{3i} = (v_{3i} - \frac{x_6}{x_5} v_{2i}) + (v_{35} - \frac{x_6}{x_5} v_{25})D_{i5} + (v_{38} - \frac{x_6}{x_5} v_{28})D_{i8},$$

$$a_{4i} = (v_{4i} - \frac{\dot{x}_4}{x_5} v_{2i}) + (v_{45} - \frac{\dot{x}_4}{x_5} v_{25})D_{i5} + (v_{48} - \frac{\dot{x}_4}{x_5} v_{28})D_{i8},$$

$$a_{6i} = (v_{6i} - \frac{\dot{x}_6}{x_5} v_{2i}) + (v_{65} - \frac{\dot{x}_6}{x_5} v_{25})D_{i5} + (v_{68} - \frac{\dot{x}_6}{x_5} v_{28})D_{i8},$$

$$(i = 1,3,4,6) \tag{22}$$

where

$$D_{i5} = -(F_{1i}F_{28} - F_{2i}F_{18})/D,$$
$$D_{i8} = -(F_{2i}F_{15} - F_{1i}F_{25})/D, \tag{23}$$
$$D = F_{15}F_{28} - F_{18}F_{25},$$

and

$$F_{1j} = \frac{\partial F_1}{\partial x_j} = \frac{\partial E}{\partial x_j}, \quad F_{2j} = \frac{\partial F_2}{\partial x_j} = \frac{\partial P}{\partial x_j}, \quad j = 1,3,4,6 \tag{24}$$

with $F_1 = E$ and $F_2 = P$ denoting respectively the energy and angular momentum integrals.

3. PRELIMINARY RESULTS

Applying the above technique, we started the computation of asymmetric periodic solutions of the general three body problem using initial conditions of such solutions of the restricted problem given by Markellos (1977) for values of the mass parameter μ in the interval (0,0.5). We chose as starting point an orbit belonging to the bifurcation series A_{20} for $\mu = 0.25$ with initial conditions $x_0 = -2.3310$. $\dot{x}_0 = -0.17292$, $\dot{y} = 1.9017$ and Jacobi constant C = 2.67054. The periodic solutions obtained are members of a continuous series formed by gradual increase of the mass of the third body m_3, in the interval (0,0.2), while the value of the mass parameter μ is kept constant: μ = 0.25. Sample numerical results are given in Table I. As can be seen in the last column of the Table, all orbits are unstable, in the linear "isoenergetic" sence.

4. REFERENCES

Hadjidemetriou, J.D. 1975, *Celes. Mech.* 12, 255

Markellos, V.V.: 1977, *Mon. Not. R. Astr. Soc.* 180, 103

CONSTRUCTION OF PERIODIC ORBITS, PROBLEMS OF STABILITY AND PERIOD DETERMINATION, IN THE ELLIPTICAL NON-PLANAR RESTRICTED PROBLEM

Colette Edelman
Bureau des Longitudes
Paris, France

ABSTRACT

Periodic orbits in a fixed frame are constructed in the vicinity of non-periodic solutions of the non perturbed problem. In a first phase, approximate initial conditions are found and in a second phase more accurate initial conditions obtained are used in order to check the periodic orbit by numerical integration of the three-body problem. Some peculiar solutions are found, for example, orbit with nearly zero angular momentum. A study of stability of periodic solutions is proposed with an approximation of the monodromy matrix Φ (T,o), not requiring numerical integration of the 6x6 variational linear system. Finally, some numerical problems of period determination are outlined.

INTRODUCTION

This paper refers to the construction of solutions of period T of a non-linear integrable system disturbed by T-periodic or autonomous perturbation, leading to an implicit system for the initial state vector. The non-integrable and non-linear perturbed system may be related to a quasi-linear one by Taylor expansion near a reference orbit. Then we can obtain rigorous periodicity conditions (Roseau 1966). Suppose now that the generating solution verifies the unperturbed integrable system, thus it is a polyparametric family where time occurs explicitely. But if this orbit is T-periodic, then periodicity conditions are singular. It is the critical case as it has been studied by I. Stellmacher (1976, 1977, 1979, 1981). Here a non-T-periodic solution is chosen to avoid singularities. It is the non critical case. The periodicity conditions lead to an implicit system for the initial state vector. For such a system solution series, relative to a small parameter of perturbation, exist if certain conditions are satisfied as demonstrated by Poincaré (1892). The method is semi-numerical; in a first step approximate initial conditions are found, in a second step more accurate conditions

V. V. Markellos and Y. Kozai (eds.), Dynamical Trapping and Evolution in the Solar System, 257–261.

are obtained by an iterative differential corrector. Finally to each particular non-T-periodic solution of the unperturbed system corresponds a T-periodic one of the perturbed motion. Therefore, a "polyparametric" family of T-periodic perturbed orbits is obtained.

This method is applicable to various perturbed problems with simplification for Hamiltonian systems. Here, numerical investigations are made to the elliptical spatial restricted three-body problem for the Sun-Jupiter system in a heliocentric (or planetocentric for lunar case) inertial frame. The generating orbit is an hyperbolic of parabolic comet.Later we will study capture orbits for these comets by this method. Several "curious" solutions are obtained: orbits almost rectilinear, near collisions, orthogonal to the orbit of the perturbing body, orbit of period T/k where k is an integer. A study of linear stability is proposed; the monodromy matrix is analytically approximated to avoid numerical integration of variational equations. Finally, some problems of period determination are outlined.

We need analytical solutions for the variational system:

1) of the unperturbed motion for a non T-periodic generating unperturbed orbit, to construct the periodic perturbed solution

2) of the perturbed system with a T-periodic generating perturbed orbit for the linear stability analysis.

We have analytical solutions for the unperturbed integrable system while for the perturbed non integrable system only an approximation is available.

2. EQUATIONS

The perturbed system has the form:

$$\dot{x} = X\ (x,t,\varepsilon), \qquad (1)$$

and the unperturbed integrable system is

$$\dot{z} = X\ (z,t,o), \qquad (2)$$

where $X(x,t,\varepsilon)$ is a T-periodic function of time t satisfying certain regularity conditions with respect to the state vector x of R^n and to the small parameter ε describing the perturbation (Roseau 1966). $x\ (x_o,t,\varepsilon)$ represents the solution of (1) with initial state vector $x(o) = x_o$; thus $z(t) = x(z_o,t,o)$ is the solution of the integrable unperturbed system (2) with starting vector $z(o)=z_o$; we put $x = z+\Theta$, Θ is the solution of the system written in the form:

$$\dot{\Theta} = \overline{X}'_x(t)\Theta + \overline{X}'_\varepsilon(t)\varepsilon + F(\Theta, t, \varepsilon) \tag{3}$$

where the surlining symbol means that the derivatives are evaluated for $(\Theta,\varepsilon) = (o,o)$ and where $F(\Theta,t,\varepsilon)$ are the terms of order ≥ 2 in the Taylor expansion of $X(z+\Theta,t,\varepsilon)$ with respect to (Θ,ε) near (o,o); we want to study now the existence and the characterization of T-periodic solutions of (1).

3. PERIODICITY CONDITIONS

We have to solve an implicit system for the starting vector x_o of the T-periodic solution of (1):

$$x_o = x(x_o, T, \varepsilon) \tag{4}$$

that leads to the implicit system, for $\Theta(o) = \Theta_o$,

$$\Delta\Psi^* \quad \Theta_o = \int_o^T \Psi^*(s) \left\{ \varepsilon \overline{X}'_\varepsilon(s) + F\left[\Theta(\Theta_o, s,\varepsilon), s,\varepsilon\right] \right\} ds \tag{5}$$

where $\Psi(t)$ is a nxn fundamental matrix solution of the joint variational unperturbed system relative to a non T-periodic generating solution $z(t)$ of the unperturbed system (2). The symbol * means matricial transposition and ΔU is $U(T)-U(o)$; if we can solve the implicit system (5) for Θ_o, then we obtain the initial state vector x_o of the T-periodic solution of (1) by : $x_o = z_o + \Theta_o$, z_o being a known vector; this system (5) will be solved semi-numerically.

4. APPROXIMATE INITIAL CONDITIONS

We put $\Theta_o = \Theta_o + \beta$ (Roseau 1966), where Θ_o satisfies the explixit system

$$\Delta\Psi^* \quad \Theta_o = \int_o^T \Psi^*(s) \ \varepsilon \overline{X}'_\varepsilon(s)\, ds + \Psi^*(T) \ \Delta z. \tag{6}$$

As $z(t)$ is non T-periodic, $\Delta\Psi^*$ is non singular, thus Θ_o is easily obtained from (6); the associated initial state vector x_o is $z_o + \Theta_o$; if $z(t)$ is T-periodic (critical case) then $\Delta\Psi^*$ is singular (Szebehely 1967).

5. MORE ACCURATE INITIAL CONDITIONS

We now have to solve an implicit system for β:

$$\Delta\Psi^* \quad \beta = \int_o^T \Psi^*(s) \quad F\left[\Theta\left(\Theta_o+\beta, s, \varepsilon\right), s,\varepsilon\right] ds. \tag{7}$$

This system has solutions provided that F satisfies certain conditions (Poincaré 1892). β is obtained by an iterative differential corrector. For a hamiltonian system Ψ^* (t) is replaced by Φ^* (t) E, where Φ (t) is a fundamental matrix solution of the unperturbed variational system relative to the generating orbit z(t), and E is the nxn matrix such that $E^2 = -I$, I being the nxn unit matrix.

6. LINEAR STABILITY ANALYSIS

The linear stability of the periodic solution x(t) is related to the eigen values of the monodromy matrix, (Hennawi 1980, Wiesel 1980). This matrix is approximated by a time averaging; the approximation is an analytical expression of the components of x(o) and x(T) (for the 3-body restricted problem); therefore we dont need numerical integration of the variational system of the perturbed motion.

7. APPLICATION TO THE RESTRICTED THREE-BODY PROBLEM

For this particular case: n=6, (2) is the two-body problem, ε is the mass of the disturbing body for the planetary case, and the mean motions ratio for the lunar case. The two primaries are the Sun and Jupiter, therefore T is Jupiter's period.

Finally, to each hyperbolic or parabolic cometary orbit z(t) is associated a T-periodic solution x(t) of the three-body problem. It should be a capture orbit for only quasiperiodic solution x(t) if:
1) $x(-\infty) = z(-\infty)$, 2) x(T)-x(o) is small, 3) x(t) is linearly stable. Several kinds of periodic solutions are found: orbits of period T/k where k is an integer, with various inclinations and eccenctricities. For these orbits, the osculating elements change slowly with time near the mean elements. Other orbits are almost rectilinear and orthogonal to Jupiter's orbit. Their osculating elements drastically vary; but for these orbits the periodicity of the first approximation is good.

Period determination

Solutions of period slightly different from T can be obtained by the procedure below: by (6) we have an approximated state vector x_o and the corresponding osculating elements give a period T_o. Now, we apply (6) with this period T_o+T and so on; we have a series of periods: T_o, T_1, ..., T_n; this series is convergent for some solution and we get by (7) an initial state vector x_o for a perturbed solution of the three-body restricted problem of period T_n + T.

Numerical investigations

Many numerical investigations have been done for various generating non T-periodic orbits **z**(t). The numerical integration procedure of the

equations of motion have been performed by P. Rocher (Rocher, 1981). The differential corrector used to solve (8) is similar to that of Markellos (Markellos, 1980). For parabolic orbit we utilize Subbotin's formulae (Subbotin, 1968) to obtain an analytical expression for a fundamental solution of the variational equations of the two-body problem.
The numerical results and further details will be published elsewhere.

8. REFERENCES

Edelman, C. 1982, *Astron. Astrophys.* 111, 220

Hennawi, A. 1980, *Celes. Mech.* 22, 237

Markellos, V.V. 1980, *Celes. Mech.* 21, 291

Poincaré, H. 1892, 1893, 1899, *Les Méthodes Nouvelles de la Mécanique Céleste, Dover Pub. Inc.* (1957)

Rocher, P. 1981, *Ephémérides des satellites faibles de Jupiter et de Saturne pour 1981 - 1982 - 1983 . Supplément de la Connaissance des Temps pour 1982,* Bureau des Longitudes

Roseau, M. 1966, *Vibrations non linéaires et théorie de la stabilité,* Springer, Berlin

Stellmacher, I. 1976, *Astron. Astrophys.* 51, 117

Stellmacher, I. 1977, *Astron. Astrophys.* 59, 337

Stellmacher, I. 1979, *Astron. Astrophys.* 80, 301

Stellmacher, I. 1981, *Celes. Mech.* 23, 145

Subbotin, M.F. 1968, *Vvedenic v teoreticeskuju astronomiju,* Moskva

Szebehely, V. 1967, *Theory of Orbits,* Academic Press

Wiesel, W. 1980, *Celes. Mech.* 21, 265

SYMMETRIC PERIODIC ORBITS IN THE ANISOTROPIC KEPLER PROBLEM

Josefina Casasayas* and Jaume Llibre**
* Facultat de Matemàtiques, Universitat de Barcelona, Barcelona 7, Spain.
**Secció de Matemàtiques, Facultat de Ciències, Universitat Autònoma de Barcelona, Bellaterra, Barcelona, Spain.

ABSTRACT. The anisotropic Kepler problem has a group of symmetries with three generators; they are symmetries respect to zero velocity curve and the two axes of motion's plane. For a fixed negative energy level it has four homothetic orbits. We describe the symmetric periodic orbits near these homothetic orbits. Full details and proofs will appear elsewhere (Casasayas-Llibre).

1. INTRODUCTION AND EQUATIONS OF MOTION.

The anisotropic Kepler problem was introduced by Gutzwiller (1973) to model certain quantum mechanical systems. But for us it has a mathematical interest because it is an easy model in order to study usual tools in the analysis of the n-body problem as non-integrability, collision manifold,... (Devaney, 1981).

This problem deals with the motion of a body which is attracted by a gravitational potential and has an anisotropic mass. It is described by the Hamiltonian system

$$\begin{aligned} \dot{q} &= M^{-1}p, \\ \dot{p} &= -\nabla V(q), \end{aligned} \tag{1}$$

where

$$q = (q_1,q_2) \in R^2-\{(0,0)\} \text{ and } p = (p_1,p_2) \in R^2$$

are the position and momentum coordinates of the body,

$$M^{-1} = \begin{pmatrix} \mu & 0 \\ 0 & 1 \end{pmatrix},$$

is the masses matrix and μ, $1 \leq \mu \leq +\infty$, is the mass parameter and

V. V. Markellos and Y. Kozai (eds.), Dynamical Trapping and Evolution in the Solar System, 263–270.

$V(q) = -|q|^{-1}$ is the potential energy. The total energy function is given by the Hamiltonian

$$H(q,p) = \tfrac{1}{2}\, p^t M^{-1} p + V(q).$$

System (1) is actually a one parameter family of Hamiltonian systems with two degrees of freedom depending analytically on the parameter μ. When $\mu=1$, we have the Kepler problem and $\mu > 1$ introduces the anisotropic matrix M which means that q_2 is the "heavy axis".

Equations (1) have a singularity of collision when $q=0$ which can be studied using the "blow up" technique of McGehee (Devaney, 1981).

Thereafter we will assume that the energy level $H=h$ is fixed and takes a prescribed negative value. Otherwise, if $h \geq 0$ there are no periodic orbits.

2. SYMMETRIES OF THE PROBLEM

The anisotropic Kepler problem has the following symmetries:

$$S_1(q_1,q_2,p_1,p_2,t) = (q_1,-q_2,-p_1,p_2,-t),$$

$$S_2(q_1,q_2,p_1,p_2,t) = (-q_1,q_2,p_1,-p_2,-t),$$

$$S_3(q_1,q_2,p_1,p_2,t) = (q_1,q_2,-p_1,-p_2,-t),$$

which can be interpreted in the following way.

Let $\gamma(t) = (q_1(t),q_2(t),p_1(t),p_2(t))$ be a solution of (1), then

$$S_1(\gamma(t)) = (q_1(-t),-q_2(-t),-p_1(-t),p_2(-t))$$

is another solution, see Figure 1. In a similar way $S_2(\gamma(t))$ and $S_3(\gamma(t))$ are solutions of (1), see Figures 2 and 3.

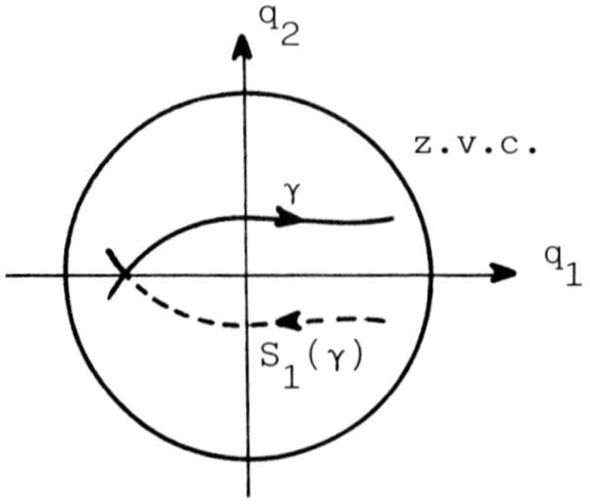

Figure 1.

Orbits which cross orthogonally the q_1-axis (resp. q_2-axis) are the symmetric orbits respect to S_1 (resp. S_2), that is $S_1(\gamma) = \gamma$ (resp. $S_2(\gamma)=\gamma$). Orbits which have some point on the zero velocity curve (z.v.c.) are the symmetric orbits respect to S_3.

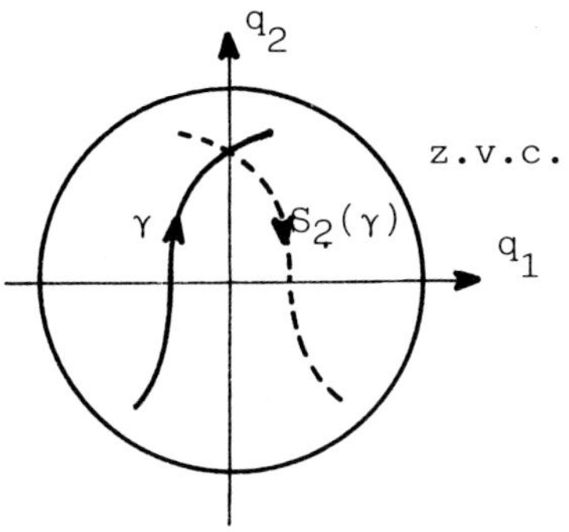

Figure 2.

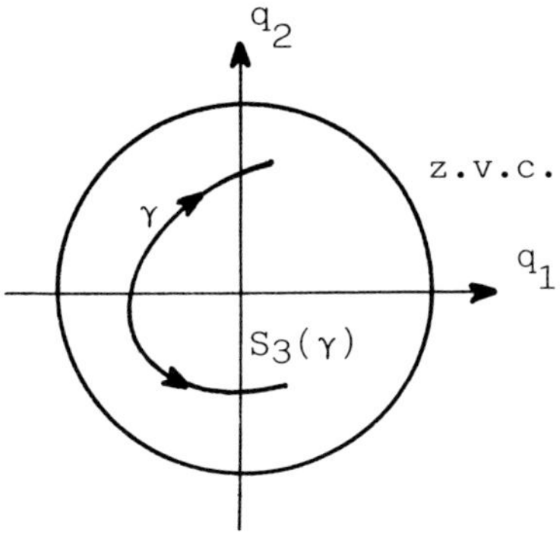

Figure 3.

If an orbit crosses two times orthogonally the q_1-axis (resp. q_2-axis) then it is a symmetric periodic orbit for S_1 (resp. S_2). Similarly, if an orbit has two points on the zero velocity curve then it is a symmetric periodic orbit for S_3, for more details see Devaney (1976). This fact is essential for our study of the symmetric periodic orbits.

3. SYMMETRIC PERIODIC ORBITS FOR THE KEPLER PROBLEM ($\mu=1$).

For the Kepler problem is not difficult to prove the following.

(1) There is a bijection between the symmetric periodic orbits respect to S_i, for each i=1,2, and two copies of the segment $(0,-1/h)$. One copy corresponds to the direct ellipses and the other one to the retrograde ellipses, see Figures 4 and 5.

(2) There is a bijection between the symmetric periodic orbits respect to S_1 and S_2 and the two points $\pm(2h)^{-1}$. They correspond to the circular orbits, see Figure 6.

(3) There is a bijection between the symmetric orbits (but not periodic) respect to S_3 and the circle. They are orbits of elliptic collision. If we regularize the equations then the orbits become periodic, see figure 7.

So when $\mu=1$ we have a complete description for the symmetric periodic orbits (note that the system is integrable).

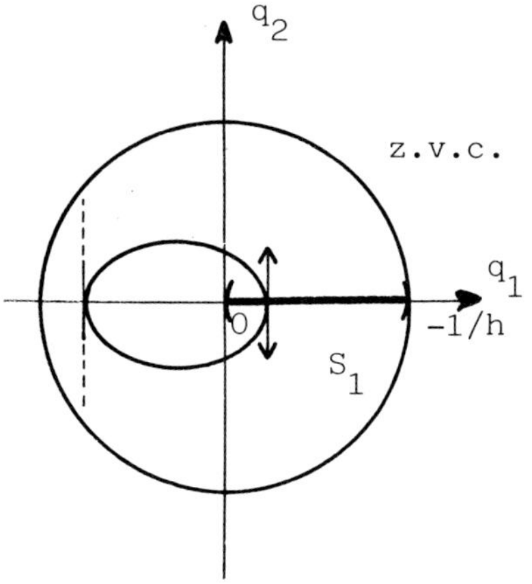

Figure 4.

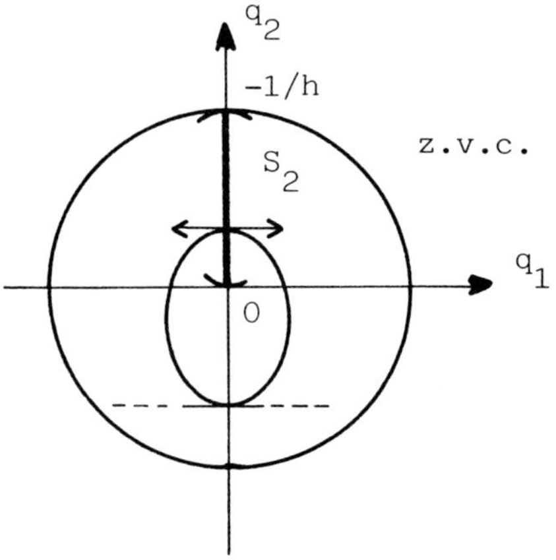

Figure 5.

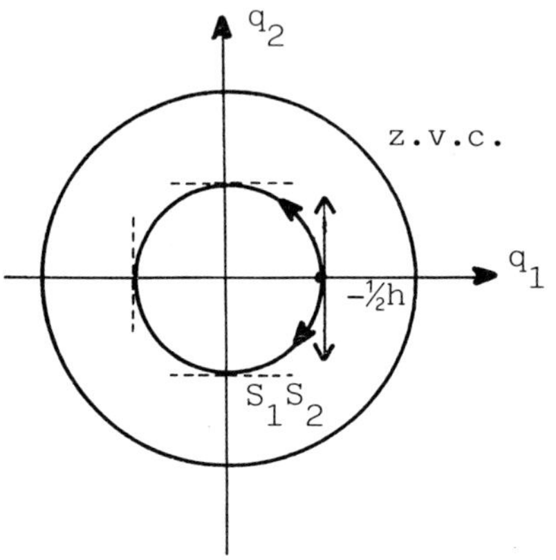

Figure 6.

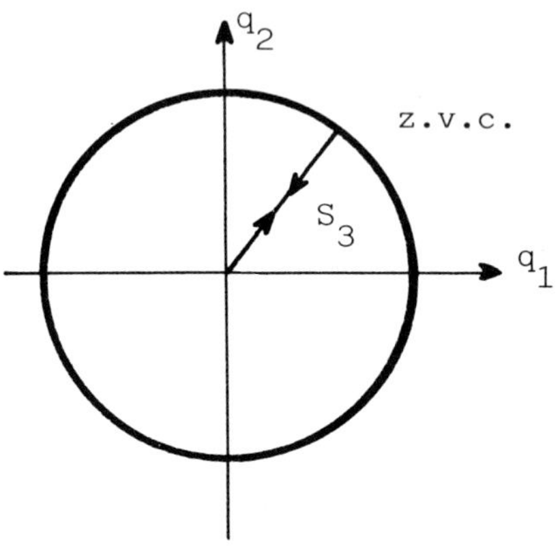

Figure 7.

4. SYMMETRIC PERIODIC ORBITS FOR $\mu > 9/8$.

A solution $(q(t),p(t))$ of the anisotropic Kepler problem is called homothetic if we can obtain $(q(t_1),p(t_1))$ from $(q(t_2),p(t_2))$ through a dilation, for every t_1, t_2 where the solution is defined. An orbit $(q(t),p(t))$ of system (1) is called a collision (resp. ejection) orbit if there exists t_o such that $q(t) \to 0$ as $t \uparrow t_o$ (resp. $t \downarrow t_o$).

It is known that the anisotropic Kepler problem has only four homothetic orbits π_i for $i = 1,2,3,4$, which are also of ejection-collision type, see Figure 8 (Devaney, 1981). We have studied the neighborhood of these orbits in order to prove the following theorem (see Casasayas-Llibre).

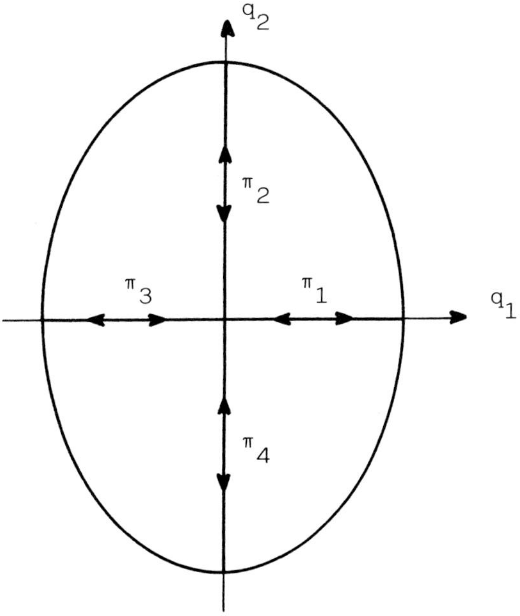

Figure 8.

THEOREM. There exists a positive integer number n_o such that for every $n,m \geq n_o$ the following statements hold.

(i) There are four symmetric periodic ejection-collision orbits respect to S_3 (resp. S_2) such that the number of crossings with the q_2-axis is $2n$ (resp. $2n+1$), see Figures 9 and 10 (resp. Figures 11 and 12). There are similar figures for the region $q_2 \leq 0$.

(ii) There are two symmetric periodic orbits respect to S_2 (resp. S_2 and S_3) such that the qualitative behaviour is given in Figure 13 (resp. Figure 14). There are similar figures for the region $q_2 \geq 0$ obtained by the change $\theta = \theta + \pi$.

(iii) There is one symmetric periodic orbit respect to S_2 (resp. S_3) such that the qualitative behaviour is shown in Figure 15 (resp. Figure 16). When m=n the orbit is also symmetric respect to S_1.

(iv) There are four (resp. two) symmetric periodic orbits respect to S_3 (resp. S_2 and S_3) such that the qualitative behaviour is given in Figures 17 and 18 (resp. Figure 19). There are similar figures for the region $q_2 \leq 0$ obtained by the change $\theta = \theta + \pi$.

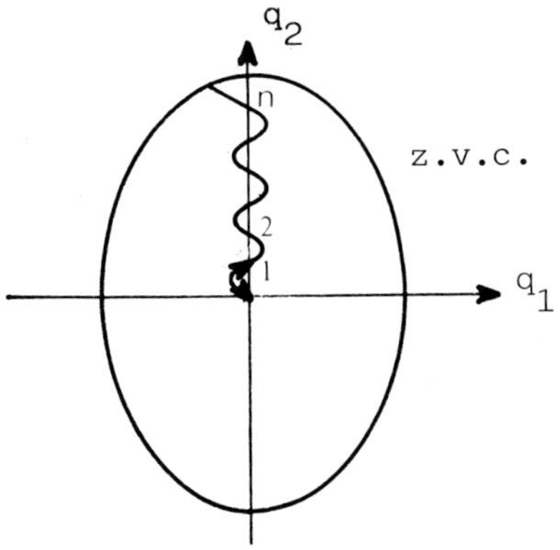

Figure 9.

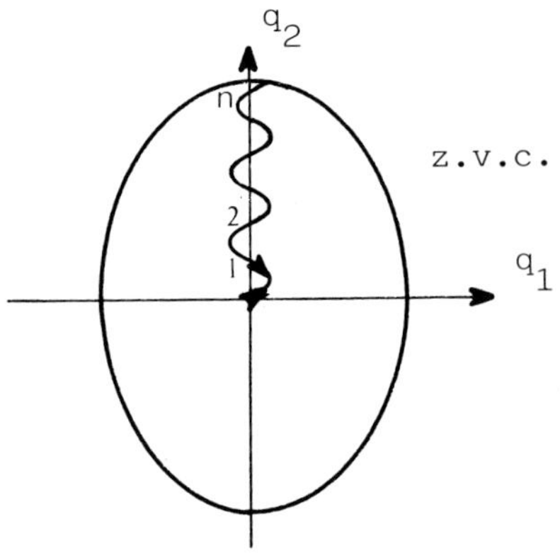

Figure 10.

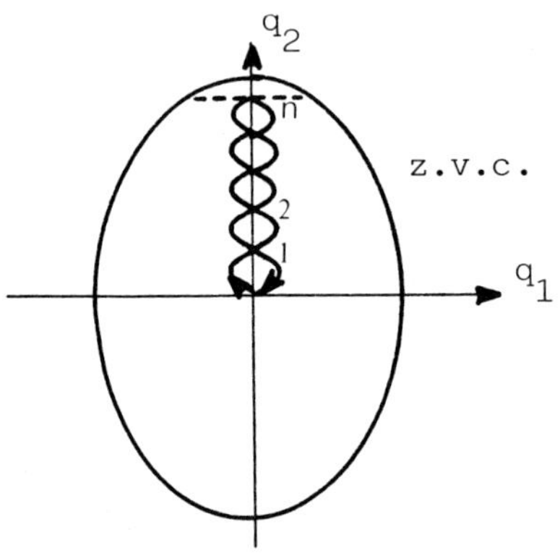

Figure 11.

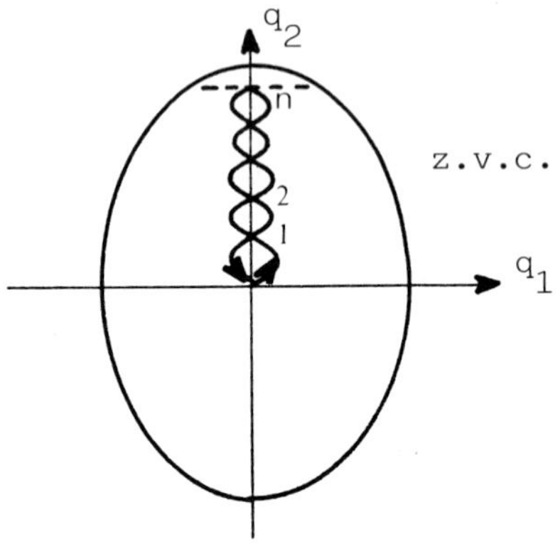

Figure 12.

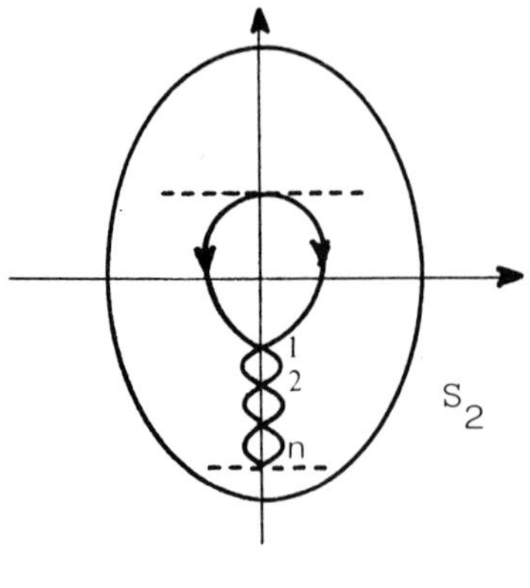

Figure 13.

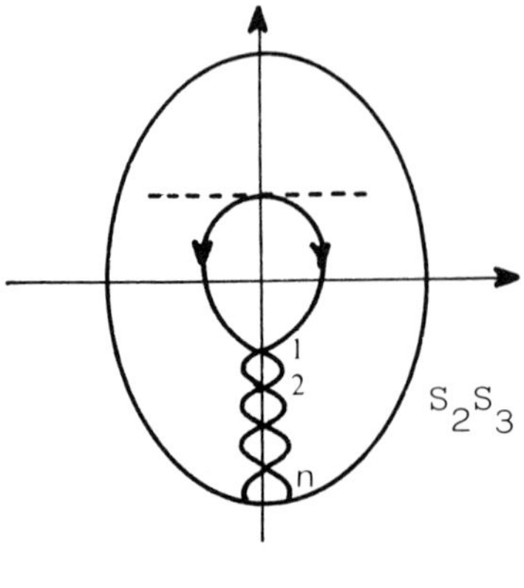

Figure 14.

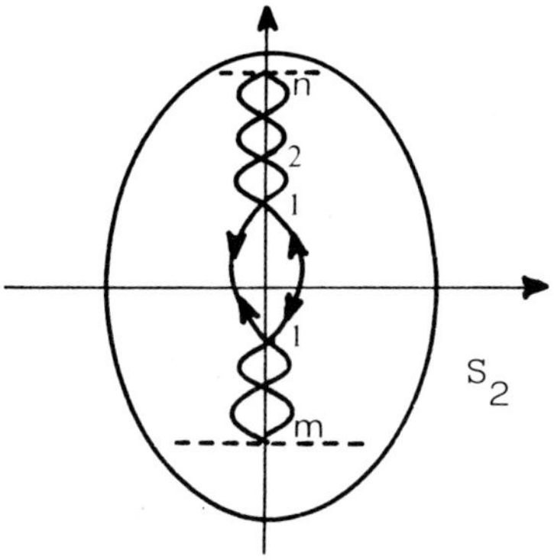

Figure 15.

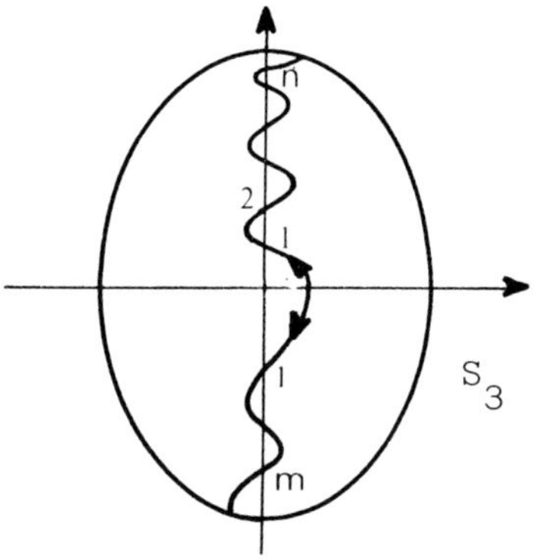

Figure 16.

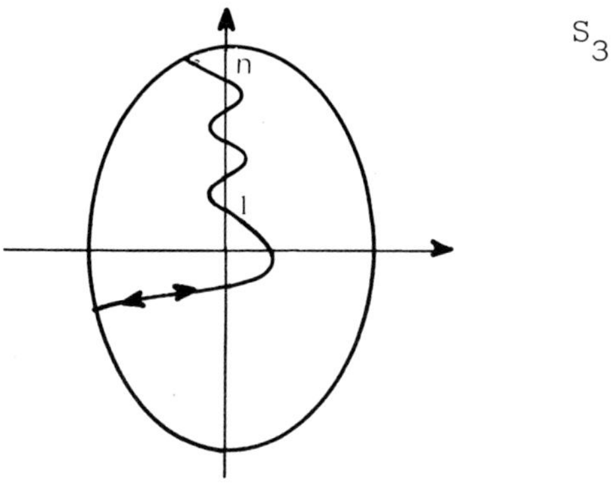

Figure 17.

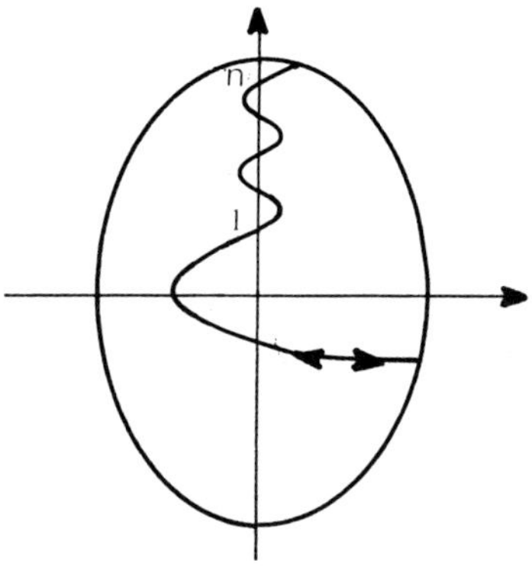

Figure 18.

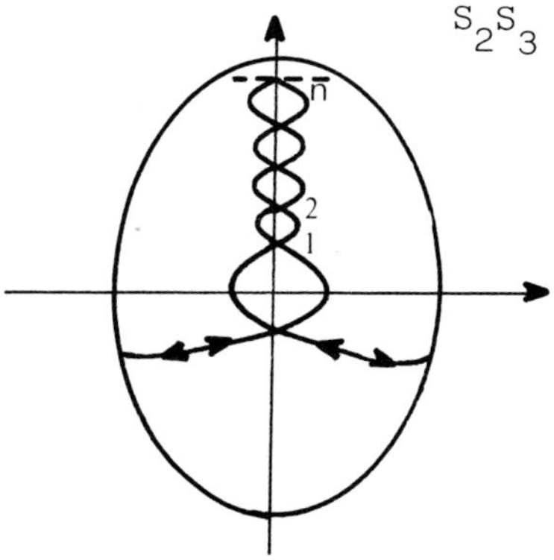

Figure 19.

REFERENCES

Casasayas, J. and Llibre, J., The global flow of the anisotropic Kepler problem (to appear).

Devaney, R. (1976), Reversible diffeomorphisms and flows, Trans. Amer. Math. Soc. 218, pp. 89-113.

Devaney, R. (1981), Singularities in Classical Mechanical Systems, Progress in Mathematics vol. 10, Birkhäuser, Basel, pp. 211-333.

Gutzwiller, M. (1973), The anisotropic Kepler problem in two dimensions, J. Math. Phys. 14, pp. 139-152.

CHARACTERISTICS OF PERIODIC ORBITS IN ELLIPTICAL GALAXIES

N. CARANICOLAS
Astronomy Department, University of Thessaloniki
Thessaloniki, Greece

Abstract. The properties of the characteristic curves of several families of periodic orbits, in a conservative dynamical system of two degrees of freedom, symmetric with respect to both axes, are reviewed. The two main types of families are presented. One sees that the pattern of the characteristics in the exact resonance case is similar to that of the near resonance case except for the basic characteristic. The form of the characteristics can be found theoretically by means of the second integral.

1. INTRODUCTION

The motion of a star on the plane of symmetry of a non axisymmetric galaxy can be described by the Hamiltonian

$$H = \frac{1}{2}(\dot{x}^2 + \dot{y}^2 + \omega_1^2 x^2 + \omega_2^2 y^2) + \text{higher order terms} \qquad (1)$$

Several investigators have used the Hamiltonian (1) in order to study the families of periodic orbits and successive bifurcations, considering only one mixed third order term (see, e.g., Contopoulos, 1970a; Contopoulos and Michaelides, 1980; Contopoulos and Zikides, 1980; Contopoulos, 1981).

In the following we shall deal with the Hamiltonian

$$H = \frac{1}{2}(\dot{x}^2 + \dot{y}^2 + \omega_1^2 x^2 + \omega_2^2 y^2) - \varepsilon(a_1 x^4 + 2a_2 x^2 y^2 + a_3 y^4) \qquad (2)$$

where $\omega_1, \omega_2, a_1, a_2, a_3$ are positive constants and ε is the perturbation. The Hamiltonian (2) may be considered to describe the motion of a star on the plane of symmetry of a non rotating elliptical galaxy.

This choice is justified by the fact that the dynamics of elliptical galaxies have gained a considerable interest in the last few years. Now our understanding of the shapes and ways to form elliptical galaxies is clearly different from that of a few years ago. Elliptical galaxies were then thought to be rotationally flattened oblate systems with isotropic velocity dispersions (e.g., Freeman 1975). Observations and detailed dynamical studies have recently shown that most elliptical galaxies are not flattened by rotation but instead they appear to be triaxial systems with anisotropic velocity dispersions (Binney 1978, Illingworth 1981).

V. V. Markellos and Y. Kozai (eds.), Dynamical Trapping and Evolution in the Solar System, 271–274.

As a consequence of the above thoughts a study of periodic orbits, in a galaxy described by the Hamiltonian (2) seems to be of interest because these orbits are the main landmarks in exploring the totality of orbits.

2. DESCRIPTION OF THE FAMILIES

It is well known that, for any given value of ε one can find periodic orbits making n oscillations along the x-axis and m oscillations along the y-axis whenever the ratio of the unperturbed frequencies ω_1/ω_2 is near a rational number n/m. We study periodic orbits intersecting the x-axis perpendicularly at a point x for a constant value of the energy h. For a given value of n and m the value of x varies by varying the perturbation ε. Then the curve $x = x(\varepsilon)$ is called the characteristic of the corresponding family of periodic orbits.

In general, two types of families are present: a) Regular families which are connected with the main families directly or through bifurcations. b) Irregular families not connected with the above families. These families appear at relatively large perturbations and seem to be connected with the dissolution of the invariant curves of non periodic orbits (Contopoulos 1970a).

Figure 1 shows the characteristic curves of several families of periodic orbits when $\omega_1^2 = 0.4$, $\omega_2^2 = 0.1$ $(\omega_1/\omega_2 = 2/1)$ $a_1 = 0.1$, $a_2 = 0.5$, $a_3 = 0.02$. Solid line indicate stable orbits while dashed lines indicate unstable orbits. We can see three groups of regular families:

i) In the first group belong the basic characteristic and its bifurcations. The basic characteristic starts from the $\bar{x}$-axis ($\bar{x} = \omega_1 x$) in the exact resonance cases while in the near resonance cases starts always from the boundary line given by the equation

$$2\varepsilon a_1' \bar{x}^{-4} - \bar{x}^{-2} + 2h = 0, \tag{3}$$

where $a_1' = a_1/\omega_1^4$.

ii) In the second group belong the characteristics which bifurcate from the boundary line.

iii) Finally, in the third group belong the characteristic which bifurcate from the $\bar{x} = 0$ axis.

We can find approximately the form of the characteristics for small values of ε using the formula

$$r = -\frac{\omega_1}{\omega_2}\{1 - \varepsilon[(\frac{3a_1}{\omega_1^4} - \frac{4a_2}{\omega_1^2\omega_2^2} + \frac{3a_3}{\omega_2^4})\Phi_{10} + (\frac{2a_2}{\omega_1^2\omega_2^2} - \frac{3a_3}{\omega_2^4})h]\} \tag{4}$$

where r is the rotation number, h is the total energy and Φ_{10} is the second integral (Caranicolas 1982). This formula is not valid when ω_1/ω_2 is near 1. In this case we can find the form of the basic characteristic using a special form of the second integral (Caranicolas and Barbanis 1982).

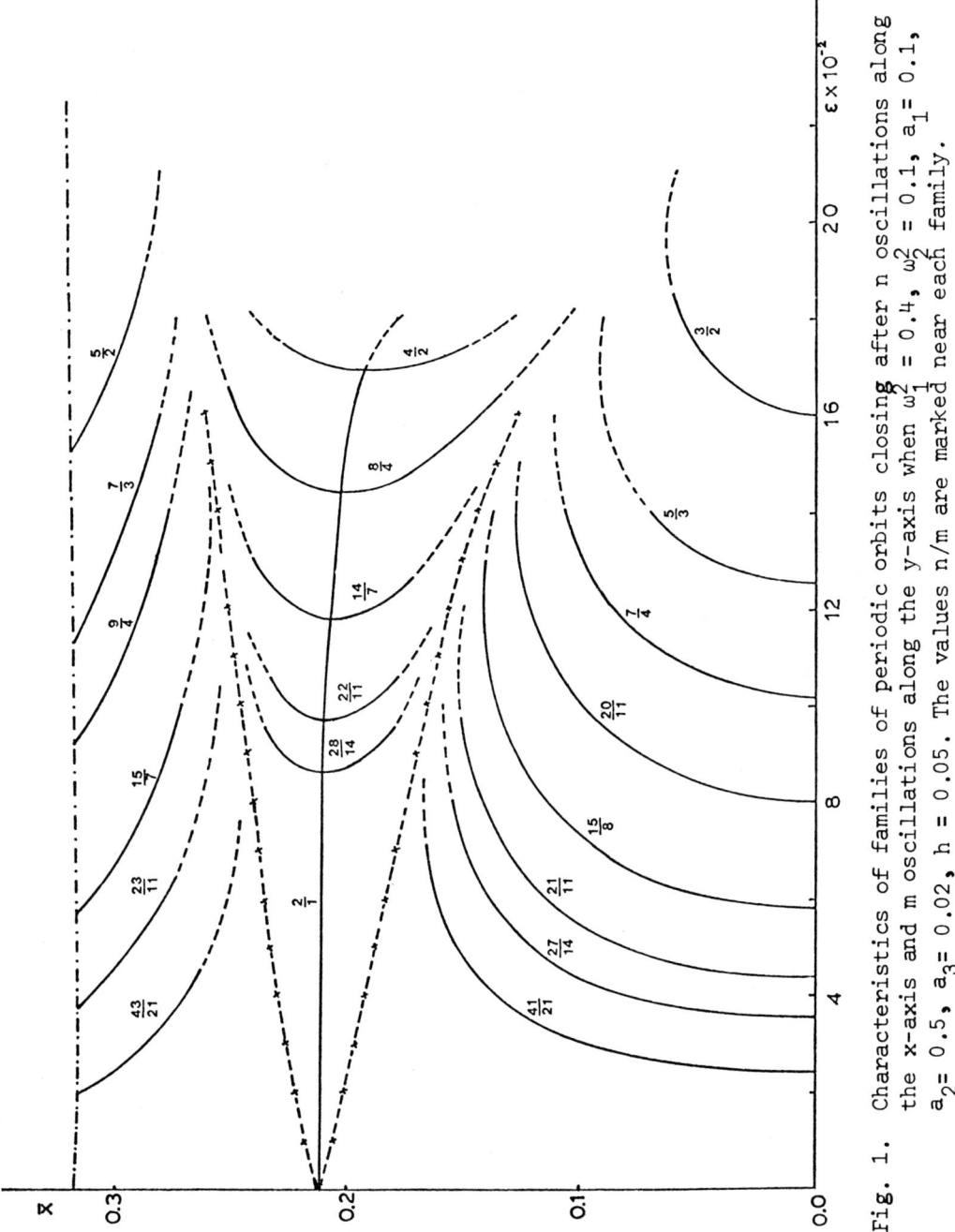

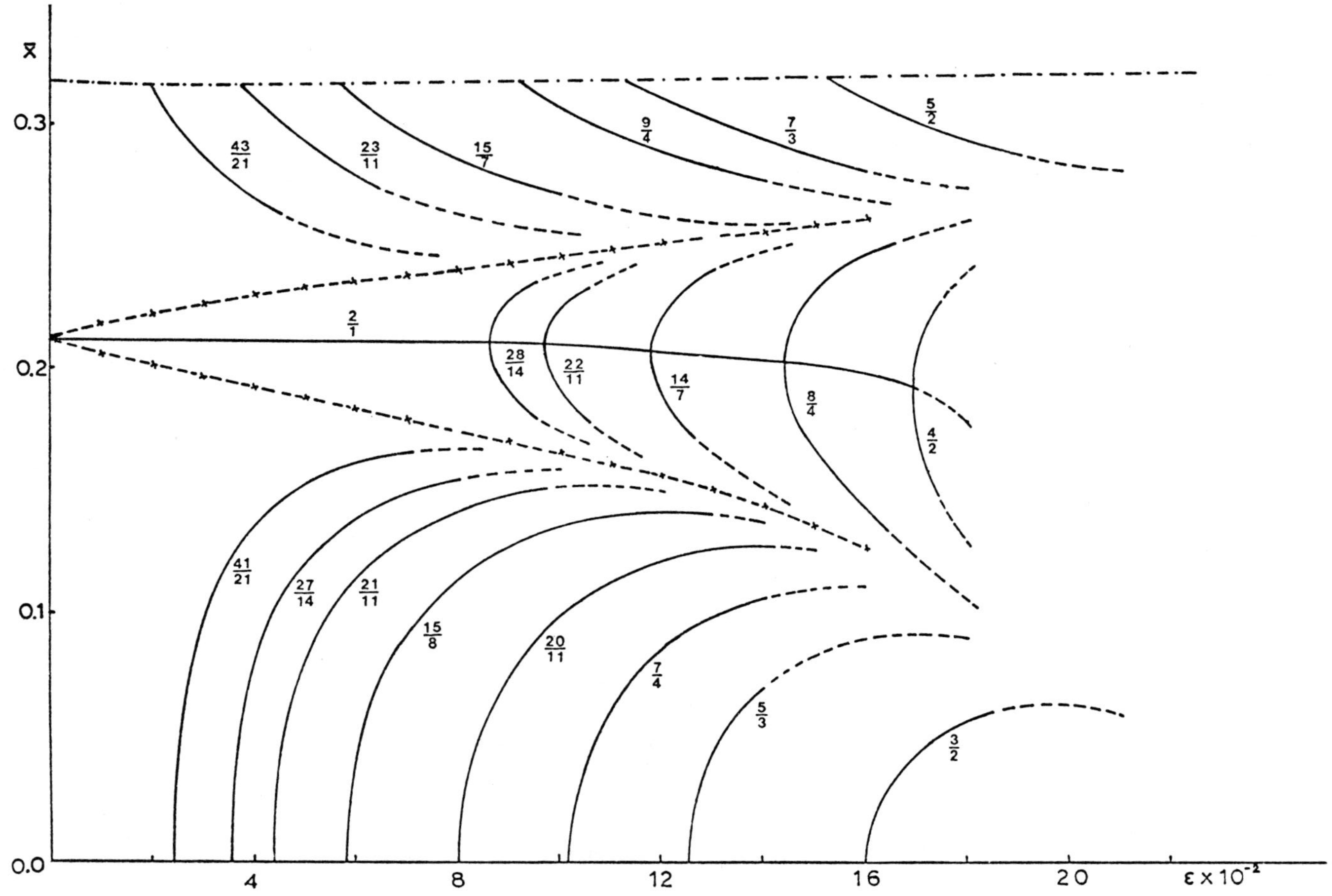

Fig. 1. Characteristics of families of periodic orbits closing after n oscillations along the x-axis and m oscillations along the y-axis when $\omega_1^2 = 0.4$, $\omega_2^2 = 0.1$, $a_1 = 0.1$, $a_2 = 0.5$, $a_3 = 0.02$, $h = 0.05$. The values n/m are marked near each family.

REFERENCES

1. Binney, J.: 1978, Comment Ap., 8, pp. 2-
2. Caranicolas, N.: 1982, Astrophys. Space Sci. 87, 237.
3. Caranicolas, N., and Barbanis, B.: 1982, Astron. Astrophys. (in press).
4. Contopoulos, G.: 1970a, Astron. J. 75, pp. 96-107.
5. Contopoulos, G., and Zikides, M.: 1980, Astron. Astrophys. 90, pp. 198-205.
6. Contopoulos, G., and Michaelidis, P.: 1980, Celest. Mech. 22, pp. 403-413.
7. Contopoulos, G.: 1981, Lett. Nuovo Cimento 28, pp. 498-502.
8. Freeman K.C.: 1975, in Galaxies and the Universe, Stars and Stellar Systems, Vol. IX, ed. A. Sandage and J. Kristian (Chicago,University of Chicago Press), pp. 406-507.
9. Illingworth, G.: 1981, in the Structure and Evolution of Normal Galaxies, ed. S.M. Fall and D. Lynden-Bell (London: Cambridge University), pp. 27-41.

PART V

TRAPPED MOTION
IN THE THREE-BODY PROBLEM

ASYMPTOTIC APPROACH TO MIRROR CONDITIONS AS A TRAPPING MECHANISM IN N-BODY HIERARCHICAL DYNAMICAL SYSTEMS

Archie E. Roy

Department of Astronomy,
Glasgow University,
Glasgow, U.K.

ABSTRACT

The consequences for hierarchical stability of almost circular almost coplanar, low perturbation orbits in an n-body hierarchical dynamical system is discussed. It is shown that frequent close approaches to mirror conditions with subsequent reversing of perturbations is ensured by such properties. The part played by near commensurabilities in mean motion is also discussed, the Sun-Jupiter-Saturn case being taken as an example.

1. INTRODUCTION

The Solar System, comprised of the planetary system and the satellite systems, exhibits a hierarchical structure in that the orbits can be ordered in size in each system. The main bodies in each system also exhibit hierarchical stability in that the ordering in size does not change in a time long compared with the longest period of revolution in that system. In addition we see that for the most part the orbits are almost circular and almost coplanar.

A question worth examining is: how did the solar system come to be trapped into such a situation and what maintains it in the trap? Indeed a further related question is: what is the nature of the trap?

With respect to n-body hierarchical systems with $n > 3$, there would appear to be no analytical criterion of stability analogous to the criterion based on the product of the square of the angular momentum and the energy of a hierarchical three-body system, drawn attention to in recent years by a number of authors (Easton 1971; Marchal 1971; Marchal and Saari 1975; Smale 1970; Saari 1974; Zare 1976, 1977).

V. V. Markellos and Y. Kozai (eds.), Dynamical Trapping and Evolution in the Solar System, 277–287.

Nevertheless, although in what follows we are interested in systems with $n > 3$, we first of all consider the general three-body hierarchical problem before turning our attention to cases where $n > 3$. As a particular example of this problem we will consider the Sun-Jupiter-Saturn case though it will become evident that many examples within the solar system could have been taken.

2. THE MIRROR THEOREM

Consider a general three-body system of the hierarchical type with bodies P_1, P_2, and P_3 of masses m_1, m_2 and m_3 such that P_1 and P_2 form a binary and P_3 is in orbit about the centre of mass C_{12} of P_1 and P_2. Let $m_2 < m_1$.

Let the osculating semimajor axes and eccentricities of the binary and the third body orbits be a_2, a_3, e_2 and e_3, suffix two referring to the binary orbit. Let $a_3 > a_2$ and let the system for simplicity be coplanar.

Then the mutual attractions of the bodies will perturb the values of a_2, a_3, e_2 and e_3.

$$\text{If} \quad a_2\,(1+e_2)\left(\frac{m_1}{m_1+m_2}\right) < a_3(1-e_3)\,, \tag{1}$$

the pericentre distance of P_3 from C_{12} will be greater than the apocentre distance of P_2 from C_{12} and the orbits will not cross.

Let us define two synodic periods S_{23} and Π_{23}. The first is the synodic period of the bodies P_2 and P_3, being the time between successive similar configurations of P_1, P_2 and P_3, for example between conjunctions $P_1\ P_2\ P_3$. If n_2 and n_3 are the mean motions of P_2 and P_3 about C_{12} while T_2 and T_3 are their periods of revolution, then

$$n_2 = \frac{2\Pi}{T_2}\ ; \quad n_3 = \frac{2\Pi}{T_3} \tag{2}$$

and

$$\frac{1}{S_{23}} = \frac{1}{T_2} - \frac{1}{T_3}\,. \tag{3}$$

The second synodic period Π_{23} is the synodic period of the apses of the orbits. It is the time between successive similar configurations of C_{12} with Π_2 and Π_3, the pericentres of the inner and outer orbits respectively. For example such a configuration may be a conjunction $C_{12}\ \Pi_2\ \Pi_3$. If $\dot{\omega}_2$ and $\dot{\omega}_3$ are the mean secular motions of Π_2 and Π_3 about C_{12}, while τ_2 and τ_3 are their periods of revolution, then

$$\dot{\omega}_2 = \frac{2\Pi}{\tau_2}\ ; \quad \dot{\omega}_3 = \frac{2\Pi}{\tau_3} \tag{4}$$

and

$$\frac{1}{\Pi_{23}} = \frac{1}{\tau_2} - \frac{1}{\tau_3} \qquad (5)$$

Consider for example the system Sun-Jupiter-Saturn, neglecting the inclinations of the orbits. Then $m_1 = 1$; $m_2 = 0.0009551$; $m_3 = 0.0002680$ in solar mass units.

Also, at the present time,

$a_2 = 5.203$ AU	$a_3 = 9.539$ AU
$e_2 = 0.048$	$e_3 = 0.056$
$n_2 = 0.2649$ rad/yr	$n_3 = 0.1066$ rad/yr
$T_2 = 11.86$ yr	$T_3 = 29.46$ yr
$\dot{\omega}_2 = 2.053 \times 10^{-5}$ rad/yr	$\dot{\omega}_3 = 1.366 \times 10^{-4}$ rad/yr
$\tau_2 = 306{,}000$ yr	$\tau_3 = 46{,}000$ yr.

Then

$S_{23} = 19.852$ years.

and

$\Pi_{23} = 54{,}138$ years.

In all reasonably durable systems $S_{23} << \Pi_{23}$.

The Roy-Ovenden mirror theorem (1955) states that if, in an n-body dynamical system, the mutual radius vectors are all perpendicular to the mutual velocity vectors at any time, then the behaviour of the system after that time is a mirror image of the behaviour before that time.

In the coplanar three-body system under discussion, if the bodies' configuration satisfies any of the following eight cases, a mirror configuration occurs.

$P_1P_2P_3$ in line; P_2 and P_3 at pericentre
$P_1P_2P_3$ in line; P_2 at pericentre, P_3 at apocentre
$P_1P_2P_3$ in line; P_2 at apocentre, P_2 at pericentre
$P_1P_2P_3$ in line; P_2 and P_3 at apocentre
$P_2P_1P_3$ in line; P_2 and P_3 at pericentre
$P_2P_1P_3$ in line; P_2 at pericentre, P_3 at apocentre
$P_2P_1P_3$ in line; P_2 at apocentre, P_3 at pericentre
$P_2P_1P_3$ in line; P_2 and P_3 at apocentre.

3. APPROXIMATE MIRROR CONFIGURATIONS

We now consider the conditions under which approximations of particular degree of accuracy to mirror configurations occur.

Let the apse lines of the orbits coincide at time t_0. It is extremely unlikely that a conjunction of the bodies takes place at this moment and even more unlikely that if it does, the bodies will be on the common apse line.

The apse lines separate at a rate of $2\pi/\pi_{23}$. Let the first conjunction of the bodies occur at time $t_1 > t_0$. Then the angle θ between the apses at t_1 will be given by

$$\theta = \frac{2\pi}{\pi_{23}}(t_1 - t_0) = \theta_0, \quad \text{say.}$$

Now $t_1 - t_0 \lesssim S_{23}$ otherwise the previous conjunction must have been nearer the apse line. This point will be re-examined later more carefully.

Hence

$$\theta_0 \lesssim \frac{2\Pi\ S_{23}}{\Pi_{23}} \quad . \tag{6}$$

The angle advanced through by the conjunction line of the bodies in one synodic period S_{23} is Φ, given by

$$\Phi = n_3\ S_{23} \tag{7}$$

since the radius vector of the body P_2 has to advance 2π radians with respect to the radius vector of the body P_3.

Let the angle between the common apse line at t_0 and the conjunction line at t_1, be α_0 radians. Then at t_1, the angles between the conjunction line at t_1 and the positions of the apses at t_1 will be given by

$$\beta_2 = \alpha_0 - \dot{\omega}_2(t_1 - t_0),\ \beta_3 = \alpha_0 - \dot{\omega}_3(t_1 - t_0) = \beta_2 + \theta_0 \ . \tag{8}$$

The angles β_2 and β_3 are the true anomalies of P_2 and P_3 respectively when the conjunction takes place.

For a mirror configuration to take place, the angles between the velocity vectors and the mutual radius vectors should be $\pi/2$ radians. We now consider what is the size of the angle γ between the velocity vector in an elliptical orbit and the radius vector for a given pair of values of the true anomaly f and eccentricity e.

The angle γ may easily be shown to be given by the relations:

$$\sin\gamma = \frac{1+e\cos f}{(1+2e\cos f+e^2)^{\frac{1}{2}}}\ ; \quad \cos\gamma = \frac{-e\sin f}{(1+2e\cos f+e^2)^{\frac{1}{2}}} \tag{9}$$

It may be noted that for e equal to zero, $\gamma = \pi/2$ while for $f = 0$ or π, $\gamma = \pi/2$.

From (9) we have

$$\frac{\partial\gamma}{\partial f} = \frac{e(\cos f+e)}{(1+2e\cos f+e^2)^{\frac{1}{2}}} \tag{10}$$

so that for a given value of e, the maximum value of γ , namely γ_{max} is given by putting

$$\cos f = -e \tag{11}$$

in (9). If we do so, we find that

$$\sin\gamma_{max} = \sqrt{1-e^2}\ ; \quad \cos\gamma_{max} = -e. \tag{12}$$

As $e \to 1$, $\gamma_{max} \to \pi$.

Also, remembering that the radius vector r is given by

$$r = \frac{a(1-e^2)}{1+e\cos f},$$

we have, when $\cos f = -e$,

$$r = a.$$

In other words, maximum γ occurs when the orbiting body lies at the end of the semiminor axis.

In the case of Jupiter, γ_{max} is found to be $92^\circ.75$ while in the case of Saturn, γ_{max} has the value of $93^\circ.21$.

These values occur at true anomalies $92^\circ.75$ and $93^\circ.21$ respectively. The approach of γ towards 90° for the orbit of Jupiter and Saturn as the true anomaly f decreases from f given by $\cos f = -e$ is shown in Table 1 below:

v / f	Jupiter e = .048	Saturn e = .056
90°	92°.7	93°.2
80	92.7	93.1
50	92.0	92.4
20	90.9	91.0
10	90.5	90.5
5	90.2	90.3
1	90.05	90.05

Table 1

Returning to the three-body system of Sun-Jupiter-Saturn we see that by (6), the maximum separation θ_o of the apses (after they have coincided at t_o) before the first conjunction of the planets occurs at t_1 is only about 4 arc minutes at most.

The poorest approximation to a satisfaction of a mirror condition will therefore occur if the first conjunction of the bodies after apse coincidence occurs with α of order 90°. If however, a conjunction takes place thereafter, with true anomalies much nearer zero, before the apses have separated appreciably, then a much better approximation to a mirror configuration will occur. Even in the case of Jupiter and Saturn, it is seen that for $f < 20^\circ$, the value of γ is within one degree of 90°. The implications of this for stability are worth listing.

For systems with perturbations of the Keplerian orbits small enough to ensure that $\pi_{23} >> S_{23}$, each apse conjunction epoch will be preceded and followed by a time interval in which the apse separation angle θ is very small; this time will itself be large compared with the bodies' synodic period S_{23} so that a number of conjunctions scattered round the orbits can take place in this time interval. A good chance will exist that one of them will occur at small true anomalies so providing a good approximation to a mirror condition with almost complete reversal of the previous build-up of perturbations.

If the perturbations are large, however, this time interval will not only be much smaller because Π_{23} will be smaller, but the orbital eccentricities will be larger. In order to produce as good an approximation to a mirror condition as before, the system will have to find a conjunction in that smaller time interval at much smaller true anomalies than before.

For example, if the eccentricities of Jupiter and Saturn's orbits were as large as 0.2, say, then for γ to be within one degree of 90°, the true anomaly would have to be less than 6° instead of less than 20°. And if the masses of Jupiter and Saturn were increased by a factor of 20, say, then Π_{23} would be decreased by a corresponding factor.

4. THE EFFECT OF NEAR COMMENSURABILITY IN MEAN MOTION

Now let us consider the effect of the well-known commensurability in mean motion in theSun-Jupiter-Saturn case.
We have:

$$n_2 = 0.2649 \text{ rad./yr} \quad ; \quad n_3 = 0.1066 \text{ rad./yr},$$

so that

$$2n_2 - 5n_3 = - 0.0032 \text{ rad./yr}.$$

The conjunction system of lines may therefore be looked upon as a three-spoke wheel slowly rotating. The angle between two consecutive conjunction lines or spokes is given by (7); viz.

$$\Phi = n_3 S_{23} = 4.23244 \text{ rad.} = 242^\circ.5$$

so that in this case $3\Phi = 727^\circ.5$ or $7^\circ.5$.

The fourth conjunction line therefore lies $7^\circ.5$ ahead of the first and the wheel may be looked upon as rotating at an angular speed of $7^\circ.5 / 3 S_{23}$ or $2^\circ.5$ /synodic period of the bodies.

Let us suppose that the apse lines come together at t_o and that the first conjunction of the bodies occurs at time t_1 so that by (6)

$$\theta_o = \frac{2\pi}{\Pi_{23}} (t_1 - t_o) < \frac{2\pi}{\Pi_{23}} S_{23}$$

which as we have seen is less than 4 arc minutes.

The angle θ between the apse lines after a further number k of synodic periods has elapsed is given by

$$\theta = \frac{2\pi}{\Pi_{23}} (t_1 - t_o + kS_{23}) = \theta_o + \frac{2k\pi S_{23}}{\Pi_{23}} \tag{13}$$

or

$$\theta < \frac{2\pi(k+1)\, S_{23}}{\Pi_{23}} \, . \tag{14}$$

Suppose the angle α between the directions of the apse line conjunction at t_o and the bodies' conjunction at t_1 , is α_o . Suppose further that it is the worst possible case, in that the conjunction occurs where $\alpha_o \sim 90^\circ$, so making the true anomalies f at the time of conjunction also $\sim 90^{\circ}$. Then considering only this particular conjunction line spoke, we know that since the conjunction line wheel rotates at a rate of $2^{\circ}.5$ /synodic period, there must have been a conjunction within $7^{\circ}.5$ of the apse line conjunction direction at t_o at a time$(90^{\circ}/2^{\circ}.5)S_{23}$ before t_1 that is 715 years before t_1. In this time interval the apse lines would have separated by about $5^{\circ}.15$ the faster moving apse of Saturn's orbit having contributed most of this $5^{\circ}.15$. The true anomalies at this time must therefore be of order 5° at most. From Table 1 we see that there is a mirror condition to within $0^{\circ}.2$ accuracy.

Some considerations not yet taken into account should be mentioned here which improve the situation considerably. The wheel of conjunctions has 3 spokes. If we now include oppositions, which from the point of view of the mirror condition are just as good at reversing perturbations, it has six spokes. Consecutive spokes are therefore separated by 120° and therefore the worst possible case is not $\alpha_o \sim 90$ but $\alpha_o \sim 60^{\circ}$. Then the maximum time before or after t_o at which a conjunction <u>or opposition</u> occurred or will occur within $7^{\circ}.5$ of the apse line conjunction direction at t_o is ~ 470 years. In this time the apse lines would have separated by about $3^{\circ}.4$. We are now therefore considering true anomalies of order 7° or smaller and for the orbits of Jupiter and Saturn the departure from a perfect mirror condition is certainly within $0^{\circ}.2$ of 90°

In the Sun-Jupiter-Saturn system, therefore, whenever the apse lines coincide, a very close approximation to a mirror configuration will take or has taken place within a short time interval of the apse line coincidence event. This event occurs every $\Pi_{23}/2=25000$ years since it is only necessary that the apse lines coincide and not that there be a conjunction of perihelia.

A general three-body system of almost circular orbits, almost commensurable in mean motions, and with small perturbations is therefore in a stable, or trapped mode. Before perturbations can build up disastrously, an apse line coincidence event will take place ensuring that a close approximation to a mirror condition event occurs or has occurred. This latter event almost completely ensures that perturbations produced in the system will be reversed.

For three-body systems of higher eccentricity and large perturbations, the chance of finding an efficient reversal of perturbation changes is decreased by the necessity(because of the higher eccentricities) of finding a conjunction of much smaller true anomalies in a much smaller time interval (because of the increased secular speeds of the apse lines).

5. FOUR- OR MORE- BODY HIERARCHICAL SYSTEMS

Such systems can be looked upon as a 'nested' mirror-seeking set. It may be remarked that in the development of the disturbing function, first order perturbations are additive. Each sub-set of three bodies in the hierarchical system will have its orbital characteristics perturbed by the other members of the system. Nevertheless, the two apses of the sub-set must still coincide at regular intervals ensuring that the close approximation to a mirror configuration that must occur will still effectively reverse the first-order three-body perturnations. If, moreover, at much longer intervals of time, there happens to occur, with four bodies of the system, a conjunction of the three apse lines, and shortly thereafter, or before, a conjunction of all four bodies, a more complete reversal of perturbations will take place. For this higher order event, however, to occur within a reasonably short time it would appear that some commensurablē locking mechanism would have to be provided such as the Laplace relation for the inner three Galilean satellites of Jupiter, Io, Europa and Ganỳmede. Their mean longitudes ℓ and mean motions n are related such that, in order from the planet,

$$n_1 - 3n_2 + 2n_3 = 0$$
$$\ell_1 - 3\ell_2 + 2\ell_3 = 180^{\circ},$$

thus ensuring that frequent mirror reversals of mutual perturbations take place. In addition,

$$n_1 - 2n_1 \sim 0$$
$$n_2 - 2n_3 \sim 0.$$

It seems more likely, therefore, that the frequent occurrence of the three-body sub-set near mirror configurations is the main mechanism by which perturbations are squashed before they destroy the stability of the system. In doing so, they provide the system with a durability long enough to ensure that 4 and higher mirror events can occur in which more complete cancelling of perturbations occurs.

6. EFFECT OF ORBITAL INCLINATIONS

We now consider the effect of the mutual inclination of the orbits in a general hierarchical three-body system. For the sake of simplicity we consider the orbits to be circular, of radius a_2 and a_3 , $a_2 < a_3$. Let the mutual inclination be i. Then for a mirror configuration to occur, it is obvious that the conjunction or opposition of the bodies has to take place at the common node or 90° from the common node. At any other position, each of the velocities is not at right angles to the radius vector of the other orbit. It should be remembered that now, it is only at the common node that a conjunction or opposition results in the radius vectors being collinear. In this context a conjunction is defined as a configuration where the longitudes ℓ of the bodies are equal, where the longitude of each body is defined to be measured from the common node along the body's orbital plane to the body's radius vector. An opposition occurs when $\ell_2 = \ell_3 + \pi$. Thus perfect mirror configurations occur at

(i) $\ell_2 = \ell_3 = 0, \frac{\pi}{2}, \pi, \frac{3\pi}{2}$

(ii) $\ell_2 = \ell_3 + \pi, \ \ell_3 = 0, \frac{\pi}{2}, \pi, \frac{3\pi}{2}$

(iii) $\ell_3 = \ell_2 + \pi, \ \ell_2 = 0, \frac{\pi}{2}, \pi, \frac{3\pi}{2}$.

It is easy to show that for any other longitude of conjunction ℓ , the angle ϕ between the radius vector of one body and the velocity vector of the other is given by

$$\cos \phi = \tfrac{1}{2} (\cos i - 1) \sin 2\ell \tag{15}$$

For a given value i of the mutual inclination, ϕ departs from 90° most when

$$\ell = \frac{\pi}{4}, \frac{3\pi}{4}, \frac{5\pi}{4}, \frac{7\pi}{4} .$$

Then for any of these values it is obvious that maximum departure of ϕ from 90° occurs when $i = 90^\circ$, the angle ϕ being 60° or 120°.

If we again consider the Sun-Jupiter-Saturn case, we find that $i \sim 1^\circ$. For $\ell = \frac{\pi}{4}$, $\phi = 90\overset{\circ}{.}0044$, a value so close to 90° that it is obvious that in this system the effect of true anomalies and eccentricities on departure of conjunctions or oppositions from perfect mirror configurations is more important than any inclination effect. Even in the case of asteroids, or of the Sun-Jupiter-Pluto case, where inclinations may be of order 20°, for $\ell = \frac{\pi}{4}$ and $i = 20^\circ$, $\phi = 91\overset{\circ}{.}73$. Eccentricities of order 0.2 are common in such systems. For such a value, the departure of the angle between radius and velocity vectors from 90° can be as high as $11\overset{\circ}{.}5$ for a true anomaly of $78\overset{\circ}{.}5$ and we have seen that to cut this departure to 1°, the true anomaly has to be less than 6°.

7. CONCLUSIONS

It is deduced that as far as a hierarchical n-body dynamical system of the kind found in the Solar System is concerned, it maximises its survival chances if its orbits are so spaced and shaped that inclinations and eccentricities are minimised and perturbations are such that the rates of rotation of apses are kept low. In such a system, frequent and efficient cancellation of perturbations by the occurrence of close approximations to mirror configurations is realised.

The problem of the origin of the low inclinations and eccentricities remains.

In the early days of the Solar System, the formation of the planets by accretion from the disc of dust and gas must have given rise to bodies in orbits with a large distribution in eccentricities and inclinations. Collisions, near-collisions and expulsions must have been common. It may be concluded that survival would have favoured the more massive bodies in orbits of smaller eccentricities and inclinations so that for that reason alone the system would have evolved towards the almost circular, almost coplanar system we observe today, a system of survivors.

In addition, however, in the system's early days, the dissipative and smoothing power of the remaining dust and gas of the disc the protoplanets ploughed their way through must have tended to reduce eccentricities and possibly inclinations. This power of course diminished sharply with the growth of the protoplanets as they accreted the remaining dust and gas. This process must therefore have aided the occurrence of even closer approaches to the ideal but unattainable perfect mirror condition that reverses completely the system's perturbations.

The trapping of the hierarchical systems in the Solar System could then be said to be an example of the old explanation put forward to explain the stage conjurer's tricks - "It's all done by mirrors"!

REFERENCES

Easton, R. 1971. J. Diff. Eq. 10, 371.
Easton, R. 1975. J. Diff. Eq. 19, 258.
Marchal, C. 1971. Astron. Astrophys., 10, 278.
Marchal, C. and Saari, D. 1975. Cel. Mech., 12, 115.
Roy, A.E. and Ovenden, M.W. 1955. Mon.Not.Soc.Astron.Soc.115, 3.
Saari, D.G. 1974. SIAM J. Appl. Math., 26, 806.
Smale, S. 1970. Inventiones Math. 11, 45.
Zare, K. 1976. Cel. Mech., 14, 73.
Zare, K. 1977. Cel. Mech., 16, 35.

NEW RESULTS FOR THE LINEAR STABILITY OF THE TRIANGULAR POINTS IN THE ELLIPTIC RESTRICTED PROBLEM

R. Meire

Astronomical Observatory, Ghent State University,
Ghent, Belgium.

ABSTRACT. New results are obtained for the linear stability of the triangular points in the elliptic restricted problem using the Hill equations which describe the infinitesimal motion around L_4,L_5. Also the shape of the 4π-periodic solutions along the transition curves in the μ-e plane is investigated .

1. INTRODUCTION

The stability of the triangular points in the elliptic restricted three-body problem has been the subject of many papers in the sixties and seventies (Danby, Bennett, Tschauner among others). In these papers the eigenvalues of the dynamical system were investigated to determine the transition curves which separate stable and non-stable regions in the μ-e plane. Here the approach is somewhat different. The equations of motion around L_4,L_5 can be written as a set of two Hill equations and we will use some theorems for this type of equations in order to obtain analytical stability conditions.

In a second part of this paper, we will investigate the 4π-periodic solutions which exist along the transition curves. It will be shown that the shape of these curves can be completely explained by analytical means.

2. EQUATIONS OF MOTION

The plane motion around the triangular points L_4,L_5 in the elliptic restricted problem is given by the equations (see Tschauner) :

$$\begin{cases} x'' - 2\,y' = r\,c_1\,x \\ y'' + 2\,x' = r\,c_2\,y \end{cases} \qquad (2.1)$$

V. V. Markellos and Y. Kozai (eds.), Dynamical Trapping and Evolution in the Solar System, 289–299.

where

$$' = \frac{d}{dv} \quad (v = \text{true anomaly})$$

$$r = (1 + e \cos v)^{-1}$$

$$c_i = \frac{3}{2} [1 + (-1)^i . \sqrt{1 - g}] \qquad (i=1,2)$$

$$g = 3\mu(1 - \mu)$$

The two parameters e and μ determine the stability or instability of the equations (2.1). Stability investigations on this 4-th order system have been done by Danby and Bennett among others.

Tschauner succeeded in separating the 4-th order system (2.1) into two independent 2-nd order systems :

$$\begin{bmatrix} y_1 \\ y_2 \\ y_1^\star \\ y_2^\star \end{bmatrix}' = \begin{bmatrix} P_1 & 0 \\ 0 & P_2 \end{bmatrix} . \begin{bmatrix} y_1 \\ y_2 \\ y_1^\star \\ y_2^\star \end{bmatrix} \qquad (2.2)$$

or

$$\begin{cases} Y' = P_1 \, Y & (2.3a) \\ Y^{\star\prime} = P_2 \, Y^\star & (2.3b) \end{cases}$$

with

$$Y = \begin{bmatrix} y_1 \\ y_2 \end{bmatrix} \quad \text{and} \quad Y^\star = \begin{bmatrix} y_1^\star \\ y_2^\star \end{bmatrix}$$

Furthermore, we have the following relations between the old and the new variables

$$\begin{cases} x = y_1 + y_1^\star \\ y = y_2 + y_2^\star \end{cases}$$

and the (2 x 2)-matrices P_l (l=1,2) are given by their elements $p_{ij}^{(l)}$

$$p_{11}^{(1)} = p_{11}^{(2)} = -\frac{1}{2} r e \sin v (1 + k e \cos v) \qquad (2.4a)$$

$$p_{22}^{(1)} = p_{22}^{(2)} = -\frac{1}{2} r e \sin v (1 - k e \cos v) \qquad (2.4b)$$

$$p_{12}^{(1)} = p_{12}^{(2)} - \frac{rc}{2} = r (a_2 + e \cos v - \frac{1}{4} k e^2 \cos 2v) \qquad (2.4c)$$

$$p_{21}^{(1)} = p_{21}^{(2)} + \frac{rc}{2} = - r (a_1 + e \cos v + \frac{1}{4} k e^2 \cos 2v) \qquad (2.4d)$$

where

$$k^2 = (1 - g)^{-1}$$

$$c^2 = 1 - 9 g + 2 e^2 + k^2 e^4$$

$$a_i = \frac{1}{4} (2 c_i + 1 - c) \qquad (i=1,2)$$

Another important aspect of Tschauner's work is that the curve c = 0 corresponds to one of the transition curves obtained (numerically) by Danby and Bennett using the equations (2.1).

Instead of using the 2-nd order equations (2.3) we can also transform them into a set of two Hill equations.
We obtain (see Meire 1980) :

$$\begin{cases} \xi_1'' + J_1(v).\xi_1 = 0 & (2.5a) \\ \xi_2'' + J_2(v).\xi_2 = 0 & (2.5b) \end{cases}$$

with

$$J_l = - \left\{ r c_l + 2 - \frac{3 \det Q_l + c_2}{q_{12}^{(l)}} + 3 \left[\frac{q_{22}^{(l)}}{q_{12}^{(l)}} \right]^2 \right\} \quad (l=1,2) \qquad (2.6)$$

and

$$q_{ij}^{(l)} = r^{-1} p_{ij}^{(l)}$$

$$\det Q_1 = r^{-1}.\det P_1 = r^{-1}.[p_{11}^{(1)}.p_{22}^{(1)} - p_{12}^{(1)}.p_{21}^{(1)}]$$

$$= \frac{1}{2}[1 + (-1)^1.c + 3\, e \cos v] \tag{2.7}$$

3. STABILITY ANALYSIS

A lot of work has been done on the stability of Hill equations. Here we will use some theorems which give analytical conditions for the stability of the general Hill equation

$$\xi'' + J(v).\xi = 0 \tag{3.1}$$

where $J(v)$ is a 2π-periodic function.

a. Theorem I (Lyapunov)

If

$$J(v) \geqslant 0 \tag{3.2}$$

and

$$\int_0^{2\pi} J(v).dv \leqslant \frac{2}{\pi} \tag{3.3}$$

then all solutions of (3.1) are stable.

b. Theorem II (Krein)

All solutions of (3.1) are stable if there exists a positive integer n for which

$$J(v) \geqslant \frac{n^2}{4} \tag{3.4}$$

and

$$\int_0^{2\pi} J(v).dv < \frac{n^2\pi}{2} + n(n+1)\,\mathrm{tg}\,\frac{\pi}{2(n+1)} \tag{3.5}$$

Both theorems involve the integral

$$Z = \int_0^{2\pi} J(v).dv$$

For the equations (2.5a) and (2.5b) one obtains after some calculations

$$Z = \int_0^{2\pi} J_1(v).dv = \frac{\pi}{2}.\Big\{ 4 - 4\, c_1\, (1-e^2)^{-1/2}$$

$$+ 3\,\{[\,2 + (-1)^1 kc\,].A_1^{-1} - 3\,\}.[\,(A_1 - 1)^2 - k^2e^2\,]^{-1/2}$$

$$+ 3\,\{[\,2 + (-1)^1 kc\,].A_1^{-1} + 3\,\}.[\,(A_1 + 1)^2 - k^2e^2\,]^{-1/2}\Big\} \qquad (3.6)$$

with

$$A_1 = [\,1 + 2\,k\,q_{12}^{(1)}(\tfrac{\pi}{2})\,]^{1/2} \qquad (1=1,2)$$

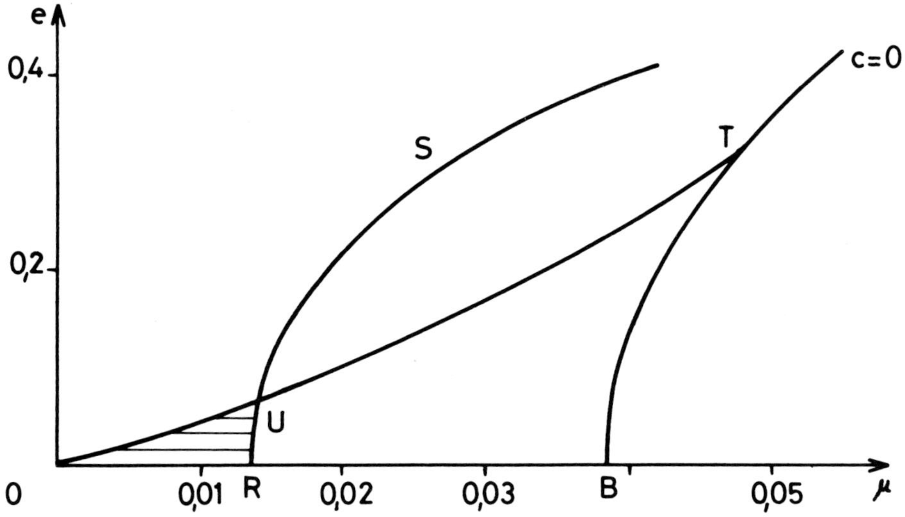

Figure 1. Stable region from Lyapunov's theorem (2.5a)

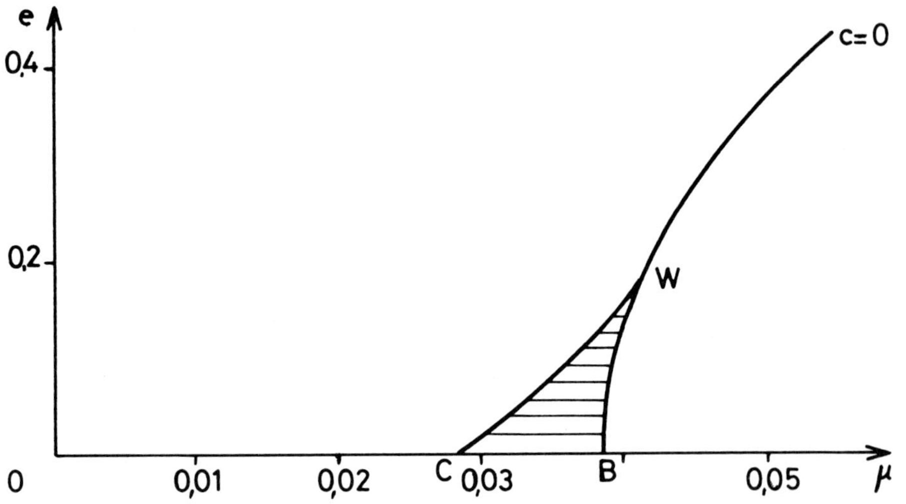

Figure 2. Stable region from Krein's theorem (n=1) for (2.5a)

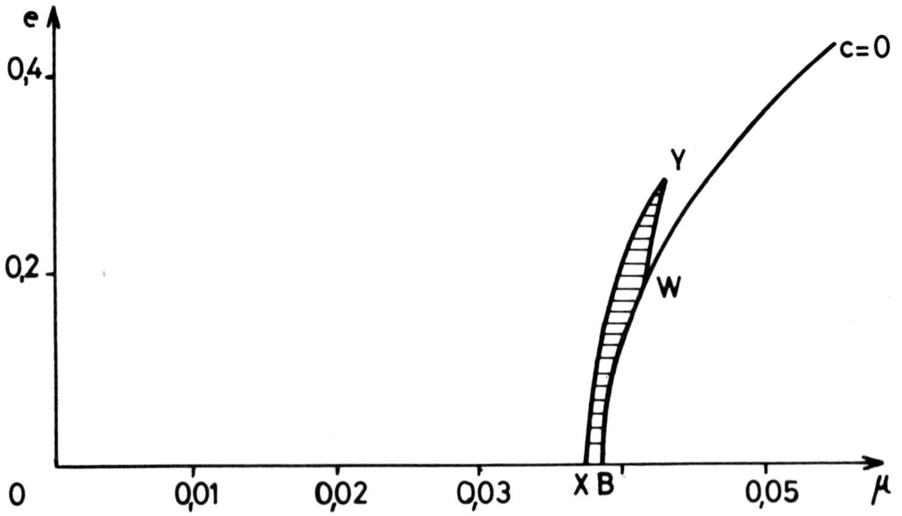

Figure 3. Stable region from Krein's theorem (n=1) for (2.5b)

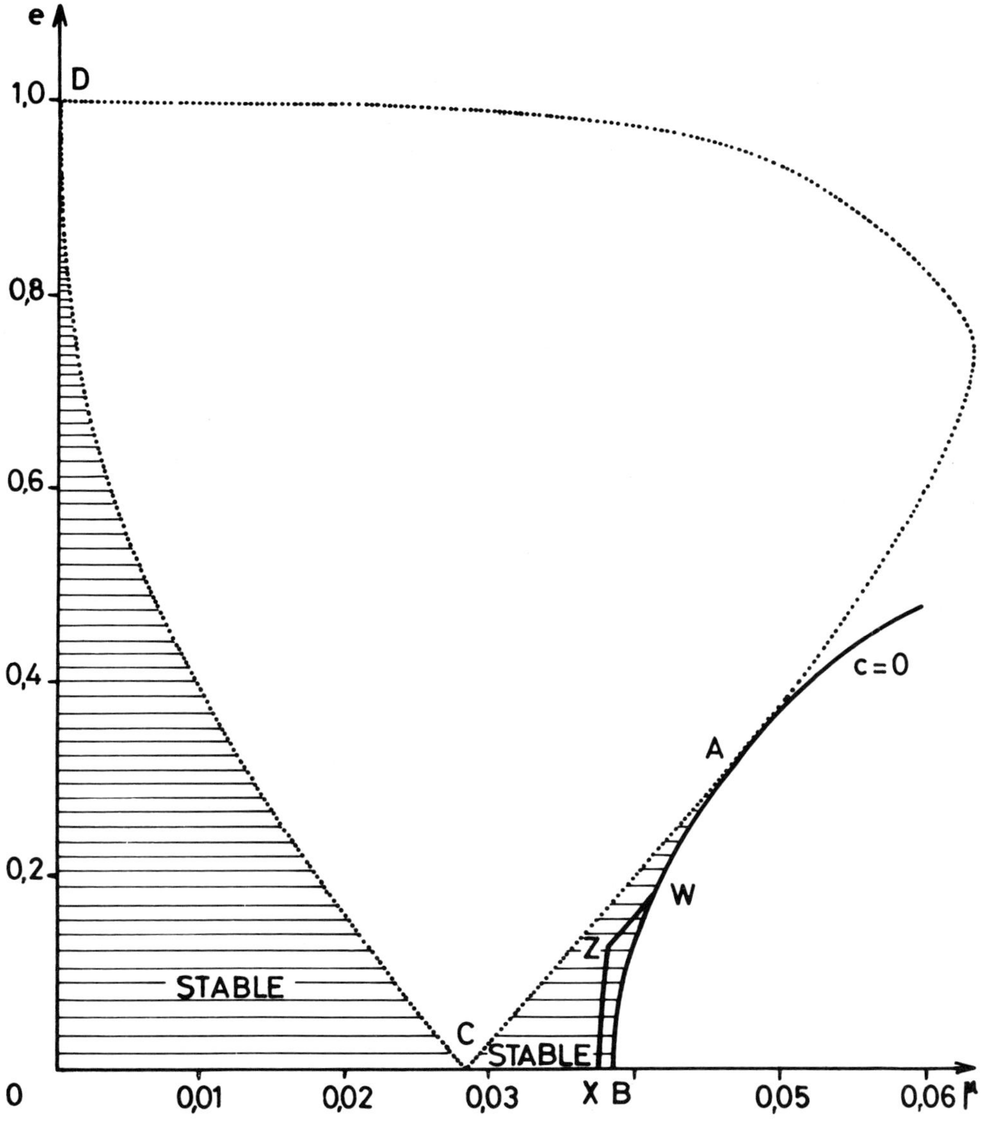

Figure 4. Combined stability region from Krein's theorem

So the conditions for these theorems can be checked analytically for both equations (2.5a) and (2.5b).

Figure 1 shows the results of theorem I, obtained from (2.5a). So we may conclude that all solutions of (2.5a) are stable in the ORU-region of the μ-e plane.

However no results were found for equation (2.5b) using theorem I.
So no general conclusion can be made on the complete motion. Yet we should remark that theorem I (and also theorem II) only states sufficient stability conditions.
More interesting results are obtained from theorem II.
Fig. 2 shows a stable region CBW obtained from (2.5a) and fig. 3 shows a stable region XYWB obtained from (2.5b). (For both cases the integer value n = 1).
A very interesting point is that there is a common region XZWB of fig. 2 and fig. 3 which is shown in fig. 4.
So we may conclude :

> "The triangular points in the elliptic problem are stable for all μ-e values which belong to the XZWB-region of the parameterplane."

For the first time, a stable region is found on a purely analytical basis. Indeed, all previous results were accomplished by numerical integration procedures or analytical approximations for the transition curves. Note however that these results cover only a small part of the numerical results (shaded regions in fig. 4).

4. PERIODIC MOTIONS

Along the transition curves CD, CA and AD of fig. 4 one of the eigenvalues is -1 (see Meire 1981). This means that there exists 4π-periodic solutions along these curves.
We were interested how those 4π-periodic solutions look like.
The curves CD and CA were obtained from the equation (2.3a) and the curve AD from (2.3b) and the curve BA represents c = 0.
From equations (2.4) it is clear that (2.3a) and (2.3b) are very similar so we will only consider the explicit form of the equations (2.3a)

$$\begin{cases} y_1' = p_{11}\, y_1 + p_{12}\, y_2 & (4.1a) \\ y_2' = p_{21}\, y_1 + p_{22}\, y_2 & (4.1b) \end{cases}$$

where the upper indices $^{(1)}$ for the p_{ij} have been omitted.
Now taking

$$\begin{cases} y_1 = \exp[\int p_{11} dv]\,.u_1 = r^{\frac{k-1}{2}}\,.\,\exp(\frac{k}{2r})\,.\,u_1 & (4.2a) \\ y_2 = \exp[\int p_{22} dv]\,.u_2 = r^{\frac{-(k+1)}{2}}\,.\,\exp(\frac{-k}{2r})\,.\,u_2 & (4.2b) \end{cases}$$

the equations (4.1) are transformed into

$$\begin{cases} u_1' = m_{12}\, u_2 & (4.3a) \\ u_2' = m_{21}\, u_1 & (4.3b) \end{cases}$$

where

$$m_{12} = p_{12}\, r^{-k} \exp\left(-\frac{k}{r}\right) \qquad (4.4a)$$

$$m_{21} = p_{21}\, r^{k} \exp\left(\frac{k}{r}\right) \qquad (4.4b)$$

From the equations (4.2) it follows that if y_1 becomes zero then so does u_1 and vice versa (the same holds for y_2 and u_2).
Equations (4.3) show that every crossing of the axes in the u_1-u_2 plane is perpendicular to these axes.
It can be shown (see Meire 1980) that $p_{12}(v)$ and thus $m_{12}(v) \geqslant 0$ for $0 \leqslant e \leqslant 1$, $0 \leqslant \mu \leqslant 1/3$.
Also it is easy to prove that $p_{21}(0)$ or $m_{21}(0) < 0$ for all possible μ- and e-values.
For certain regions in the μ-e plane p_{21} and consequently m_{21} will remain negative for all values of v whereas for other regions, they can become positive. In the former case the extremum values of u_1 and u_2 can only be reached when crossing the u_1- or u_2-axis, while in the latter case u_2 will reach additional extremum values (where $u_1 \neq 0$ but $m_{21} = 0$).

Fig. 5a, fig. 6a and fig. 7a show three possible types of 4π-periodic orbits in the u_1-u_2 plane and fig. 5b, fig. 6b and fig. 7b show the corresponding orbits in the y_1-y_2 plane.
In the case of fig. 5a the coefficient $m_{21}(v)$ remains negative for every value of the true anomaly, while for fig. 6a and fig. 7a $m_{21}(v)$ also becomes positive, so that u_2 reaches additional extremum values.

Further research on the shape of these curves has been done and will be published in another paper.
This however is a first attempt to explain the shape of periodic orbits around the triangular points in the elliptic restricted problem and it has become clear that the work of Tschauner is very important for the further study of this interesting problem.

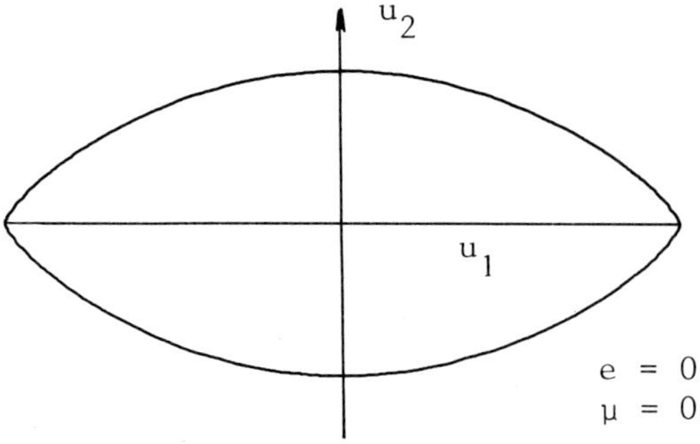

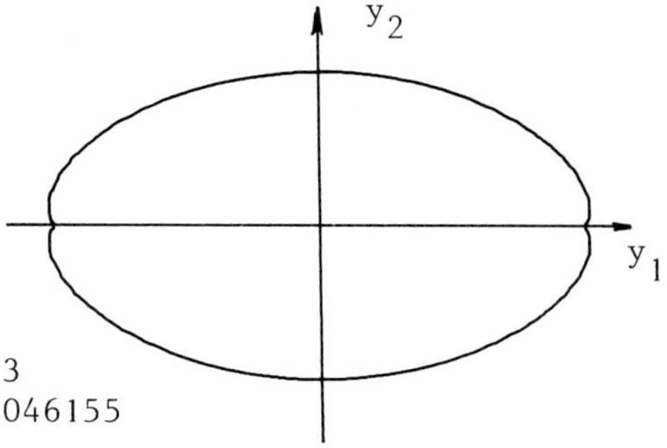

e = 0.3
μ = 0.046155

Fig. 5a

Fig. 5b

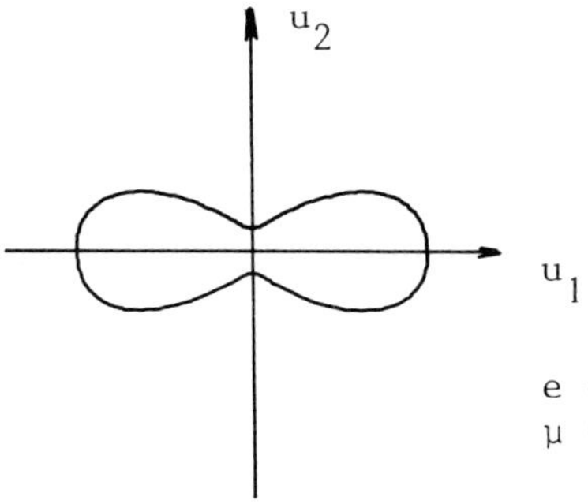

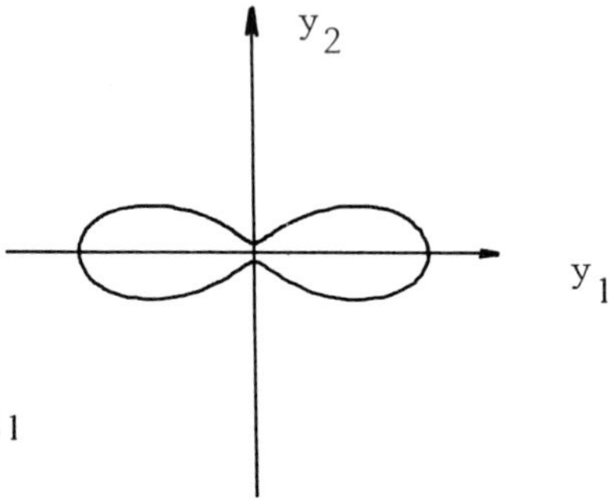

e = 0.6
μ = 0.003781

Fig. 6a

Fig. 6b

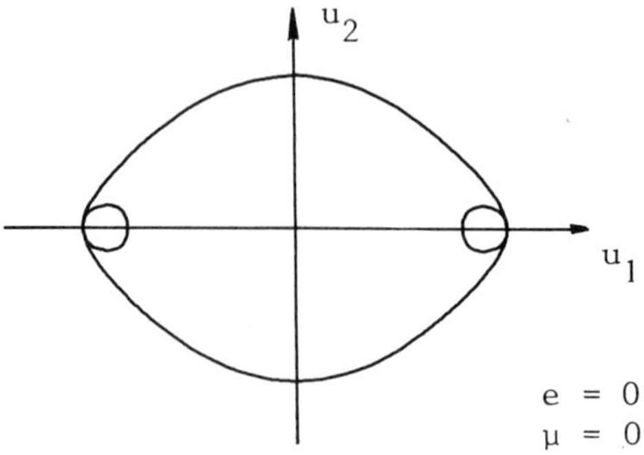

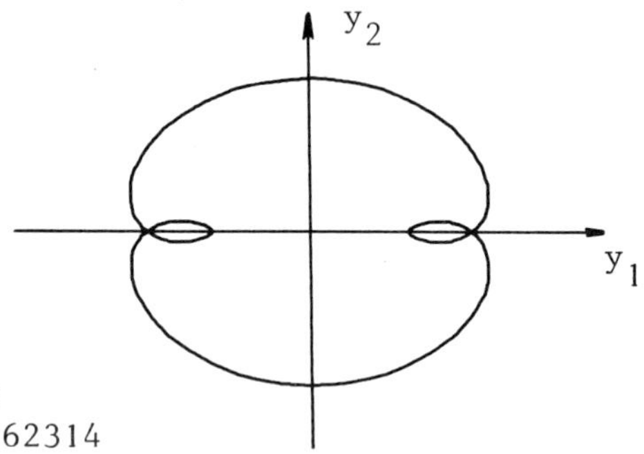

e = 0.8
μ = 0.062314

Fig. 7a

Fig. 7b

5. REFERENCES

Bennett A. (1965), Icarus 4, p 177

Danby J.M.A. (1964), Astron. J. 69, p 165

Hale J.K. (1969), "Ordinary Differential Equations", Wiley, New York

Krein M.G. (1955), Prikl. Mat. Mech. 15 (1951), English translation in Am. Math. Soc. Trans. Series 2, 1, p 163

Matas V. (1973) Bull. Astron. Inst. Czech. 24, p 249

Meire R. (1980), Bull. Astron. Inst. Czech. 31, p 312

Meire R. (1981), Cel. Mech. 23, p 89

Meire R. (1982), Astron. Astroph. 110, p 152

Szebehely V. (1967), "Theory of Orbits", Academic Press, New York

Tschauner J. (1971), Cel. Mech. 3, p 189

ON TOPOLOGICAL STABILITY IN THE GENERAL THREE AND FOUR-BODY PROBLEM

A. Milani and A.M. Nobili

Department of Astronomy, Glasgow University, U.K.
Permanent address: Department of Mathematics,
University of Pisa, Italy.

1. INTRODUCTION

By simple symmetry and change-of-scale considerations the topology of the level manifolds of the classical integrals of the N-body problem is shown to depend only on the value of the integral $z = c^2h$ (total angular momentum squared times total energy). For every hierarchical structure given to the N bodies the problem can be described as a set of N-1 perturbed two-body problems by means of a fitted Jacobian coordinate system; in this setting the Easton inequality, relating potential, momentum of inertia and the z integral, is easily rederived. For N=3 the confinement conditions due to this inequality can be described, in a pulsating synodic reference system, as level lines of a modified potential function on a plane.

For the small parameter $\varepsilon = m_3/m_2$ (mass of the smallest body divided by mass of the secondary body in the main binary) going to zero these level lines reduce to the zero velocity curves of the restricted circular 3-body problem; however, if the two larger masses have an eccentricity $e_2 > 0$, the difference between the actual value of z and its critical value corresponding to the Lagrangian point L_2 contains a "destabilizing" term porportional to e^2_2. By neglecting terms of the order of ε^2 an approximate, and very easy to check, stability criterium is established. Moreover, since it contains a zero order term proportional to e^2_2 it allows also an order-of-magnitude-estimate of the minimum mass m_3 below which no stability at all can be guaranteed on the basis of ten classical integrals only. The minimum mass is given by the reduced mass of the main binary times $e^2_2/2$ and in the Sun-Jupiter-third body system it turns out to be about one half of the mass of the Earth. This means that no stability can be guaranteed in this way for Mercury, Mars, Pluto and, of course, the asteroids.

For $N \geqslant 4$ every hierarchy can be broken: more the hierarchy is strong more easily a close approach of two bodies can be obtained without violating Easton inequality, then the connectedness of the

V. V. Markellos and Y. Kozai (eds.), Dynamical Trapping and Evolution in the Solar System, 301–315.

collision subset allows any exchange of bodies. However, the time needed to change the z functions of the 3-body subsystems enough to allow such exchange is very long, as can be estimated by a perturbation theory approach using as small parameters not only the mass ratios but also the "scale ratios" among the subsystems.

This paper contains only the statements, with some sketches of the proofs; for a full account the reader should refer to (Milani and Nobili, 1982 and Milani and Nobili, 1983).

2. THE LEVEL MANIFOLDS OF THE CLASSICAL INTEGRALS

The N-body problem, with masses m_i, position $\underline{r}_i$ and velocities $\dot{\underline{r}}_i$ is defined by its kinetic energy T and potential U:

$$T = \frac{1}{2} \sum_{i=1}^{N} m_i \dot{r}_i^2 \qquad U = G \sum_{i<j} \frac{m_i m_j}{r_{ij}} \qquad \underline{r}_{ij} = \underline{r}_i - \underline{r}_j . \tag{1}$$

The 10 classical integrals will be denoted by $\underline{\alpha}$ (linear momentum), $\underline{\beta}$ (position of the centre of mass at t=0), $\underline{J}$ (angular momentum), E=T-U (energy). In the 6N-dimensional phase space, if we impose $\underline{\alpha}=\underline{\beta}=0$, $\underline{J}(\underline{r},\dot{\underline{r}})=c$ and $E(\underline{r},\dot{\underline{r}})=h$ we define a manifold $V_{\underline{c},h}$ (generically smooth and 6N-10 dimensional). Then the problem of "topological stability" can be stated in this way: for which N, $\underline{c}$, h does $V_{\underline{c},h}$ have more than 1 connected component (Birkhoff,1927, pp 287-288)? In this case, is the projection of $V_{\underline{c},h}$ on the configuration space also disconnected? A second and more difficult part of the problem is the following: given two open subsets in the same connected component $V_{\underline{c},h}$, is there a solution of the dynamical equations of the N-body problem going from the one to the other? How long does it take?

A major breakthrough occurred in Celestial Mechanics in the seventies with the reply to Birkhoff's old question on the topology of the level manifolds of the classical integrals in the general 3-body problem. The result was that for some values of the energy and of the angular momentum the level manifolds of the classical integrals are topologically disconnected as subsets of the phase-space; moreover, the projections of these disconnected components on the configuration space are also disconnected. Hence forbidden configurations do form a boundary that separates regions of "trapped" motions (Golubev, 1968; Smale 1970a; Marchal 1971; Smale 1970b; Easton, 1971; Tung,1974; Easton, 1975; Marchal and Saari, 1975; Zare,1976 and 1977; Bozis, 1976). Although the relevance of this result for the stability of planetary systems was perceived (Szebehely and Zare,1977; Szebehely and McKenzie,1977; Roy,1979; Walker et al.,1980; Walker and Roy, 1981) there have been some problems in fully exploiting this discovery in assessing the stability of such systems. One of the difficulties is that the proofs of the topological stability criterium use relatively difficult mathematical tools and the computational procedures for actually checking whether it is satisfied or not are long, so that in the process the physical intuition of the meaning of the topological criterium is easily lost.

As it often happens in scientific research, after a new general result has been obtained and it has been definitely assessed, it comes out that there is the possibility to get the same result in an easier way that sometimes gives also some hints for getting further interesting results. That is why we started our work on this subject by giving a new proof of the topological stability criterium for 3-body systems (also in 3 dimensions) that uses only Lagrangian mechanics and elementary calculus (Milani and Nobili, 1982).

If the symmetries of the N-body problem are taken into account it can be shown that the problem of the topological type of $V_{\underline{c},h}$ does not depend really on the 4 parameters $\underline{c}$,h but only on a scalar bifurcation parameter $z = c^2h$. Since T and U are invariant under rotations $R \in SO(3)$ every rotation R maps diffeomorphically $V_{\underline{c},h}$ onto $V_{R\underline{c},h}$ and the topology of $V_{\underline{c},h}$ depends only on c and h. Moreover, since U is a homogeneous function, every change of scale that multiplies all the r_i by a factor $\alpha > 0$ and the time by a factor $\tau > 0$ maps orbits onto orbits provided that α and τ satisfy the "third Kepler law": $\alpha^3/\tau^2 = 1$. (We could also say that the universal constant of gravitation, G, must be invariant under this change of scale). Then $V_{\underline{c},h}$ is mapped diffeomorphically onto $V_{\underline{c}',h'}$ with $c' = c(\alpha^2/\tau)$ and $h' = h/\alpha$. We conclude that there is only one function of $\underline{c}$,h which is invariant under rotations and changes of scale (apart from others functionally dependent) and it is the integral $z = c^2h$.

The number of connected components of $V_{\underline{c},h}$ will thus change only when $z = hc^2$ crosses some "critical value"; at the critical value $V_{\underline{c},h}$ is not smooth, the singular points corresponding to the relative equilibrium configurations $\nabla(E - \langle\underline{\omega}, \underline{J}\rangle) = 0$, stationary in a reference frame rotating with angular velocity $\underline{\omega}$ (Smale, 1970b).

3. HIERARCHICAL DYNAMICAL SYSTEMS

The most formal definition of a hierarchy for an N-body system can be given as follows: a hierarchy A is a symbol constructed by using the masses $m_1, m_2, \ldots. m_N$, each once and only once, and the operation of forming couples. As an example, for N=4 all the hierarchies are equivalent, by relabelling of the masses or by changing the order in some couples, to one of the following two:

$$B = ((m_1, m_2),(m_3, m_4)) \qquad P = (((m_1, m_2),m_3),m_4) \qquad (2)$$

where B is a double-binary hierarchy and P is a planetary hierarchy (see Figure 1).

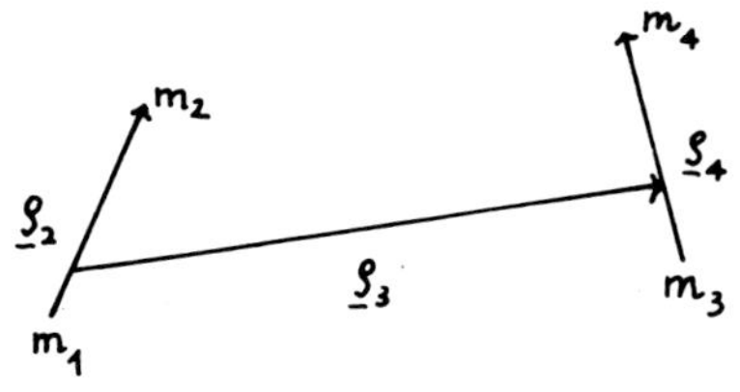

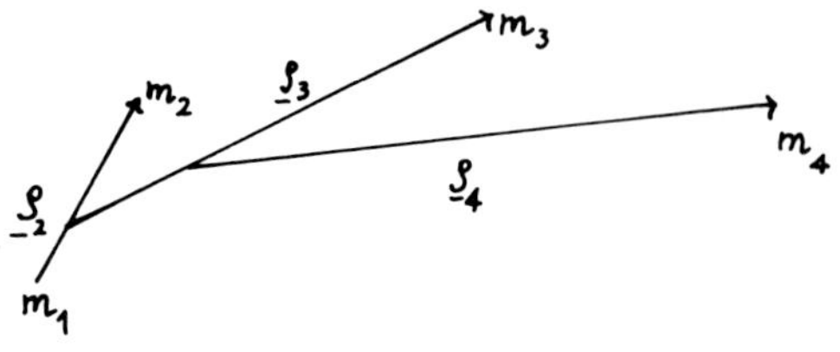

B = double-binary hierarchy P = planetary hierarchy

Figure 1

Perhaps a better understanding of the structure of a hierarchy can be obtained by representing it with graphs of the kind introduced by Evans (1968): whenever a couple is formed, an oriented segment is introduced in the graph (Roy, 1982). To write the dynamical equations and the integrals of motion in a way that shows the physical significance of a hierarchy as a set of perturbed 2-body subsystems we need to use Jacobian coordinates (Walker, 1982). They are defined by the reduced masses M_j and by the Jacobian vectors $\underline{\rho}_j$, j=1,...,N. The $\underline{\rho}_j$ are obtained as linear combinations of the $\underline{r}_i$:

$$\underline{\rho}_j = \sum_{i=1}^{N} a_{ji}\, \underline{r}_i \tag{3}$$

and must satisfy the following four conditions:

(i) the first vector $\underline{\rho}_1$ is the center of mass vector, M_1 is the total mass:

$$M_1 \underline{\rho}_1 = \sum_{i=1}^{N} m_i r_i \qquad M_1 = \sum_{i=1}^{N} m_i \tag{4}$$

(ii) the kinetic energy as a quadratic form in the $\dot{\underline{\rho}}_j$ is diagonal with eigenvalues M_j:

$$2T = \sum_{j=1}^{N} M_j \dot{\underline{\rho}}_j^{\,2} \tag{5}$$

(iii) the product of the reduced masses M_j is equal to the product of the masses m_j:

$$\prod_{i=1}^{N} M_i = \prod_{i=1}^{N} m_i \tag{6}$$

(iv) the map $\underline{r} \to \underline{\rho}$ preserves orientation; together with (6) this means:

$$\det \left[a_{ij} \right] = +1 \tag{7}$$

A Jacobian coordinate system can be chosen in different ways. We will say that a Jacobian system is fitted to a hierarchy A if it is constructed according to the following recursion rule: whenever in A a couple B = (B', B") is formed, the Jacobian vectors of B are in this order: the center of mass of the subsystem B; the Jacobian vectors of B', excluding the first one because it is the center of mass of B'; a new added vector; and the Jacobian vectors of B", excluding the first one again. The same rule applies to the reduced masses, only one of which must be determined. Then the following existence and uniqueness result holds: for every hierarchy A there is one and only one fitted Jacobian coordinate system and it is defined by adding, when the couple (B',B") is formed, a new vector $\underline{\rho}_i$ going from the center of mass of B' to the center of mass of B", and a new reduced mass $M_i = M_i' M_i''/(M_i' + M_i'')$ where M_i', M_i'' are the total masses of B',B".

In the Jacobian coordinates angular momentum and moment of inertia with respect to an axis $\underline{e}$ are of the same form as in the usual coordinates:

$$\underline{c} = \sum_{i=1}^{N} m_i \, \underline{r}_i \times \dot{\underline{r}}_i = \sum_{i=1}^{N} M_i \, \underline{\rho}_i \times \dot{\underline{\rho}}_i \tag{8}$$

$$I_{\underline{e}} = \sum_{i=1}^{N} m_i \, |\underline{r}_i \times \underline{e}|^2 = \sum_{i=1}^{N} M_i \, |\underline{\rho}_i \times \underline{e}|^2 \tag{9}$$

Let us suppose that the cyclic coordinates $\underline{\rho}_1$ are ignored, or that $\underline{\rho}_1 = \dot{\underline{\rho}}_1 = 0$; the energy integral is:

$$h = T - U(\underline{\rho}) \tag{10}$$

where U is considered as a function of the configuration $\underline{\rho} = (\underline{\rho}_2, \ldots, \underline{\rho}_N)$. The angular momentum c (referred to the center of mass) satisfies the inequality

$$c^2 \leq I_{\underline{e}} \, .2T \tag{11}$$

involving the kinetic energy T and the moment of inertia $I_{\underline{e}}$ referred to the unit angular momentum vector $\underline{e} = \underline{c}/c$.

By using the property of U of being a homogeneous function of degree (-1) we can combine (10) and (11) in an inequality to be satisfied by the configuration $\underline{\rho}$ with given h and $\underline{c}$. Let $\lambda = \sqrt{I_e}$ be the "scale" of the configuration $\underline{\rho}$ and $\underline{u} = \underline{\rho}/\lambda$ the configuration independent from scale; since $U(\underline{u}) = \lambda\, U(\underline{\rho})$ we have

$$2h\lambda^2 + 2U(\underline{u})\,\lambda - c^2 \geq 0 \qquad (12)$$

and the reality condition for the scale λ gives the Easton (1971) inequality

$$U^2(\underline{u}) + 2hc^2 \geq 0 \qquad (13)$$

4. TRAPPING MECHANISMS IN THE GENERAL AND IN THE RESTRICTED THREE-BODY PROBLEM.

Easton inequality (13) reduces the bifurcation problem to the "constant scale configuration manifold" $I_{\underline{e}} = 1$. Since by the projection π on the invariable plane $U(\pi(\underline{\rho})) \geq U(\underline{\rho})$, and the relative equilibria are planar, the planar case always gives all the relevant information on the connected components of the level manifolds. Moreover we can choose a "pulsating synodic" reference system in the invariable plane such that $\pi(\underline{\rho_2}) = (-1,0)$; then the sphere $I_{\underline{e}} = 1$ is parametrized by $\pi(\underline{\rho_3}) = (x,y)$ and the potential U is:

$$U(x,y) = G(M_2 + r^2 M_3)^{\frac{1}{2}} \left[m_1 m_2 + m_3\left(\frac{m_2}{r_2} + \frac{m_1}{r_1}\right)\right] \qquad (14)$$

where $r = (x^2 + y^2)^{\frac{1}{2}}$, $r_1 = ((x-\mu)^2 + y^2)^{\frac{1}{2}}$ and $r_2 = ((x-\mu+1)^2 + y^2)^{\frac{1}{2}}$ are the distances of the projection of m_3 from, respectively, the origin, the projection of m_1 and the projection of m_2 and μ is, as usual, $m_2/(m_1+m_2)$. $M_2 = m_1 m_2/(m_1+m_2)$ and $M_3 = m_3(m_1+m_2)/(m_1+m_2+m_3)$ are the reduced masses. Hence the computation of the number of connected components of $U(x,y) \geq (-2z)^{1/2}$ = constant can be done with the same techniques used to study the zero-velocity curves in the restricted 3-body problem (see Figure 2).

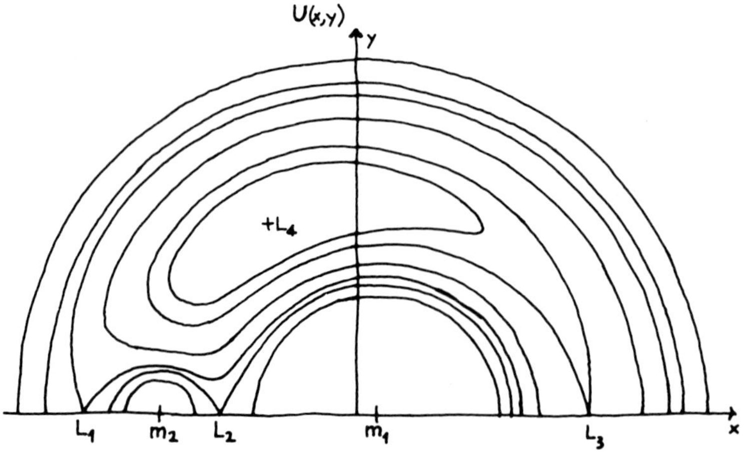

Figure 2

This similarity can be better understood by expanding $U(x,y)$ in power series of the small parameter $\varepsilon = m_3/m_2$, where m_3 is the smallest mass in the system and m_2 the mass of the secondary body in the main binary. We get

$$U(x,y) = A\left[1 + \frac{\varepsilon}{1-\mu} \, \Omega(x,y) + O(\varepsilon^2) \right] \qquad (15)$$

where $\Omega(x,y) = r^2/2 + \mu/r_2 + (1-\mu)/r_1$ is the well known function defining the zero-velocity curves in the restricted case and A is a constant, $A = GM_2^3/2\,(m_1 + m_2)$.

The topological stability criterium for a general 3-body problem, completely similar to the Hill stability criterium for the restricted problem, requires the computation of the difference Δz between the actual value z of the c^2h integral and its value z_2 at the Lagrangian point L_2. By expanding z in power series of ε we obtain

$$\frac{z}{M_2^3} = G^2(m_1+m_2)^2 \frac{e_2^2-1}{2} + \varepsilon \frac{G(m_1+m_2)a_2}{1-\mu} J + O(\varepsilon^2) + O(\varepsilon^2 e_2^2) \qquad (16)$$

where a_2, e_2, n_2 are the osculating semimajor axis, eccentricity and mean motion of the main binary (obviously changing in time) and J is the "Jacobi" function defined as

$$J = \bar{h}_3 - n_2 \frac{<\underline{c}_3 \cdot \underline{c}_2>}{e_2} \qquad (17)$$

$\bar{h}_3$ and $\underline{c}_3$ being the energy and angular momentum of M_3, and $\underline{c}_2$ the angular momentum of M_2 (per unit mass) in the usual Jacobian coordinates. If we expand also z_2 in power series of ε the resulting formula is

$$\frac{z_2}{M_2^3} = -\frac{1}{2} G^2(m_1+m_2)^2 + \varepsilon \frac{G(m_1+m_2)a_2}{1-\mu} J_2 + O(\varepsilon^2) \qquad (18)$$

where J_2 is the Jacobi constant computed at the equilibrium point L_2 of the corresponding restricted problem with the same masses m_1, m_2 and a distance between them equal to a_2. We stress that, with this definition of J_2, formula (18) is correct because it can be proved (Milani and Nobili, 1982) that for a given general 3-body problem the Jacobi function J_2 corresponding to the Lagrangian point L_2 is equal to the Jacobi integral computed at the equilibrium point L_2 of the corresponding restricted 3-body problem apart from terms of the order of ε.

From (16) and (18) we compute now $\Delta z = z - z_2$. By using the usual units of the restricted problem (such that $G = 1$, $m_1 + m_2 = 1$, $a_2 = 1$) we have:

$$\frac{\Delta z}{\mu^3(1-\mu)^3} = \frac{e_2^2}{2} + \varepsilon \frac{\Delta J}{1-\mu} + O(\varepsilon^2) + O(\varepsilon\, e_2^2) \qquad (19)$$

where $\Delta J = J - J_2$ must be less than zero in order to guarantee the stability of the restricted problem according to Hill's criterium. The analogous topological stability criterium in the general 3-body problem requires $\Delta \mathbf{z} < 0$. Neglecting terms of the order of ε^2 and terms of the order of εe_2^2 we can give an approximate stability criterium in the general 3-body problem requiring that

$$\frac{\Delta \mathbf{z}}{\mu^3(1-\mu)^3} = \frac{e_2^2}{2} + \varepsilon \frac{\Delta J}{1-\mu} < 0. \tag{20}$$

According to the approximate criterium (20) a Hill "unstable" 3-body system (i.e. $\Delta J > 0$) will still be "unstable"-in the sense that we cannot guarantee its stability-in the general case; on the other hand, a Hill stable one (i.e. J of the order of -1 in these units) can be stable in the general case only if the mass of the smallest body satisfies the inequality

$$m_3 \gtrsim \frac{m_1 m_2}{m_1 + m_2} \cdot \frac{e_2^2}{2} \tag{21}$$

(see also Marchal and Bozis, 1982) where the destabilizing effect of the eccentricity of the binary is quantified. For the Sun-Jupiter-third body system the (21) means: $m_3 \gtrsim 0.4\ m_E$, m_E being the mass of the Earth, so that no stability at all can be guaranteed, on the basis of the ten classical integrals only, for small objects like Mercury, Mars, Pluto and the asteroids even if it can be easily shown that they are all stable according to Hill's criterium (see Table 1). One way of understanding why no tiny body can be proved to be topologically stable in the general 3-body problem is that in this case the bifurcation parameter is c^2h, i.e. total angular momentum squared times total energy, and a tiny body contributes very little to it. In other words, Jupiter does not care very much where Mercury, Pluto, Mars or any asteroid is. This does not mean, of course, that they will actually be so much perturbed by Jupiter to cross its orbit. We simply cannot guarantee their stability by using a criterium based on the classical integrals. On the contrary, in the restricted problem the bifurcation parameter is the Jacobi integral, which contains energy and angular momentum of the third body only (per unit mass) so that Hill's criterium is meaningful no matter how small the third body is.

We note that the meaning of the connected components is different for the general and for the restricted 3-body problem. In the restricted problem a zero velocity curve enclosing a bounded region of allowed motion means that the test particle cannot escape; as an example, in the restricted case all the asteroids up to Thule cannot cross the critical 8-shaped Hill's curve, which is fixed in the rotating frame, so that they can never escape (see also Farinella and Nobili, 1978). On the other hand, even if a Hill stable 3-body system can be proved to be stable in the general case too, this does not exclude the escape of one of the three bodies. The reason is that the (x,y) plot must be multiplied by a variable scale factor because it is drawn in a pulsating synodic reference system. But if the topological criterium is satisfied (i.e. $\Delta z < 0$) the hierarchy will never be broken, e.g. in the sense that the distance of m_3 from the primary is constrained forever to be smaller than the distance between the primary and the secondary body in the binary.

Table 1 summarizes the results of the exact and the approximate criterium applied to 3-body subsystems of the Solar System and shows the usefulness of the simple approximate criterium.

Table 1: Exact and Approximate Computations of the Stability Parameter in the General 3-Body Problem

3-Body Subsystem: Sun+	Δz	δz	ε	ΔJ	"Stable"
Jupiter-Mercury	+0.000192	+0.000190	1.7×10^{-4}	-5.463	No
Jupiter-Venus	-0.005127	-0.005142	2.6×10^{-3}	-2.448	Yes
Jupiter-Earth	-0.003634	-0.003646	3.1×10^{-3}	-1.520	Yes
Jupiter-EM Center of Mass	-0.003691	-0.003703	3.2×10^{-3}	-1.519	Yes
Jupiter-Mars	+0.000896	+0.000895	3.4×10^{-4}	-0.726	No
Jupiter-Saturn	-0.040128	-0.030410	3.0×10^{-1}	-0.105	Yes
Jupiter-Uranus	-0.025013	-0.023389	4.6×10^{-2}	-0.535	Yes
Jupiter-Neptune	-0.056640	-0.051579	5.4×10^{-2}	-0.971	Yes
Jupiter-Pluto	+0.000752	+0.000751	3.5×10^{-4}	-1.115	No
Earth-Moon	+0.000142	+0.000138	1.2×10^{-2}	+0.0002	No

As far as the Sun-Jupiter-exterior planet case is concerned, the relevant critical value of the c^2h integral is z_1, corresponding to the L_1 relative equilibrium point (see Figure 2). But whenever $m_1 >> m_2 >> m_3$ the computation of the critical value z_1 is not needed if the approximate criterium (20) is used because the difference $z_1 - z_2$ turns out to be zero apart from terms of the order of $\varepsilon\mu$ or terms of the order of ε^2 .

5. BREAKING FOUR-BODY HIERARCHIES

We will show now that, as stated by Marchal (1971), no topological stability criterium based on the ten classical integrals only can be formulated for a 4-body system, i.e. every 4-body hierarchy could in principle be broken. For a fixed invariable plane, i.e. for a fixed $\underline{e}$, and for fixed values of h,c a given configuration $\underline{\rho}(0)$ can be changed along a continuous path $\underline{\rho}(s)$ to a new configuration $\underline{\rho}(1)$ provided that the continuous function $K(s) = I_{\underline{e}} U^2(\underline{\rho}(s))$ never falls below its initial value $K(0) \geq -2hc^2$, so that inequality (13) is always fulfilled. This does not necessarily mean that there is a solution of the dynamical equations connecting $\underline{\rho}(0)$ to $\underline{\rho}(1)$; it simply means that the ten classical integrals do not exclude the existence of such a solution.

Let us define $' \equiv d/ds$ and the Jacobian vectors $\underline{\rho}_2$ $\underline{\rho}_3$ and $\underline{\rho}_4$ as in Figure 1. Let us then keep $\underline{\rho}_4$ constant and change the length of $\underline{\rho}_2$ and $\underline{\rho}_3$ in such a way that the moment of inertia remains constant, i.e. $\dot{I}_{\underline{e}} = 0$. Now the question is: can we have $U'(\underline{\rho}(s)) \geq 0$ in such a way that Easton inequality (13) is always fulfilled and nevertheless the hierarchy of the 4-body system is finally broken? $I_{\underline{e}} = 0$ means:

$$\rho_3' = - \frac{M_2}{M_3} \frac{\rho_2}{\rho_3} \rho_2' \tag{22}$$

and we require

$$U'(\underline{\rho}(s)) = \langle \underline{\rho}_2', \nabla_2 U \rangle + \langle \underline{\rho}_3', \nabla_3 U \rangle > 0 \,. \tag{23}$$

In a 4-body system with a double-binary B hierarchy given by (2) (see also Figure 1), the potential U is easily computed as a multipole expansion of the gravitationsl effect of each binary on the center of mass of the other; the mixed terms, that means terms containing both ρ_2/ρ_3 and ρ_4/ρ_3, do not appear until the fourth order in the ratios ρ_2/ρ_3 and ρ_4/ρ_3 is reached (Milani and Nobili, 1983):

$$U = \frac{G\, m_1 m_2}{\rho_2} + \frac{G\, m_3 m_4}{\rho_4} +$$

$$+ \frac{G(m_1+m_2)(m_3+m_4)}{\rho_3} \Big[1+\mu_2(1-\mu_2) \frac{\rho_2^2}{\rho_3^2} P_2(\cos\theta_{23}) + \mu_4(1-\mu_4) \frac{\rho_4^2}{\rho_3^2} P_2(\cos\theta_{43}) +$$

$$+\mu_2(1-\mu_2)(1-2\mu_2) \frac{\rho_2^3}{\rho_3^3} P_3(\cos\theta_{23}) + \mu_4(1-\mu_4)(1-2\mu_4) \frac{\rho_4^3}{\rho_3^3} P_3(\cos(\pi-\theta_{43})) +$$

$$+ \text{4th order terms} \Big] \tag{24}$$

where $\mu_2 = m_2/(m_1+m_2) = \mu$, $\mu_4 = m_4/(m_3+m_4)$; θ_{23} is the angle between $\underline{\rho}_2$ and $\underline{\rho}_3$, θ_{43} the angle between $\underline{\rho}_4$ and $\underline{\rho}_3$; P_2, P_3 are the usual Legendre polynomials. Inequality (23) becomes:

$$U' = <\underline{\rho}_2' \, , \, \frac{-Gm_1m_2}{\rho_2^3}\underline{\rho}_2> \left\{ 1+0\left(\frac{m_3+m_4}{m_1+m_2}\,\frac{\rho_2^3}{\rho_3^3}\right) - \frac{\rho_2^3}{\rho_3^3} \cdot \frac{m_1+m_2+m_3+m_4}{m_1+m_3}\left[1 + +0\left(\mu_2(1-\mu_2)\frac{\rho_2^2}{\rho_3^2}\right) +0\left(\mu_4(1-\mu_4)\frac{\rho_4^2}{\rho_3^2}\right)\right]\right\} > 0 \qquad (25)$$

so that for any "sufficiently" hierarchical 4-body system all the terms inside the curly brackets are small compared to the first one and this inequality can be satisfied by shortening ρ_2 (that means, because of (22), by lengthening ρ_3). As far as a 4-body system with a planetary hierarchy P is considered, the gravitational potential expansion is given by Walker et al (1980) and an inequality similar to (25) can be written, containing different small parameters depending on the different hierarchy.

When ρ_2 is so small that

$$\frac{Gm_1m_2}{\rho_2} > U(\underline{\rho}(0)) \qquad (26)$$

$U' > 0$ is no more required provided $\rho_2' = 0$; $\underline{\rho}_3, \underline{\rho}_4$ can be rotated at will, with $\rho_2 =$ constant; ρ_3 and ρ_4 can be changed with the condition $M_3\rho_3\rho_3' + M_4\rho_4\rho_4' \geq 0$ (i.e. $I' = 0$ with $\rho_2' = 0$) until when

$$\frac{Gm_4m_2}{\mathbf{r}_{24}} > U(\underline{\rho}(0)) \; ;$$

then, with m_2 and m_4 fixed, m_1 and m_3 can be moved and the hierarchy is definitely broken.

This hierarchy-breaking procedure can be easily understood by considering that the set of collisions on $I_e =$ constant $\neq 0$ is connected for $N > 3$; therefore the sets $U_e >$(very large constant) are also connected.

6. SECULAR PERTURBATIONS ON THE c^2h INTEGRAL DUE TO THE FOURTH BODY

Breaking a 4-body hierarchy requires breaking also the hierarchy of a 3-body subsystem stable according to the topological stability criterium. This means that the c^2h function of the subsystem, which is no more an integral of the motion because of the fourth body perturbation, does change by a significant amount. However, if the hierarchy is very strong, the perturbations will surely act very **slowly**. After all, the Solar System is a hierarchical system and 70% of all the observed 3 and 4-body multiple stellar systems are a close pair with a distant companion or two close pairs at a large distance (Voigt, 1974), i.e. they are strongly hierarchical dynamical systems.

So the relevant question is: how slowly do the perturbations of a fourth body act on a topologically stable 3-body subsystem? Can we estimate the lifetime of the 3-body system against the perturbations of a given fourth body?

Let us consider the case of a double-binary B hierarchy given by (2) and let us restrict for simplicity to the planar case (computations in the planetary P case are similar, although more involved). We want to study the secular time variation $\dot{z}_{23sec}$ of the c^2h "integral" z_{23} of the 3-body subsystem $((m_1, m_2), m_3+m_4)$ (see Figure 3) due to the fact that (m_3+m_4) is not actually a point-mass but a binary.

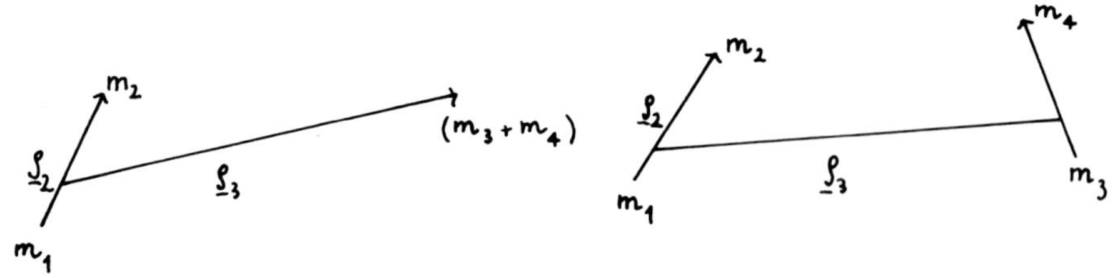

3-body subsystem 4-body system

Figure 3

$z_{23} = c_{23}^2 h_{23}$ can be computed from the angular momentum and energy of the 3-body subsystem, which are given by:

$$h_{23} = M_2 h_2 + M_3 h_3 - R_{23}$$
$$c_{23} = M_2 c_2 + M_3 c_3 \qquad (27)$$

where h_2, c_2 and h_3, c_3 are the energy and angular momentum (per unit mass) of the two binaries (m_1, m_2) with Jacobian vector $\underline{\rho_2}$ and $(m_1 + m_2, m_3 + m_4)$ with Jacobian vector $\boldsymbol{\rho}_3$ assumed as unperturbing each other, while R_{23} is the interaction potential:

$$R_{23} = \frac{G(m_1+m_2)(m_3+m_4)}{\rho_3} \mu_2(1-\mu_2) \left[\frac{\rho_2^2}{\rho_3^2} P_2(\cos\theta_{23}) + 0\left(\frac{\rho_2^3}{\rho_3^3}\right)\right] \qquad (28)$$

so that the relationship between h_3, R_{23} and $\bar{h}_3$ as defined in section 4 is simply

$$\bar{h}_3 = h_3 + R_{23} \qquad (29)$$

To compute $\dot{z}_{23}$ we would like to apply the usual techniques of perturbation theory, but one more difficulty arises because the small parameters with respect to which the perturbing functions can be developed as in (28) should be independent from the dynamical variables $\underline{\rho}_i$, $\dot{\underline{\rho}}_i$. To overcome this difficulty we consider an auxiliary de-hierarchized system in which the initial $\underline{\rho}_3$ has been shortened by a factor $\lambda (0 < \lambda < 1)$ and the masses have been changed in such a way that the ratios $\mu_2(1-\mu_2)$ and $\mu_4(1-\mu_4)$ are divided by a factor $\sigma (0 < \sigma < 1)$. If we assume that the initial 4-body system (i.e. without shortening) is the actual hierarchical dynamical system whose stability we are investigating, and that the de-hierarchized system has a ratio between its Jacobian vectors of the order of 1 and very large μ (i.e. $\mu_2 = \mu_4 \simeq 1/2$ in the de-hierarchized system), then λ is of the order of $\mathrm{Max}(\rho_2/\rho_3, \rho_4/\rho_3)$ and σ is of the order of $\mathrm{Max}\,(\mu_2(1-\mu_2), \mu_4(1-\mu_4))$, ρ_2, ρ_3, ρ_4 and μ_2, μ_4 being the actual ones. Let us call S_{j3} (j=2 or 4) the perturbing functions of the two de-hierarchized 3-body systems with Jacobian radius vectors $\underline{\rho}_2$, $\lambda\underline{\rho}_3$ and $\underline{\rho}_4$ $\lambda\underline{\rho}_3$. Let us also call R_{j3} (j=2 or 4) the perturbing functions of the corresponding 3-body systems where $\underline{\rho}_3$ has not been shortened; R_{23} is given by (28) and R_{43} is obviously analogous. Then, the relationship between R_{j3} and S_{j3} is simply

$$R_{j3} = \lambda^3 \sigma S_{j3} + O(\lambda^4 \sigma) \tag{30}$$

The potential function U of the 4-body system, given by (24), contains also a "mixed term", i.e.:

$$U = \frac{G\, m_1 m_2}{\rho_2} + \frac{G\, m_3 m_4}{\rho_4} + \frac{G(m_1+m_2)(m_3+m_4)}{\rho_3} + R_{23} + R_{43} + R_{24} \tag{31}$$

where R_{24} contains the mixed terms but is of higher order, as can be seen from (24), that is, comparing with the de-hierarchized system:

$$R_{24} = \sigma^2 \lambda^5 S_{24} + O(\sigma^2 \lambda^6) \ . \tag{32}$$

We now compute the time derivative of z_{23}, using e.g. a Poisson bracket formalism with the Hamiltonian of the full 4-body system:

$$h = M_2 h_2 + M_3 h_3 + M_4 h_4 - R_{23} - R_{34} - R_{24}$$

where h_4 is the energy, per unit mass, of the binary (m_3, m_4) with Jacobian radius vector $\underline{\rho}_4$ and reduced mass $M_4 = m_3 m_4/(m_3 + m_4)$. For $\dot{z}_{23}$ (in the planar case) we get:

$$\dot{z}_{23} = \{z_{23}, h\} = c_{23}^2 \{h_{23}, h\} + 2c_{23} h_{23} \{c_{23}, h\} =$$

$$= c_{23}^2 \left\{ n_3 \frac{\partial (R_{43} + R_{24})}{\partial \ell_3} + n_2 \frac{\partial R_{24}}{\partial \ell_2} + \{R_{23}, R_{43} + R_{24}\} \right\} - \tag{33}$$

$$-2h_{23}\, c_{23} \frac{\partial (R_{43} + R_{24})}{\partial g_4}$$

where ℓ_j, g_j ($j=2,3,4$) are the usual angular variables of the binary system with Jacobian radius vector $\underline{\rho}_j$, i.e. ℓ_j are the mean anomalies and g_j the arguments of the pericenters.

Before exploiting (30) and (32) to estimate the orders of magnitude, let us remember that we are interested only in secular effects because we assumed that the initial configuration is such that z_{23} is smaller than the critical value , and therefore short term perturbations do not affect stability (close approaches are impossible before a large enough change in z_{23} takes place). We can therefore apply the usual technique of averaging so that terms containing $\partial/\partial\ell_2$ give zero when averaged over ℓ_2 and terms containing $\partial/\partial\ell_3$ give zero when averaged with respect to ℓ_3 (this is the so-called "Lagrange theorem on the stability of the Solar System"), and therefore only second order terms are left in the long-term evolution of z_{23} . Also the $\partial/\partial g_4$ terms average out whenever $e_4 = 0$ or $e_2 = e_2 = 0$ (e_j, $j=2,3,4$, being the osculating eccentricity of the binary with Jacobian radius vector $\underline{\rho}_j$), i.e. the long term evolution of z_{23} can be described as

$$\dot{z}_{23} \text{ (long term)} \simeq \text{2nd order terms} + O(e_2 e_4) + O(e_3 e_4) \qquad (34)$$

By using the estimates (30) and (32) of the perturbing functions we finally get (see Milani and Nobili, 1983):

$$\dot{z}_{23} \text{ (long term)} \simeq O(\lambda^6\sigma^2) + O(\lambda^3\sigma e_3 e_4) + O(\lambda^5\sigma^2 e_2 e_4). \qquad (35)$$

We can comment formula (35) by saying that it provides a significant order-of-magnitude upper estimate of the time needed to break the stability of the $((m_1,m_2),m_3 + m_4)$ subsystem:

$$\Delta t \gtrsim - \Delta z_{23}/\dot{z}_{23}\text{(long term)} \qquad (36)$$

(where Δz_{23} is z_{23} minus the critical value corresponding to the L_2 equilibrium for the same masses m_1, m_2 and $m_3 + m_4$), provided that also the analogous z_{43} for the $(m_1 + m_2, (m_3, m_4))$ system is controlled in a similar way. However, as usual in perturbation theory, the order of magnitude estimates that depend upon the "principle of the averages" as (34) and (35) hold only in the assumption that no significant reasonance occurs between the three mean motions n_2, n_3 and n_4 .

REFERENCES:

Birkhoff, G.D. : 1927, "Dynamical Systems", AMS, Providence.

Bozis, G.: 1976, "Zero Velocity Surfaces for the General Planar 3-Body Problem", Astroph. and Space Sci., 43, 355-368.

Easton, R.: 1971, "Some Topology of the 3-Body Problem", Journal of Diff. Equations, 10, 371-377.

Easton, R.: 1975, "Some Topology of the n-Body Problems", Journal of Diff. Equations, 19, 258-269.

Evans, D.S.: 1968, "Stars of Higher Multiplicity", Quarterly J.Roy. Astr. Soc. 9, 388.

Farinella, P. and Nobili, A.M. : 1978, "A Simple Explanation of Some Characteristics of the Asteroidal Belt Based on the Restricted 3-Body Problem". Moon and Planets 18, 241-250.

Golubev, V.G. : 1968, "Hill Stability in the Unrestricted 3-Body Problem" Soviet Phys. Dokl. 13, 373.

Marchal, C. : 1974, "Qualitative Study of a N-Body System: a New Condition of Complete Scattering". Astron. Astrophys. 10, 278-289.

Marchal, C. and Saari, D.G.: 1975, "Hill Regions for the General 3-Body Problem", Cel. Mech. 12, 115-129.

Marchal, C. and Bozis, G.: 1982, "Hill Stability and Distance Curves for the General 3-Body Problem". Cel. Mech. 26, 311-333.

Milani, A. and Nobili, A.M.: 1982, "On Topological Stability in the General 3-Body Problem". Cel. Mech., in press.

Milani, A. and Nobili, A.M.: 1983, "On Topological Stability in the 4-Body Systems". In preparation.

Roy, A.E.: 1979, "Empirical Stability Criteria in the Many-Body Problem." In "Instabilities in Dynamical Systems". V. Szebehely, Ed., Reidel, Dordrecht, pp.177-210.

Roy, A.E.: 1982, "The Stability of N-Body Hierarchical Dynamical Systems." In "Applications of Modern Dynamics to Celestial Mechanics and Astrodynamics". V. Szebehely, Ed., Reidel, Dordrecht, pp.103-130.

Smale, S.: 1970a, "Topology and Mechanics, I." Inventiones Math. 10, 305-331.

Smale, S.: 1970b, "Topology and Mechanics, II. "Inventiones Math. 11, 45-64.

Szebehely, V. and Zare, K. : 1977, "Stability of Classical Triplets and of their Hierarchy". Astron. Astrophys. 58, 145-152.

Szebehely, V. and McKenzie, R.: 1977, "Stability of Planetary Systems with Bifurcation Theory". Astronomical Journal 82, 79-83

Tung, C.C.: 1974, "Some Properties of the Classical Integrals of the General Problem of three Bodies". Scientia Sinica, 17, 306-330.

Voigt, H.H. 1974, "Outline of Astronomy" Vol.II, Noordhoff International Publishing, Leyden.

Walker, I.W., Emslie G.A., and Roy, A.E.: 1980, "Stability Criteria in Many Body Systems, I." Celestial Mechanics, 22, 371-402.

Walker, I.W. and Roy, A.E.: 1981, "Stability Criteria in Many Body Systems, II". Celestial Mechanics, 24, 195-225.

Walker, I.W.: 1983, "Stability Criteria in Many Body Systems, IV", Celestical Mechanics, 29, 149.

Zare, K.: 1976, "The Effects of Integrals on the Totality of Solutions of Dynamical Systems". Celestial Mechanics, 14, 73-83

Zare, K.: 1977, "Bifurcation Points in the Planar Problem of Three Bodies". Celestial Mechanics, 16, 35-38.

BOUNDARIES FOR THE EQUIPOTENTIAL CURVES IN THE ELLIPTIC RESTRICTED THREE BODY PROBLEM

Magda Delva
Institut für Astronomie, Universität Graz, Austria

ABSTRACT

In the elliptic restricted three body problem an invariant relation between the velocity square of the third body and its potential is studied for long time intervals as well as for different values of the eccentricity. This relation, corresponding to the Jacobian integral in the circular problem, contains an integral expression which can be estimated if one assumes that the potential of the third body remains finite. Then upper and lower boundaries for the equipotential curves can be derived. For large eccentricities or long time intervals the upper boundary increases, while the lower decreases, which can be interpreted as shrinking respectively growing zero velocity curves around the primaries.

1. INTRODUCTION

In a paper by Szebehely and Giacaglia (1964) the equations of motion for the infinitesimal third body in the elliptic restricted three body problem are presented in the same analytical form as for the circular case. From these equations, they derived an invariant relation for the velocity square, which corresponds to the Jacobian integral in the circular problem. In our paper, using this relation, we search for boundaries for the equipotential curves, which limit the motion of the third body to well defined parts of the plane.

The problem is described in a synodic, pulsating, barycentric coordinate system (ξ,η); the primaries are situated on the ξ axis and their masses are taken to be $(1-\mu)$ und μ (Fig.1). Starting from the usual synodic barycentric coordinate system (ξ^*,η^*), the pulsating system (ξ,η) is obtained, when the variable distance r between the primaries:

$$r = \frac{a(1-e^2)}{1+e\cos f}$$

V. V. Markellos and Y. Kozai (eds.), Dynamical Trapping and Evolution in the Solar System, 317–323.

(a,e and f being the semimajor axis, the eccentricity and the true anomaly of the elliptic orbit) is chosen to be the unit of length.

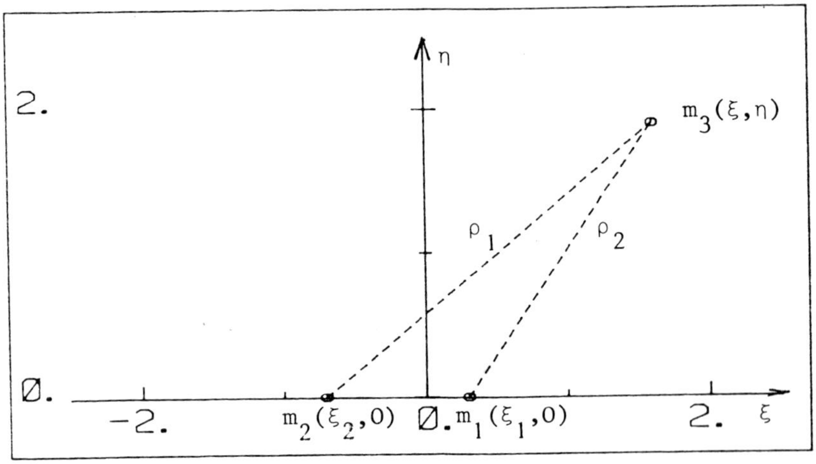

Fig.1.: Synodic, pulsating barycentric coordinate system (ξ,η).

This results in fixed abscissae for these bodies:

$$\begin{aligned} \xi_1 &= \xi_1^*/r = \mu \\ \xi_2 &= \xi_2^*/r = -(1-\mu) \end{aligned} \qquad (1.1)$$

A transformation to the true anomaly as the independent variable leads to the desired form for the equations of motion for the third body (primes denote derivatives with respect to the true anomaly f):

$$\begin{aligned} \xi'' - 2\eta' &= \frac{\partial\omega}{\partial\xi} \\ \eta'' + 2\xi' &= \frac{\partial\omega}{\partial\eta} \end{aligned} \qquad (1.2)$$

The function ω depends on the potential function Ω:

$$\Omega\ (\xi(f),\eta(f)) = \frac{1}{2}\left[(1-\mu)\,\rho_1^{\ 2} + \mu\rho_2^{\ 2}\right] + \frac{1-\mu}{\rho_1} + \frac{\mu}{\rho_2}$$

$$\omega(\xi,\eta,f) = \frac{\Omega(\xi(f),\eta(f))}{1+e\cos f} \qquad (1.3)$$

where ρ_1,ρ_2 are the normed distances from the third body to the primaries:

$$\rho_1^{\ 2} = (\xi-\xi_1)^2 + \eta^2 \qquad (1.4)$$

$$\rho_2^{\ 2} = (\xi-\xi_2)^2 + \eta^2$$

The invariant relation for the velocity square in the pulsating system is given by the formula (Szebehely and Giacaglia, 1964):

$$\xi'^2 + \eta'^2 = 2\omega(\xi,\eta,f) - 2\int\frac{\partial\omega}{\partial f}\,df - C \qquad (1.5)$$

and corresponds to the Jacobian integral for e = 0.

It is shown by Erdi (1982) that it is possible to evaluate the integral in Equation (1.5) for the case of Trojan asteroids using an asymptotic solution for their motion. In a paper by Delva and Dvorak (1979) a series expansion was used to study an equivalent of equation (1.5).

2. ESTIMATION OF THE INTEGRAL $\int\frac{\partial\omega}{\partial f}\,df$

In equation (1.5) for the velocity square, the integral term on the right hand side causes variations of the zero velocity curves, which define the regions of motion. We now estimate boundaries for its value for long intervals of the time. For this purpose it is necessary to restrict the motion of the third body in such a way, that the value of the potential function $\Omega(\xi,\eta)$ on the orbit always remains bounded by an upper limit M(M>0) for all times:

$$\Omega(\xi,\eta) \leq M \qquad (2.1)$$

or we have to exclude collisions with the primaries and very large distances from them. A lower limit for $\Omega(\xi,\eta)$ is the value in the Lagrangian points L_4 and L_5; the inequality

$$m = \Omega(\xi_{L_4},\eta_{L_4}) = \frac{3}{2} \leq \Omega(\xi,\eta) \leq M \qquad (2.2)$$

is valid on the orbit for all times. To be able to estimate the integral

$$I = \int\frac{\partial\omega}{\partial f}\,df = \int\Omega(\xi(f),\eta(f))\,\frac{e\sin f}{(1+e\cos f)^2}\,df \qquad (2.3)$$

we will use the inequality (2.2) and a theorem on integration (Smirnow, (1973):
if two functions $\Omega(\xi(f),\eta(f))$ and g(f) are integrable (and thus bounded) on an interval $[f_a,f_b]$ and if g(f) does not change its sign on the interval, the following inequalities hold for the integral of $(\Omega.g)$:

for $g(f) \geq 0$:

$$m\int_{f_a}^{f_b} g\,df \leq \int_{f_a}^{f_b} \Omega\, g\,df \leq M\int_{f_a}^{f_b} g\,df \qquad (2.4)$$

for $g(f) \leq 0$:

$$M \int_{f_a}^{f_b} g \, df \leq \int_{f_a}^{f_b} \Omega \, g \, df \leq m \int_{f_a}^{f_b} g \, df$$

Since we are interested in long time intervals or long intervals of the true anomaly, we choose the integration interval to be an integer multiple of 2π:

$$f_a = f_o = 0 \tag{2.5a}$$

$$f_b = f_{2n} = 2n\pi \quad , \quad n \in N$$

and a decomposition of the interval:

$$f_k = k\pi \quad , \qquad k = 0,\ldots,2n$$

$$f_o = 0 \leq f_1 \leq \ldots \leq f_k \leq \ldots \leq f_{2n} \tag{2.5b}$$

Defining the function $g(f)$ by the equation

$$g(f) = \frac{e \sin f}{(1+e \cos f)^2} \tag{2.6}$$

it is positive or zero on all intervals $[f_{2j},f_{2j+1}]$ and negative or zero on $[f_{2j+1},f_{2j+2}]$, $j=0,\ldots,n-1$. Calculating its integral on these intervals and using (2.4), we find the inequalities:

$$m \frac{2e}{1-e^2} \leq \int_{f_{2j}}^{f_{2j+1}} \Omega g \, df \leq M \frac{2e}{1-e^2} \tag{2.7a}$$

$$-M \frac{2e}{1-e^2} \leq \int_{f_{2j+1}}^{f_{2j+2}} \Omega \, g \, df \leq -m \frac{2e}{1-e^2} \tag{2.7b}$$

Summation then gives for the total integral I (2.3):

$$-n(M-m) \frac{2e}{1-e^2} \leq I \leq n(M-m) \frac{2e}{1-e^2} \tag{2.8}$$

3. BOUNDARIES FOR THE EQUIPOTENTIAL CURVES

The result (2.8) is used to study the possible changes of the zero velocity curves with changing eccentricity and length of the integration interval. The curves are defined by the equation

$$F(\xi,\eta,f) = 2\omega - 2I - C = 0 \tag{3.1}$$

or by the condition that

$$G(\xi,\eta,f) = 2\left\{\frac{1}{2}\left[(1-\mu)\rho_1{}^2 + \mu\rho_2{}^2\right] + \frac{1-\mu}{\rho_1} + \frac{\mu}{\rho_2}\right\} = (2I+C)\ (1+e\cos f) \tag{3.2}$$

The relation holds and is meaningful for any fixed value of f. From the inequality (2.8) and estimating the factor (1+e cos f), the curves $G(\xi,\eta,f)$ can vary within the following limits:

$$C - eC - \frac{4en(M-m)}{1+e} \leq G(\xi,\eta,f) \leq C + e\,C + \frac{4en(M-m)}{1-e} \tag{3.3}$$

For values of the lower boundary smaller than $2\Omega\,(\xi_{L_4},\eta_{L_4}) = 3$ no solutions for (3.2) are found, no equipotential curve will limit the motion of the third body. For any value of both boundaries greater than $2\Omega(\xi_{L_4},\eta_{L_4})$, zero velocity curves exist and define forbidden regions for the motion (Fig.2)

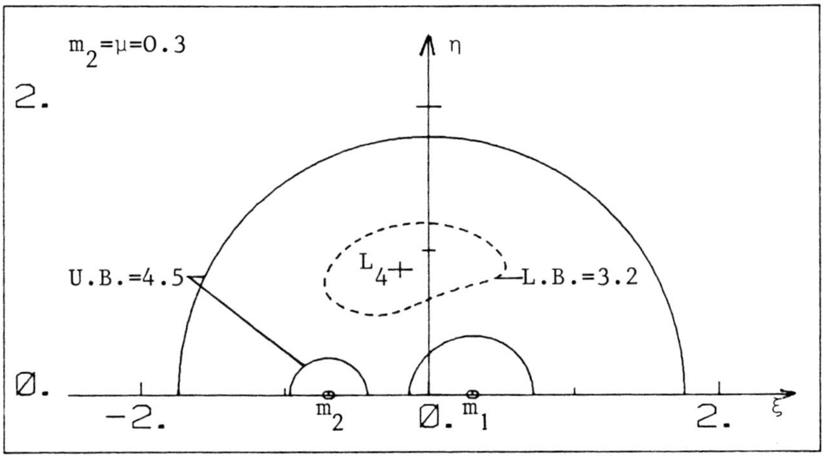

Fig.2: Equipotential curves for the lower and upper boundary of $G(\xi,\eta,f)$, if greater then $2\Omega(\xi_{L_4},\eta_{L_4})$.

The constant C is calculated from the moment with f = o:

$$C = \frac{2\Omega(\xi(0),\eta(0))}{1+e} - [\xi'(0)^2 + \eta'(0)^2] \tag{3.4}$$

3.A Boundaries for various values of the eccentricity e

In the case e=0 the inequality (3.3) reduces to the known form
$G(\xi,\eta) = C$

which defines the regions of motion once for all values of f.

In the case $e \neq 0$ and $C>0$, the lower boundary decreases with increasing values of e, while the upper one increases. For large values of $e (e<1)$ the equipotential curves $G(\xi,\eta,f)$ can vary within broad limits, defining small or no forbidden regions for the lower and large ones, becoming more and more closed around the primaries, for the upper boundary (Fig.3). It is clear that, if the body trespasses the restriction of inequality (2.1), these regions of motion are not valid and no predictions on their long time behaviour are allowed.

We conclude that from a theoretical point of view and within the restrictions (2.1) for the motion, the larger the eccentricity is, the larger variations in the regions of motion may occur, permitting more possibilities for the orbit of the third body.

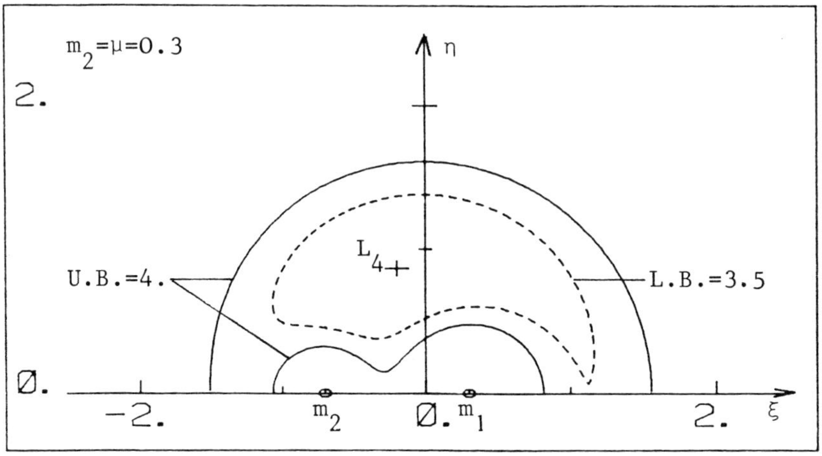

Fig.3a: Zero velocity curves for small e.

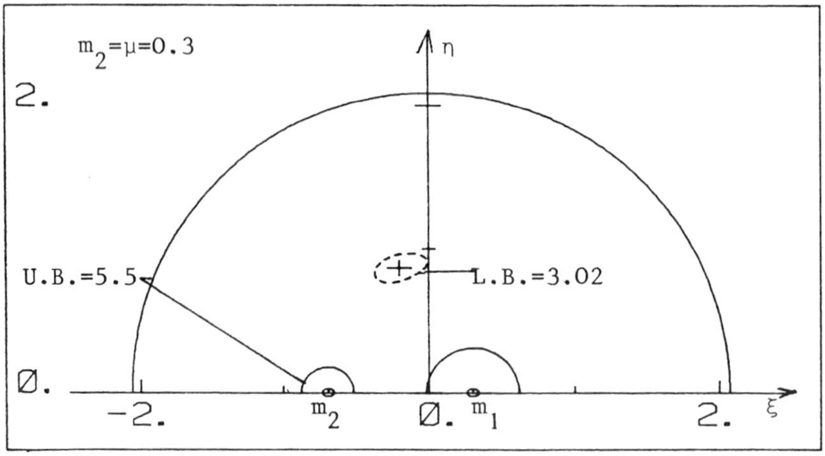

Fig.3b: Zero velocity curves for large e.

3.B Boundaries for long time intervals

Since we chose the integration interval to have the length $n(2\pi)$, we consider the boundaries (3.3) for large values of the integer n. It is easily seen, that increasing the value of n causes the same effect as increasing the eccentricity e. The situation for long time intervals corresponds therefore to the one in Fig.3b: long time intervals can allow more possibilities of variation for the equipotential curves.

REFERENCES

Delva,M. and Dvorak,R.: 1979, Astron.Astrophys. 77, pp. 252-254.

Erdi,B.: 1982, in "Dynamical Astronomy", B.A.Balázs and V.Szebehely (eds.) Budapest, pp. 17-36.

Smirnow,W.I.: 1973, Lehrgang der höheren Mathematik, V.E.B.Berlin, Vol.I, p. 247.

Szebehely,V. and Giacaglia, G.E.O.: 1964, Astron.J. 69, pp. 230-235.

CAPTURE/ESCAPE BOUNDARY IN THE COLLINEAR RESTRICTED THREE-BODY PROBLEM

Gerard Gómez and Jaume Llibre

Secció de Matemàtiques, Facultat de Ciències,
Universitat Autònoma de Barcelona, Bellaterra,
Barcelona, Spain.

ABSTRACT. We study the cantorian structure of the successive intersections of the invariant manifolds of infinity (parabolic orbits) with a certain surface of section. The first of these intersections is computed numerically. The structure of the set of orbits of capture or escape after n binary collisions is given.

1. FORMULATION OF THE PROBLEM. EQUATIONS OF MOTION.

Two bodies (called primaries) of point masses m_1 and m_2 are moving in an elliptic collision orbit under the influence of their mutual gravitational attraction and a third body of mass $m_3 \approx 0$ (attracted by the previous two but not influencing their motion) moves in the line defined by the two primaries. The collinear restricted three body problem is to describe the motion of this third body.

We select units of length, time and mass such that the length of the major axis of the collision elliptic orbit equals 2, the period is 2π and $m_1 = m$, $m_2 = 1 - m$ with $m \in (0,1)$. Units are taken in such a way that the gravitation constant equals 1. We take the center of masses at the origin.

Let $-x_1, x_2$ ($x_i \geqslant 0$) be the coordinates of m_1, m_2, respectively. Then the motion of the two primaries is given by

$$x_1 = (1-m)(1-\cos E) \;, \qquad x_2 = m(1-\cos E) \;, \tag{1}$$

with $t = E - \sin E$. The parameter E is the so called eccentric anomaly and the origin of time is taken at a collision between m_1 and m_2. We remark that t, E are defined modulus 2π and from now on this will be understood without explicit mention each time that t or E appears.

Let $x \geqslant 0$ be the coordinate of the third body. We assume $x \geqslant x_2$ (see Figure 1). The equation of motion of the third body is:

V. V. Markellos and Y. Kozai (eds.), Dynamical Trapping and Evolution in the Solar System, 325–338.

$$d^2x/dt^2 = (1-m)/(x-x_2)^2 - m/(x+x_1)^2 , \qquad (2)$$

with x_1, x_2 given by (1).

We have a singularity in (2) when $x = x_2$. Furthermore, if $x_2 = 0$ we obtain a triple collision because $x_1 = 0$, too. If $x_2 > 0$ we encounter a binary collision between m_2 and m_3. A change of variables will regularize such last collision as usual. In what follows we will not consider triple collision orbits.

Let $h_{123} = \dot{x}^2/2 - (1-m)/(x-x_2) - m/(x+x_1)$ be the energy per unit mass of the third body, and $h_{23} = (\dot{x}-\dot{x}_2)^2/2 - (1-m)/(x-x_2)$ the energy associated with the binary m_2,m_3. We relate the motion close to collision m_2,m_3 with the motion of the third body near infinity. As h_{123} does not have a finite limit when $x - x_2 \to 0$ and h_{23} is not suitable when $x \to \infty$, we define $h = (x^r - x_2^r)h_{123}/x^r + x_2^r h_{23}/x^r$ with $r \geqslant 1$. Using the behaviour of bodies near collision (Siegel-Moser, pp30), we see that h is well defined along solutions of (2) (if and only if $r \geqslant 1$) and that $h \to h_{23}$ if $x - x_2 \to 0$, and $h \to h_{123}$ if $x \to \infty$.

We scale the time in order to regularize binary collisions m_1,m_2 and m_2,m_3 introducing the s variable (Stiefel-Scheifele, pp20)

$$dt = (x - x_2)(1 - \cos E)ds .$$

Then (2) becomes

$$\frac{dx}{ds} = \left\{\frac{x_2^r\dot{x}_2}{x^r} + \left[\left(\frac{x_2^r\dot{x}_2}{x^r}\right)^2 + 2\left(\frac{x^r - x_2^r}{x^r}\,\frac{m}{x+x_1} + \frac{1-m}{x-x_2} + h - \frac{x_2^r\dot{x}_2^2}{2x^r}\right)\right]^{1/2}\right\}\frac{dt}{ds} ,$$

$$\frac{dh}{ds} = \frac{x^r - x_2^r}{x^r}\,\frac{dh_{123}}{ds} + \frac{x_2^r}{x^r}\,\frac{dh_{23}}{ds} + (h_{23} - h_{123})\,\frac{d}{ds}\left(\frac{x_2^r}{x^r}\right) , \qquad (3)$$

$$\frac{dt}{ds} = (x - x_2)(1 - \cos E) ,$$

where

$$dh_{123}/ds = (m\dot{x}_1/(x+x_1)^2 - (1-m)\dot{x}_2/(x-x_2)^2)\cdot dt/ds ,$$

$$dh_{23}/ds = (\dot{x} - \dot{x}_2)(-m/(x+x_1)^2 - \ddot{x}_2)\cdot dt/ds ,$$

$$d/ds(x_2/x)^r = r x_2^{r-1}(\dot{x}_2 x - x_2\dot{x})/x^{r+1}\cdot dt/ds ,$$

$$h_{23} - h_{123} = -\dot{x}\dot{x}_2 + \dot{x}_2^2/2 + m/(x+x_1) ,$$

$$\dot{x}_1 = (1-m)\sin E/(1-\cos E) ,$$

$$\dot{x}_2 = m\sin E/(1-\cos E) ,$$

$$\ddot{x}_1 = -(1-m)/(1-\cos E)^2 \,,$$

$$\ddot{x}_2 = -m/(1-\cos E)^2 \,.$$

We remark that $r \geqslant 1$ is enough to make dx/ds and $(x^r - x_2^r)x^{-r} dh_{123}/ds$ regular, but values $r \geqslant 2$ and $r \geqslant 3$ are necessary for $x_2^r x^{-r} dh_{23}/ds$ and $(h_{23} - h_{123})\, d/ds(x_2/x)^r$, respectively. From now on we take $r = 3$.

From (3) it follows that the motion of m_3 is determined by four initial conditions

$$s = s_0 \,, \quad (x - x_2)(s_0) = 0 \,, \quad h(s_0) = h_0 \,, \quad t(s_0) = t_0 \,.$$

Note that every orbit of m_3 has at least one collision m_2, m_3.

Due to the autonomous character of the equations the value of s_0 is irrelevant. Let us take $s_0 = 0$. Then $h(0) = h_0$, $t(0) = t_0$, or equivalently $E(0) = E_0$, are enough to determine the motion if we assume that bodies m_2, m_3 are at collision. Then we have:

LEMMA 1 (see Proposition 2.1 of Llibre-Simó, 1980a). Orbits of the third body excluding triple collision are determined by points of a cylinder K excluding one generatrix (see Figure 2).

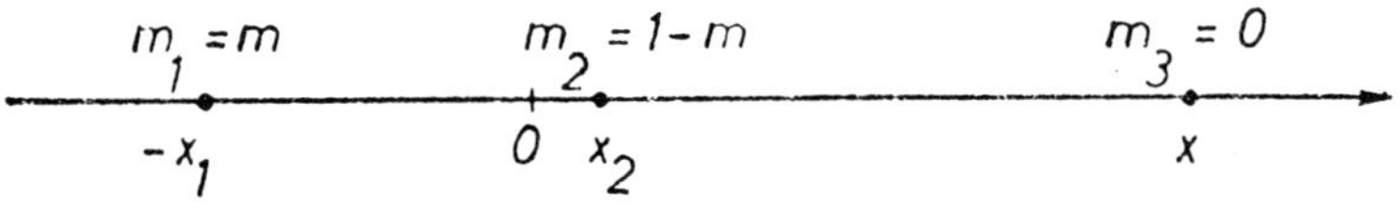

Figure 1.

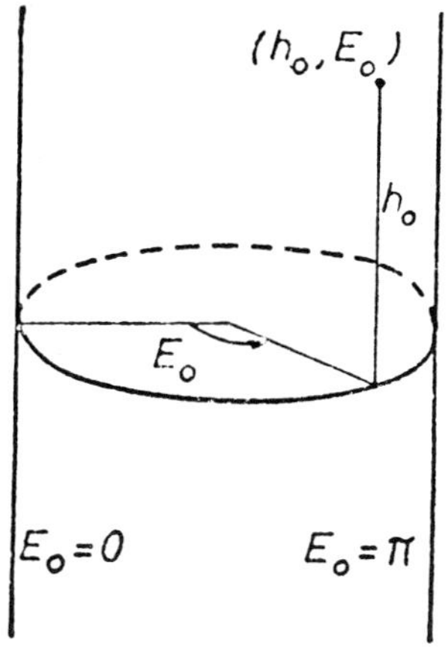

Figure 2.

We remark that more than one point can give the same orbit. The generatrix $E_0 = 0$ which is related to triple collision is excluded.

We define a mapping in a subset D_1 of K, $f: D_1 \longrightarrow K$ in the following way: Given a point $(h_0, E_0) \in K$ and $s_0 = 0$ let us consider the orbit defined by (h_0, E_0). We have that $(x - x_2)(s_0) = 0$. Let s_1 be the next zero of $x - x_2$ for increasing values of s, if it exists. Then, if $h(s_1) = h_1$ and $E(s_1) = E_1$, we define $f(h_0, E_0) = (h_1, E_1)$. We call f the *Poincaré map* of the collinear restricted three body problem associated with K.

The set D_1 is associated with the orbits of m_3 which collide again with m_2. We call D_1 the set of *elliptic orbits* for increasing time. It is clear that D_1 is open. Now we study the complementary of D_1. If f is not defined on the point (h_0, E_0) as $\ddot{x}(t) < 0$ for all $t > t_0$, the function $\dot{x}(t)$ is monotonically decreasing. There exist $\lim_{t \to \infty} \dot{x}(t) = \dot{x}(+\infty) \geq 0$ and $\lim_{t \to \infty} x(t) = +\infty$. We say (h_0, E_0) is *hyperbolic (parabolic)* for $t \to +\infty$ if $\dot{x}(+\infty) > 0$ ($\dot{x}(+\infty) = 0$). The set of hyperbolic (parabolic) initial conditions for $t \to +\infty$ in K will be called H_1 (P_1). The set H_1 is open. We call D_{-1} the subset of K where f^{-1} is defined, we have $D_{-1} = f(D_1)$. D_{-1} is called the set of elliptic orbits for decreasing time. We define H_{-1} (P_{-1}) as hyperbolic (parabolic) orbits for $t \to -\infty$ in a natural way.

We remark that equation (2) is invariant with respect to the symmetry $(x, \dot{x}, t) \to (x, -\dot{x}, -t)$. This symmetry on K is given by $S(h_0, E_0) = (h_0, 2\pi - E_0)$. Then we have $f^{-1} = S^{-1} \circ f \circ S$ and hence $D_{-1} = S(D_1)$, $P_{-1} = S(P_1)$, $H_{-1} = S(H_1)$.

2. THE INTEGRABLE CASE $m = 0$.

The system (2) for $m = 0$ reduces to a rectilinear two body problem. The energy $h = h_{123} = h_{23} = \dot{x}^2/2 - 1/x$ is constant for each orbit and the solutions such that for a given E_0 verify $(x - x_2)(E_0) = 0$, are associated with elliptic, parabolic, hyperbolic collision orbits according as the energy h is negative, zero or positive.

Furthermore the cylinder K is divided by the circle $h_0 = 0$ (parabolic orbits) in two regions: elliptic orbits ($h_0 < 0$) and hyperbolic ones ($h_0 > 0$).

Now the Poincaré map f is given by:

$$f(h_0, E_0) = (h_1, E_1) \ , \tag{4}$$

with

$$h_1 = h_0 ,$$
$$E_1 = E_0 + E(2\pi(-2h_0)^{-3/2}) ,$$

where $E(t)$ is the solution of the equation $t = E - \sin E$. It is clear that f is defined only for values of $h_0 < 0$. Therefore, the domain of definition D_1 is a semi-cylinder and $D_1 = D_{-1} = f(D_1)$. Here D_1 is the circle $h_0 = 0$ of K and $P_1 = P_{-1}$, of course the semi-cylinder $h_0 > 0$ is $H_1 = H_{-1}$.

We look at D_1 as an annulus of radius $h_0 \in (-\infty, 0)$. Then the Poincaré map on D_1 is a twist map, which means that the concentric circles $h_0 =$ constant < 0 are rotated by an angle $E(2\pi(-2h_0)^{-3/2})$ which tends increasing to ∞ as h_0 approaches 0. Thus the image of any radius is a curve spiralling infinitely about the origin as it approaches $h_0 = 0$.

3. THE NONINTEGRABLE CASE $m > 0$

In order to study P_1 and P_{-1} when $m > 0$ we need to consider the flow in the neighborhood of ∞. Following McGehee we introduce the variables

$$x = 2/q^2 , \quad \dot{x} = -p , \quad dt = (4/q^3)dw ,$$

with $0 < q < \infty$. Then (2) becomes

$$\begin{aligned} dq/dw &= p , \\ dp/dw &= [(1-m)/(1-q^2x_2/2)^2 + m/(1+q^2x_1/2)^2] q , \\ dt/dw &= 4/q^3 . \end{aligned} \tag{5}$$

In (McGehee) it was proved that in the infinity there is an hyperbolic periodic orbit whose invariant manifolds are P_1 (stable) and P_{-1} (unstable).

The symmetry of the problem in these new coordinates is given by $(q,p,t,w) \longrightarrow (q,-p,-t,-w)$. Then if the manifold P_{-1} has the expression $q = F(p,t)$, the P_1 is $q = F(-p,-t)$. In (Llibre-Simó, 1980a) it has been shown that $F(p,t)$ is 2π-periodic in t and is not analytic in t but in E. Moreover

$$F(p,t) = \sum_{n \geq 0} a_n(t)p^n , \tag{6}$$

with

$$\begin{aligned} a_0 &= a_2 = a_3 = a_4 = a_6 = 0 , \\ a_1 &= 1 , \\ a_5 &= -5m(1-m)/16 , \\ a_7 &= 35m(1-m)(1-2m)/128 , \\ a_8 &= -3m(1-m)(-5t/2 + \textstyle\int_0^t (1-\cos E)^3 dE) , \\ &\cdots\cdots \end{aligned} \tag{7}$$

Using the above results we have:

LEMMA 2 (see Proposition 3.2 of Llibre-Simó, 1980a). If $m > 0$ is sufficiently small the set P_1 of parabolic initial conditions for $t \to +\infty$ in K is a simple closed curve (if we add the generatrix $E_0 = 0$ to K) which divides K in two components D_1 and H_1.

The geometry of the curves P_1 and P_{-1} will be studied numerically in the next section.

LEMMA 3. If $m > 0$ is sufficiently small, let $\gamma = \{(h_0,E_0) : h_0 = h_0(s), E_0 = E_0(s), s \in [0,1]\}$ be an arc having in P_1 the point belonging to $s = 0$. We suppose that at this point γ and P_1 are not tangent. Then $f(\gamma) = \{(h_1,E_1) : h_1 = h_1(s), E_1 = E_1(s), s \in [0,1]\}$ approaches P_{-1} spiraling, i.e. $E_1(s) \to \infty$ when $s \to 0$.

The proof of this lemma is obtained taking into account that near the periodic orbit at infinity the equations of motion are approximately the same as in the Sitnikov problem (see Moser).

LEMMA 4 (see Theorem 4.1 of Llibre-Simó, 1980a). If $m > 0$ is small enough then $P_1 \neq P_{-1}$ and they intersect at a non-tangential homoclinic point p on $E_0 = \pi$.

Using this fact the Bernoulli shift can be embedded as a subsystem of the Poincaré map f in a neighborhood of the homoclinic point p (for more details see Llibre-Simó, 1980a).

4. NUMERICAL STUDY OF P_1 and P_{-1}

In this section we shall describe the geometry of the curve P_{-1} for values of $m \in (0,1)$, see Figures 3, 4 and 5.

Since $P_1 = S(P_{-1})$, $S(h_0,E_0) = (h_0, 2\pi - E_0)$ and the curve P_{-1} is increasing in E_1, we have that $P_1 \cap P_{-1}$ reduces to an unique point, the homoclinic point p of Lemma 4. Moreover, we compute the angle β of the intersection of P_1 and P_{-1} at p. These values together with the value of the energy at the intersection point are given in Table I.

In Figures 6 and 7 we have plotted the angle β in front of the mass ratio m. As it is clear from Figure 7 this angle varies linearly with m, if m is sufficiently small. This is due to the analytical dependence of the angle β with respect m. The coefficient of m in the development of β in power series of m was computed analytically in Llibre-Simó, 1980a. Its numerical value is $\tan\beta = 1.19...m + O(m^2)$.

In short we have numerical evidence of the following:

CONJECTURE. For all value of $m \in (0,1)$ the intersection of P_1 with P_{-1} reduces to an unique transversal homoclinic point.

In order to compute P_{-1}, for the different values of m, we start

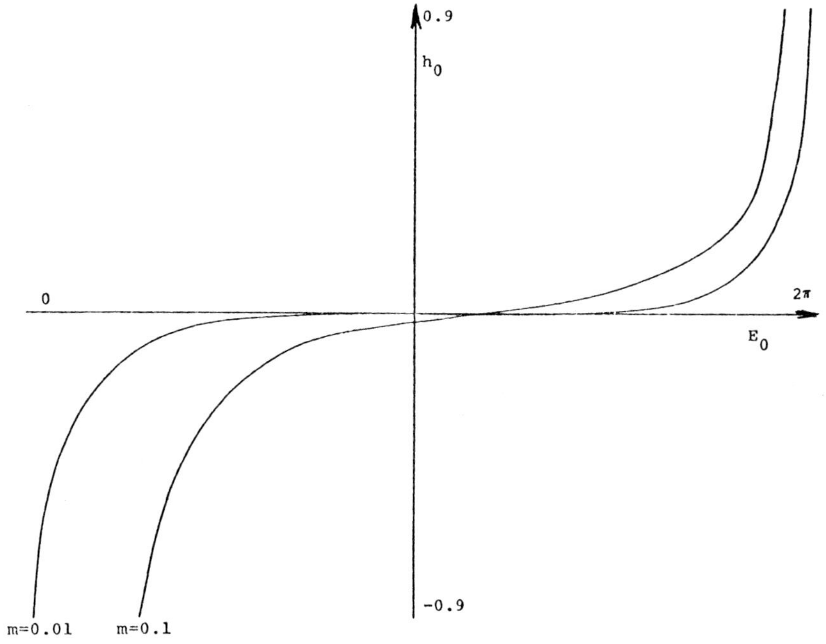

Figure 3.

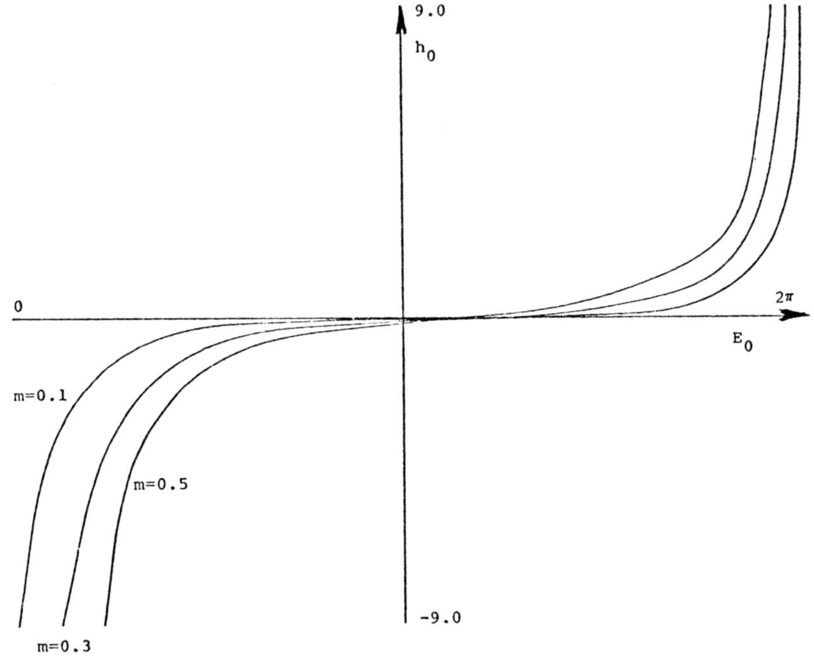

Figure 4.

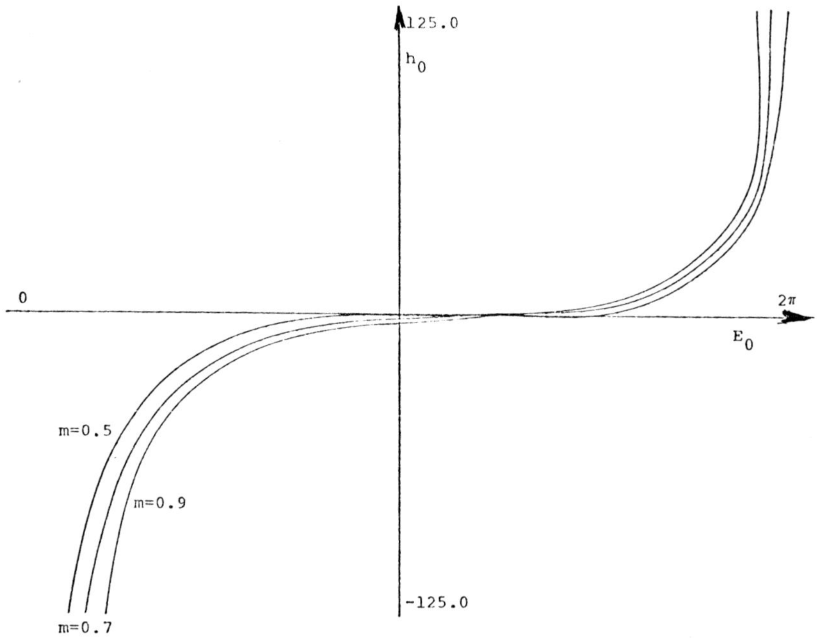

Figure 5.

m	E_0	$\beta/2$
.0001	$-2.9450 \cdot 10^{-5}$	$5.97 \cdot 10^{-5}$
.0002	$-5.8903 \cdot 10^{-5}$	$1.19 \cdot 10^{-4}$
.0003	$-8.8331 \cdot 10^{-5}$	$1.77 \cdot 10^{-4}$
.0004	$-1.1771 \cdot 10^{-4}$	$2.39 \cdot 10^{-4}$
.0005	$-1.4717 \cdot 10^{-4}$	$2.98 \cdot 10^{-4}$
.001	$-2.9500 \cdot 10^{-4}$	$5.67 \cdot 10^{-4}$
.005	$-1.4767 \cdot 10^{-3}$	$2.87 \cdot 10^{-3}$
.01	$-2.9570 \cdot 10^{-3}$	$5.67 \cdot 10^{-3}$
.05	$-1.5021 \cdot 10^{-2}$	$2.89 \cdot 10^{-2}$
.1	$-3.0661 \cdot 10^{-2}$	$5.64 \cdot 10^{-2}$
.2	$-6.3891 \cdot 10^{-2}$	0.1106
.3	$-9.9873 \cdot 10^{-2}$	0.1673
.4	-0.1388	0.2202
.5	-0.1812	0.2700
.6	-0.2276	0.3176
.7	-0.2787	0.3616
.8	-0.3364	0.4022
.9	-0.4040	0.4380

Table I.

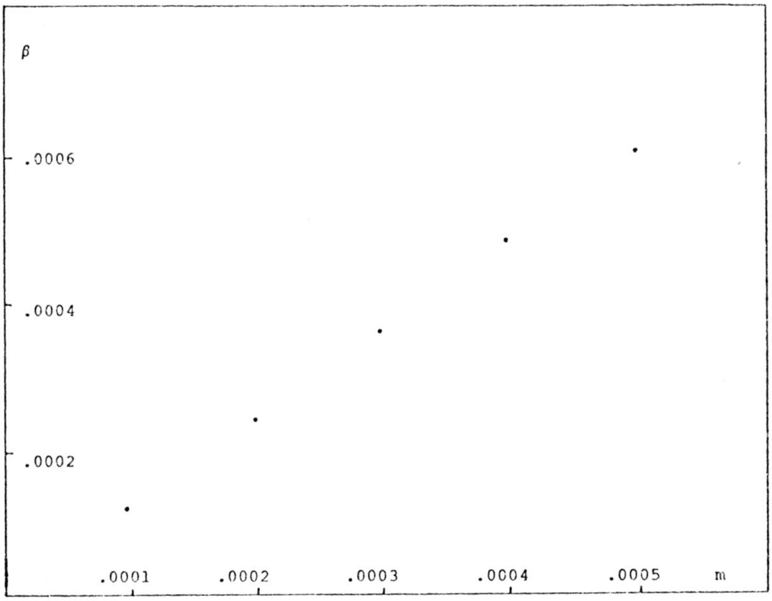

Figure 6.

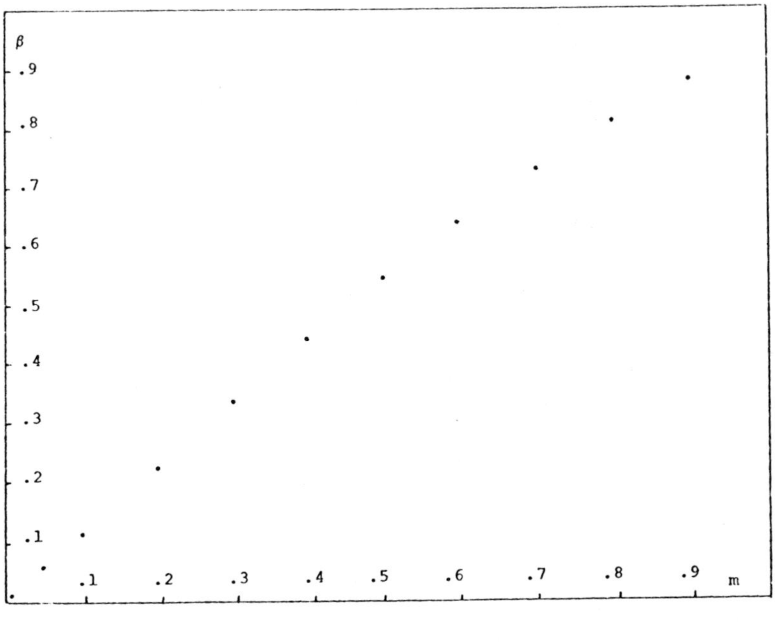

Figure 7.

with initial conditions on the manifold of infinity given by (6) and (7) retaining terms up to the seventh order in p and taking a starting value of p = 0.01 .

At the beginning, we integrate the differential system (5) until the singularity due to the binary collision m_2, m_3 makes regularization necessary. At that point we change the set of variables (q,p,t,w) to another one (u, h_{23}, t, s) which is obtained by applying a Levi-Civita transformation to the system m_2, m_3. The differential equations in this new set of variables are

$$\frac{d^2u}{ds^2} = \frac{h_{23}}{2} u - \frac{mu^3}{2(u^2+1-\cos E)^2} + \frac{mu^3}{(1-\cos E)^2},$$

$$\frac{dh_{23}}{ds} = -2u \frac{du}{ds}\left[\frac{m}{(u^2+1-\cos E)^2} - \frac{m}{(1-\cos E)^2}\right],$$

$$\frac{dt}{ds} = u^2 .$$

The integration is done with a Runge-Kutta-Fehlberg, Rk78, routine with local error of 10^{-13}, 10^{-15}.

5. FINAL EVOLUTIONS

We call final evolutions of the collinear restricted three body problem the behavior of the infinitesimal body when time tends to $\pm\infty$. There are eight types of final evolutions. Parabolic (hyperbolic) evolutions, P or P_- (H or H_-) in which the infinitesimal body reaches infinity with zero (positive) velocity as time tends to $+\infty$ or $-\infty$, respectively. Lagrange evolutions L or L_- in which the orbit of the infinitesimal body remains bounded as time tends to $+\infty$ or $-\infty$, respectively. Finally, oscillatory evolutions OS are characterized by the fact $\limsup_{t\to+\infty} x(t) = +\infty$ and $\liminf_{t\to+\infty} x(t) < +\infty$. Similarly, we have oscillatory evolutions OS_-.

When we consider the two final evolutions associated to one orbit, there are sixteen possibilities. For example, $H_- \cap P$ means that as $t\to -\infty$ the evolution is of hyperbolic type and as $t\to +\infty$ it is parabolic.

In (Llibre-Simó,1980a) it is shown that in a neighborhood of the homoclinic point p there are points with associated orbits of all the possible sixteen types of final evolutions.

We shall say that a point $q \in K$ gives rise to a parabolic orbit for f^n (f^{-n}), and we shall say that this orbit is of type P_n (P_{-n}) if it goes parabolically to infinity when $t \to +\infty$ $(-\infty)$ after $n-1$ binary collisions m_2, m_3. In a similar way we shall define hyperbolic orbits for f^n (f^{-n}) and we shall denote them by H_n (H_{-n}). We define the set

C_n (C_{-n}) as those points $x \in K$ such that $f^n(x)$ ($f^{-n}(x)$) is a point of triple collision, i.e. a point of K on the generatrix $E = 0$.

As it follows from Section 2, all the final evolutions for $m = 0$ are given in Table II. In order to study all the kinds of final evolutions for $m > 0$ we consider the successive intersections of the invariant manifolds of the infinity, formed by the orbits of P_1 and P_{-1} with the surface of section $x - x_2 = 0$.

Let γ be an arc contained in D_1 and having an ending point on P_1, see Figure 8. By using Lemma 3, the domain of definition of f^2 restricted to γ is obtained taking out of γ infinite closed intervals I_i such that $f(I_i) \not\subset D_1$ (see Figures 8 and 9). It is clear that these intervals accumulate at P_1. According with our notation we shall call D_n (D_{-n}) the domain

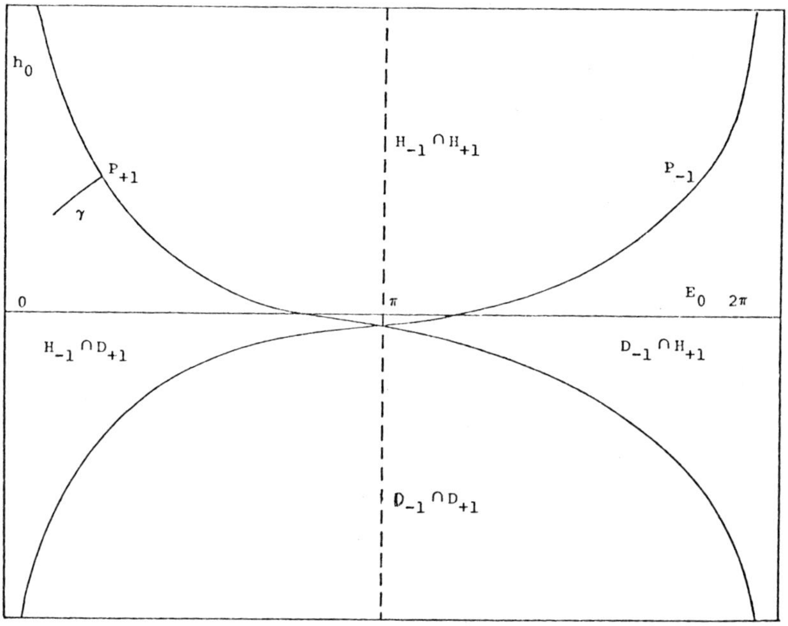

Figure 8.

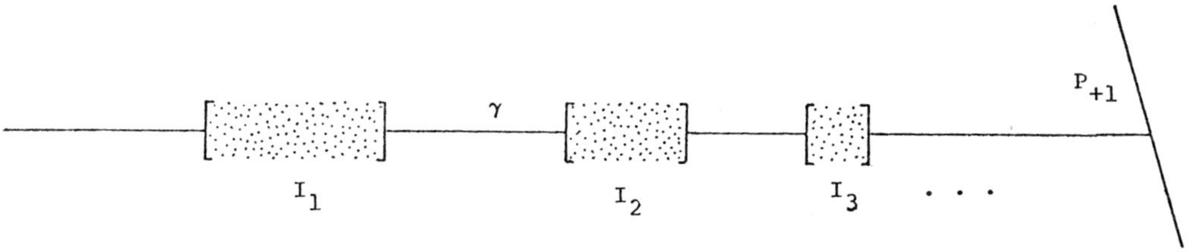

Figure 9.

of definition of f^n (f^{-n}), recall that $D_{-n} = S(D_n)$. By continuity the domain D_2 is D_1 minus an infinite set of bands B_1, B_2,... which cut the arc γ at the closed intervals I_1, I_2, ..., respectively. These bands are the antiimage by f of $D_{-1} - D_1$.

Each one of these bands has two boundary curves. The points of the one that is closer to P_1 correspond to points of C_2 and the other one correspond to points of P_2. The interior of a band is formed by points of H_2. By using the symmetry S we obtain the sets D_{-2}, C_{-2}, P_{-2} and H_{-2}.

Again, to obtain the domain of definition D_3 of f^3 on γ, we must exclude not only

$$\bigcup_{i=1}^{\infty} I_i$$

but

$$\bigcup_{i=1}^{\infty} \bigcup_{j=1}^{\infty} I_{ij}$$

where each I_{ij} is a closed interval. The intervals I_{ij} as varying j accumulate to the interval I_i as it is shown in Figure 10. By continuity the domain D_3 is D_2 minus an infinite set of bands $B_{11}, B_{12}, \ldots, B_1, B_{21}, B_{22}, \ldots, B_2, B_{31}, B_{32}, \ldots$ which intersect the arc γ at the closed intervals $I_{11}, I_{12}, \ldots, I_1, I_{21}, I_{22}, \ldots, I_2, I_{31}, I_{32}, \ldots$ It is clear that $\bigcap_{n \geqslant 0} D_n$ is a cantorian set.

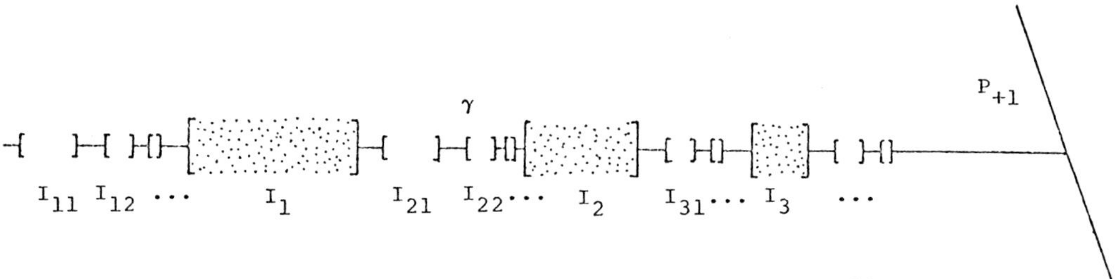

Figure 10.

In short, we have Table II for $m > 0$.

Remark. In (Serra) it has been proved: (1) that for $m = 0$ the set of initial conditions on K which give rise to triple collision orbits is dense in the set of bounded orbits, (2) for any $\delta > 0$ there exists a $m_0 > 0$ such that if the mass parameter $m \in [0, m_0]$, the set of bounded orbits which are not contained in the closure of the set of triple collision orbits has Lebesgue measure less than δ.

The set of orbits with final evolution of type	is homeomorphic to	
	$m = 0$	$m \in (0,1)$
$P_{\pm 1}$	open interval	open interval
$P_{\pm n}$, $n \geq 2$	empty	countable union of disjoint open intervals
$H_{\pm 1}$	open 2-dimensional set	open 2-dimensional set
$H_{\pm n}$, $n \geq 2$	empty	countable union of disjoint open 2-dimensional sets
$P_{-1} \cap P_{+1}$	open interval	the point p
$P_{-n} \cap P_{+m}$, $n+m \geq 3$	empty	countable set of points
$P_{-n} \cap H_{+m}$ or $H_{-n} \cap P_{+m}$ with $n,m \geq 1$	empty	countable union of disjoint open intervals
$H_{-1} \cap H_{+1}$	open 2-dimensional set	open 2-dimensional set
$H_{-n} \cap H_{+m}$, $n+m \geq 3$	empty	countable union of disjoint open 2-dimensional sets
$C_{-1} \cap P_{+1}$ or $P_{-1} \cap C_{+1}$	one point	one point
$C_{-n} \cap P_{+m}$ or $P_{-n} \cap C_{+m}$ with $n+m \geq 3$	empty	countable set of points
$C_{-1} \cap H_{+1}$ or $H_{-1} \cap C_{+1}$	open interval	open interval
$C_{-n} \cap H_{+m}$ or $H_{-n} \cap C_{+m}$ with $n+m \geq 3$	empty	countable union of disjoint open intervals

Table II.

REFERENCES

Llibre, J. and Simó, C. (1980a), J. Differential Equations 37, p.444.

Llibre, J. and Simó, C. (1980b), Math. Ann. 248, p.153.

Llibre, J. and Simó, C. (1980c), Publicacions Mat. Universitat Autònoma de Barcelona 18, p.49.

McGehee, R. (1973), J. Differential Equations 14, p.70.

Moser, J.K. (1973), "Stable and random motions in dynamical systems", Princeton Univ. Press, Princeton, N.J.

Serra, R. (1981), to appear in "Actas del IV Congreso de Ecuaciones Diferenciales y Aplicaciones",Sevilla.

Siegel, C.L. and Moser, J.K. (1971), "Lectures on Celestial Mechanics", Springer-Verlag, Berlin/New York.

Stiefel, E.L. and Scheifele, G. (1971), "Linear and regular Celestial Mechanics", Springer-Verlag, Berlin/New York.

DOUBLY ASYMPTOTIC ORBITS AT THE UNSTABLE EQUILIBRIUM IN THE ELLIPTIC RESTRICTED PROBLEM

E. Perdios
University of Patras, Patras, Greece

ABSTRACT

Outgoing asymptotic orbits at a collinear Equilibrium point of the elliptic restricted three-body problem are constructed analytically and numerically and those which intersect perpendicularly the axis of symmetry are selected by means of a numerical procedure of differential corrections. Due to the symmetry properties of the problem, the latter orbits terminate asymptotically back to the equilibrium point and therefore provide a kind of "asymptotic trapping" at the unstable equilibrium, a trapping mechanism based on "long duration of passage" rather than stability of motion.

1. LINEARIZED EQUATIONS

The linearized equations of motion of the elliptic planar restricted three-body problem, valid for infinitesimal motion near a collinear equilibrium point, are

$$\dot{\underline{x}} = A\underline{x}\ , \qquad (1)$$

with

$$\underline{x} = (x_1, x_2, x_3, x_4)^T = (\xi, \eta, \dot{\xi}, \dot{\eta})^T\ , \qquad (2)$$

ξ, η, being the coordinates of the third particle with respect to the equilibrium. The "time"-dependent coefficient matrix A is given by

$$A = \begin{bmatrix} 0 & 0 & 1 & 0 \\ 0 & 0 & 0 & 1 \\ A^*/\sigma & 0 & 0 & 2 \\ 0 & B^*/\sigma & -2 & 0 \end{bmatrix}\ , \qquad (3)$$

where :

V. V. Markellos and Y. Kozai (eds.), Dynamical Trapping and Evolution in the Solar System, 339–350.

$$\sigma = 1 + e \cos t, \tag{4}$$

$$A^* = 1-Q_0^* - x_0 Q_1^* - \mu(1-\mu)R_1^*,$$

$$B^* = 1-Q_0^* ,$$

$$Q_0^* = \frac{1-\mu}{r_{10}^3} + \frac{\mu}{r_{20}^3} , \tag{5}$$

$$Q_1^* = -\frac{3}{\alpha\, r_{10}^3} - \mu R_1^* ,$$

$$R_1^* = -3\left[\frac{1}{\alpha r_{10}^3} - \frac{1}{(\alpha-1)\, r_{20}^3}\right],$$

$$\alpha = x_0 + \mu ,$$

$$r_{10} = |\alpha|, \quad r_{20} = |\alpha-1|,$$

We use as "time" t, the independent variable φ (true anomaly) while, in the above abbreviations, e is the eccentricity of the binary orbit and x_0 is the position of the equilibrium on the x-axis.

2. ANALYTICAL DETERMINATION OF THE OUTGOING EIGENVECTOR

The coefficient matrix A is expanded in the form

$$A(t) = A_0 - e\, A_1 \cos t + e^2\, A_1 \cos^2 t - \ldots \tag{6}$$

where:

$$A_0 = \begin{bmatrix} 0 & 0 & 1 & 0 \\ 0 & 0 & 0 & 1 \\ A^* & 0 & 0 & 2 \\ 0 & B^* & -2 & 0 \end{bmatrix} , \tag{7}$$

and

$$A_1 = \begin{bmatrix} 0 & 0 & 0 & 0 \\ 0 & 0 & 0 & 0 \\ A^* & 0 & 0 & 0 \\ 0 & B^* & 0 & 0 \end{bmatrix} . \tag{8}$$

The eigen-values of A_0 are

$$\lambda_1 = \lambda_0 , \quad \lambda_2 = -\lambda_0, \quad \lambda_3 = i s_0 , \quad \lambda_4 = -i s_0 \tag{9}$$

with

$$\lambda_0 = \sqrt{w_1 + \sqrt{w_1^2 - w_2}}$$
$$s_0 = \sqrt{-w_1 + \sqrt{w_1^2 - w_2}} \tag{10}$$

and

$$w_1 = (A^* + B^* - 4)/2, \quad w_2 = A^* B^* . \tag{11}$$

The corresponding eigenvectors are:

$$\underline{y}_1 = (1, \gamma_0, \lambda_0, \gamma_0\lambda_0)^T ,$$
$$\underline{y}_2 = (1, -\gamma_0, -\lambda_0, \gamma_0\lambda_0)^T ,$$
$$\underline{y}_3 = (1, i\delta_0, is_0, -\delta_0 s_0)^T , \tag{12}$$
$$\underline{y}_4 = (1, -i\delta_0, -is_0, -\delta_0 s_0)^T ,$$

with

$$\gamma_0 = \frac{\lambda_0^2 - A^*}{2\lambda_0} = \frac{2\lambda_0}{B^* - \lambda_0^2}$$
$$\delta_0 = \frac{s_0^2 + A^*}{2s_0} = \frac{2s_0}{B^* + s_0^2} . \tag{13}$$

In the present case the eigenvalues are single and the eigenvectors are linearly independent, thus

$$T_0 = (\underline{y}_1, \underline{y}_2, \underline{y}_3, \underline{y}_4)^T , \tag{14}$$

is non-singular and A_0 is diagonalized by:

$$T_0^{-1} A_0 T_0 = \Lambda_0 , \tag{15}$$

with

$$\Lambda_0 = \begin{bmatrix} \lambda_0 & 0 & 0 & 0 \\ 0 & -\lambda_0 & 0 & 0 \\ 0 & 0 & is_0 & 0 \\ 0 & 0 & 0 & -is_0 \end{bmatrix} . \tag{16}$$

We now put

$$B = T_0^{-1} A_1 T_0 \tag{17}$$

and consider a fundamental solution matrix $\Phi(t,t_0)$ of the system, with $\Phi(t_0; t_0) = I$. Then, we shall have,

$$\Phi(t; t_0) = P(t) \exp [(t-t_o)L] \tag{18}$$

with

$$P(t + T^*) = P(t), \quad P(t_0) = I, \quad T^* = 2\pi , \tag{19}$$

L being a constant matrix, whose eigen-values are the characteristic exponents of the problem. Let T be a constant matrix such that

$$T^{-1} L T = \Lambda \tag{20}$$

where Λ is diagonal, and let

$$\Psi(t; t_0) = \Phi(t; t_0) T. \tag{21}$$

From (18) and (21) we obtain

$$\Psi(t; t_0) = P(t) T \exp [(t-t_0)\Lambda] , \tag{22}$$

while (21) gives

$$\dot{\Psi}(t; t_0) = \dot{\Phi}(t; t_0) T = A(t)\Phi(t; t_0) T = A(t)\Psi(t; t_0), \tag{23}$$

i.e. $\Psi(t; t_0)$ is a fundamental solution matrix of system (1).

We now adopt formal series expansions for the above constant matrices L, Λ, T as well as for the matrix $Q(t)$, where

$$Q(t) = P(t)T, \quad Q(t_0) = T . \tag{24}$$

We shall have:

$$\begin{aligned} L &= L_0 + eL_1 + e^2 L_2 + \dots \\ \Lambda &= \Lambda_0 + e\Lambda_1 + e^2\Lambda_2 + \dots \\ T &= T_0 + eT_1 + e^2 T_2 + \dots \end{aligned} \tag{25}$$

and

$$Q(t) = Q_0(t) + eQ_1(t) + e^2 Q_2(t) + \dots \tag{26}$$

with

$$Q_i(t_0) = T_i, \quad i = 0,1,2,\ldots$$

From (22), (23) and (24) we obtain

$$\dot{Q}(t) = A(t)Q(t) - Q(t)\Lambda , \tag{27}$$

and from (6), (26) and (27) we shall have the following differential Equations accounting for up to fourth-order terms in e:

$$\dot{Q}_0(t) = A_0 Q_0(t) - Q_0(t)\Lambda_0, \tag{28a}$$

$$\begin{aligned}\dot{Q}_1(t) &= A_0 Q_1(t) - Q_1(t)\Lambda_0 \\ &- A_1 Q_0(t)\cos t - Q_0(t)\Lambda_1,\end{aligned} \tag{28b}$$

$$\begin{aligned}\dot{Q}_2(t) &= A_0 Q_2(t) - Q_2(t)\Lambda_0 \\ &- A_1 Q_1(t)\cos t - Q_1(t)\Lambda_1 \\ &+ \frac{1}{2}(A_1 + A_1\cos 2t) Q_0(t) - Q_0(t)\Lambda_2 ,\end{aligned} \tag{28c}$$

$$\begin{aligned}\dot{Q}_3(t) &= A_0 Q_3(t) - Q_3(t)\Lambda_0 \\ &- A_1 Q_2(t)\cos t - Q_2(t)\Lambda_1 \\ &+ \frac{1}{2}(A_1 + A_1\cos 2t) Q_1(t) - Q_1(t)\Lambda_2 \\ &- \frac{1}{4}(3A_1\cos t + A_1\cos 3t) Q_0(t) - Q_0(t)\Lambda_3 ,\end{aligned} \tag{28d}$$

$$\begin{aligned}\dot{Q}_4(t) &= A_0 Q_4(t) - Q_4(t)\Lambda_0 \\ &- A_1 Q_3(t)\cos t - Q_3(t)\Lambda_1 \\ &+ \frac{1}{2}(A_1 + A_1\cos 2t) Q_2(t) - Q_2(t)\Lambda_2 \\ &- \frac{1}{4}(3A_1\cos t + A_1\cos 3t) Q_1(t) - Q_1(t)\Lambda_3 \\ &+ \frac{1}{8}(3A_1 + 4A_1\cos 2t + A_1\cos 4t) Q_0(t) - Q_0(t)\Lambda_4 .\end{aligned} \tag{28e}$$

We also put

$$R(t) = T_0^{-1} Q(t), \quad R(t_0) = T_0^{-1} T \tag{29}$$

and expand formally,

$$R(t) = R_0(t) + eR_1(t) + e^2R_2(t) + \dots \tag{30}$$

with

$$R_i(t_0) = T_0^{-1}\,T_i, \quad i = 0,1,2,\dots$$

Combining now (15), (17), (26), (28), (29) and (30), we obtain, after some reduction, the following perturbation Equations for the first five terms of expansion (30):

$$\dot{R}_0(t) = [\Lambda_0, R_0(t)] , \tag{31a}$$

$$\dot{R}_1(t) = [\Lambda_0, R_1(t)] - BR_0(t)\cos t - R_0(t)\Lambda_1, \tag{31b}$$

$$\dot{R}_2(t) = [\Lambda_0, R_2(t)] - B R_1(t)\cos t - R_1(t)\Lambda_1$$

$$+ B R_0(t) \left(\frac{1+\cos 2t}{2}\right) - R_0(t)\Lambda_2 , \tag{31c}$$

$$\dot{R}_3(t) = [\Lambda_0, R_3(t)] - B R_2(t)\cos t - R_2(t)\Lambda_1$$

$$+ B R_1(t) \left(\frac{1+\cos 2t}{2}\right) - R_1(t)\Lambda_2$$

$$- R_0(t)\Lambda_3 - B R_0(t) \left(\frac{3\cos t + \cos 3t}{4}\right), \tag{31d}$$

$$\dot{R}_4(t) = [\Lambda_0, R_4(t)] - B R_3(t)\cos t - R_3(t)\Lambda_1$$

$$+ B R_2(t) \left(\frac{1+\cos 2t}{2}\right) - R_2(t)\Lambda_2$$

$$- R_1(t)\Lambda_3 - B R_1(t) \left(\frac{3\cos t + \cos 3t}{4}\right)$$

$$- R_0(t)\Lambda_4 + B R_0(t) \left(\frac{3 + 4\cos 2t + \cos 4t}{8}\right), \tag{31e}$$

where we have used the notation

$$[C_1, C_2] = C_1C_2 - C_2C_1 . \tag{32}$$

The solution of the differential Equations (31) finally provides the matrices $R(t)$ and Λ to fourth-order terms in e:

$$R(t) = I + (C_{11} \cos t + S_{11} \sin t)e + (C_{20} + C_{22} \cos 2t + S_{22} \sin 2t)e^2 + (C_{31} \cos t + S_{31} \sin t + C_{33} \cos 3t + S_{33} \sin 3t)e^3 + (C_{40} + C_{42} \cos 2t + S_{42} \sin 2t + C_{44} \cos 4t + S_{44} \sin 4t)e^4 \qquad (33)$$

$$\Lambda = \Lambda_0 + \Lambda_2 e^2 + \Lambda_4 e^4 . \qquad (34)$$

The constant matrices C_{ij}, S_{ij}, Λ_2, Λ_4 involved in the above solution for R(t) and Λ are obtained in the solution process and are given by the following expressions:

$$(S_{11})_{ij} = -(B)_{ij} / \left[1 + (\lambda_i - \lambda_j)^2\right]$$

$$C_{11} = \left[S_{11}, \Lambda_0\right]$$

$$(C_{20})_{ij} = \begin{cases} 0, & i = j \\ \left[(B)_{ij} - \sum_k (B)_{ik} (C_{11})_{kj}\right] / 2(\lambda_j - \lambda_i), & i \neq j \end{cases}$$

$$(S_{22})_{ij} = \left\{ B(I - C_{11}) + \frac{1}{2} \left[\Lambda_0, (BS_{11})\right] \right\}_{ij} / \left[4 + (\lambda_i - \lambda_j)^2\right]$$

$$C_{22} = -\frac{1}{2} \left[\Lambda_0, S_{22}\right] + \frac{1}{4} BS_{11}$$

$$(\Lambda_2)_{ii} = \frac{1}{2} (B)_{ii} - \frac{1}{2} \sum_k (B)_{ik} (C_{11})_{ki}$$

$$(S_{31})_{ij} = \left\{ \frac{1}{2} \left[\Lambda_0, (BS_{22})\right] + \left[\Lambda_0, (S_{11}\Lambda_2)\right] - \frac{1}{4} \left[\Lambda_0, (BS_{11})\right] - B\left(C_{20} - \frac{3}{4} C_{11} + \frac{1}{2} C_{22} + \frac{3}{4} I\right) - C_{11}\Lambda_2 \right\}_{ij} / \left[1 + (\lambda_i - \lambda_j)^2\right]$$

$$C_{31} = -\left[\Lambda_0, S_{31}\right] + \frac{1}{2} B \left(S_{22} - \frac{1}{2} S_{11}\right) + S_{11}\Lambda_2$$

$$(S_{33})_{ij} = \left\{ \frac{1}{2}\left[\Lambda_0, (B S_{22})\right] - \frac{1}{4}\left[\Lambda_0, (B S_{11})\right] - \frac{3}{2} B \left(C_{22} - \frac{1}{2} C_{11} + \frac{1}{2} I\right)\right\}_{ij} \Big/ \left[9 + (\lambda_i - \lambda_j)^2\right]$$

$$C_{33} = -\frac{1}{3}\left[\Lambda_0, S_{33}\right] + \frac{1}{6} B \left(S_{22} - \frac{1}{2} S_{11}\right)$$

$$(C_{40})_{ij} = \begin{cases} 0, & i=j \\ \left\{ -\frac{3}{8} (B)_{ij} - (C_{20})_{ij} (\Lambda_2)_{ij} + \frac{1}{8} \sum_k B_{ik} \left[4C_{20} - 3C_{11} - 4C_{31} + 2C_{22}\right]_{kj} \right\} / (\lambda_j - \lambda_i), & i \neq j \end{cases}$$

$$(S_{42})_{ij} = \left\{ \frac{1}{2}\left[\Lambda_0, (B S_{31})\right] + \frac{1}{2}\left[\Lambda_0, (B S_{33})\right] - \frac{1}{2}\left[\Lambda_0, (B S_{22})\right] + \left[\Lambda_0, (S_{22}\Lambda_2)\right] + \frac{1}{4}\left[\Lambda_0, (B S_{11})\right] - B(C_{31} + C_{33} - C_{22} + C_{20} + C_{11} - I) - 2C_{22}\Lambda_2 \right\}_{ij} \Big/ \left[4 + (\lambda_i - \lambda_j)^2\right]$$

$$C_{42} = -\frac{1}{2}\left[\Lambda_0, S_{42}\right] + \frac{1}{4} B \left(S_{31} + S_{33} - S_{22} + \frac{1}{2} S_{11}\right) + \frac{1}{2} S_{22}\Lambda_2$$

$$(S_{44})_{ij} = \left\{ \frac{1}{2}\left[\Lambda_0, (B S_{33})\right] - \frac{1}{4}\left[\Lambda_0, (B S_{22})\right] + \frac{1}{8}\left[\Lambda_0, (B S_{11})\right] - B\left(2C_{33} - C_{22} + \frac{1}{2} C_{11} - \frac{1}{2} I\right)\right\}_{ij} \Big/ \left[16 + (\lambda_i - \lambda_j)^2\right]$$

$$C_{44} = -\frac{1}{4}\left[\Lambda_0, S_{44}\right] + \frac{1}{8} B \left(S_{33} - \frac{1}{2} S_{22} + \frac{1}{4} S_{11}\right)$$

$$(\Lambda_4)_{ii} = \frac{3}{8}(B)_{ii} + \frac{1}{8} \sum_k B_{ik} (4C_{20} - 3C_{11} - 4C_{31} + 2C_{22})_{ki} .$$

From Equation (22) we now obtain

$$\Psi(t_0; t_0) = T, \tag{35}$$

where T is determined to fourth-order terms in e, from the solution process described above. The column of T corresponding to the larger (real) eigenvalue of L-the real positive element of diagonal matrix Λ-provides the initial state of the outgoing asymptotic orbit leaving the equilibrium. This initial state is normalized by forcing the first component to be equal to 1, thus introducing the orbital parameter ε appearing in the heading of the table of numerical results below.

The smalness of the orbital parameter ε is obligatory since we are treating the linearized Equations. An analytical solution, based on a similar perturbation technique, to higher order terms in ε based on second-or higher-order Equations in the place of Equations (1), is also possible to obtain. In the case of the circular restricted problem (e=0) this has been done by Deprit and Henrard (1965). In the present case of the elliptic restricted problem this perturbation in ε is much more involved and is not included here. The analysis described here for the collinear equilibria is similar to that of Bennett (1966) referring to the triangular equilibria.

NUMERICAL VERIFICATION

In practice the initial state of the outgoing asymptotic orbit (for given ε) was determined both analytically (as outlined above) and numerically. The numerical determination is based on numerical integration of the linearized system (1). For details on this we refer to Bennett (1965). The results of the two procedures agreed satisfactorily.

3. SYMMETRIC DOUBLY ASYMPTOTIC ORBITS

Having obtained the appropriate initial state of the outgoing asymptotic orbit we apply a differential corrections procedure for its "correction" in order to achieve a perpendicular crossing of the axis of the primaries by the orbit. This ensures an asymptotic "return" of the particle back to the unstable equilibrium, at "infinite time" (subject to an additional requirement mentioned below).

The initial state vector depends on the three parameters :

$$\mu, e, t_0,$$

while the occurrence of a "parpendicular crossing" is equivalent to the satisfaction of the following two conditions:

$$y(\mu, e, t_0; \tau) = 0, \qquad (36)$$
$$\dot{x}(\mu, e, t_0; \tau) = 0,$$

where x and y are the usual pulsating - rotating coordinates of the elliptic restricted problem, and $\tau(=n\pi)$ must be an integer multiple of the half-period π of the primaries. In the elliptic problem this ensures that the perpendicular crossing is also a "mirror condition" as required for the above mentioned asymptotic return (to the equilibrium), i.e for the symmetry of the orbit. $\tau = n\pi$ is therefore the additional requirement mentioned above.

For given value of n (i.e of τ)we have three parameters to adjust in order to satisfy the two conditions (36). This shows that the symmetric doubly asymptotic orbits sought are members of monoparametric sets, and it is easy to set up a predictor-corrector procedure for their numerical determination.

Many doubly asymptotic orbits, belonging to a number of monoparemetric sets-series-of such orbits have been obtained by means of such a procedure and the numerical data corresponding to five such symmetric doubly asymptotic orbits of the elliptic problem (belonging to the same series) and given as examples in the Table below. One of them is shown graphically in the Figure.

Table 1. Symmetric doubly asymptotic orbits ($\varepsilon = -0.001$, $\tau = 6\pi$)

e	μ	t_0	x_τ	$\dot{y}_\tau$
0.001	0.1782153	1.919801	-2.737897	2.249880
0.006	0.1810822	1.996968	-2.744719	2.257073
0.01	0.1834156	2.057626	-2.750137	2.262752
0.06	0.2149984	2.705028	-2.808980	2.321029
0.1	0.2403649	3.043285	-2.820386	2.322685

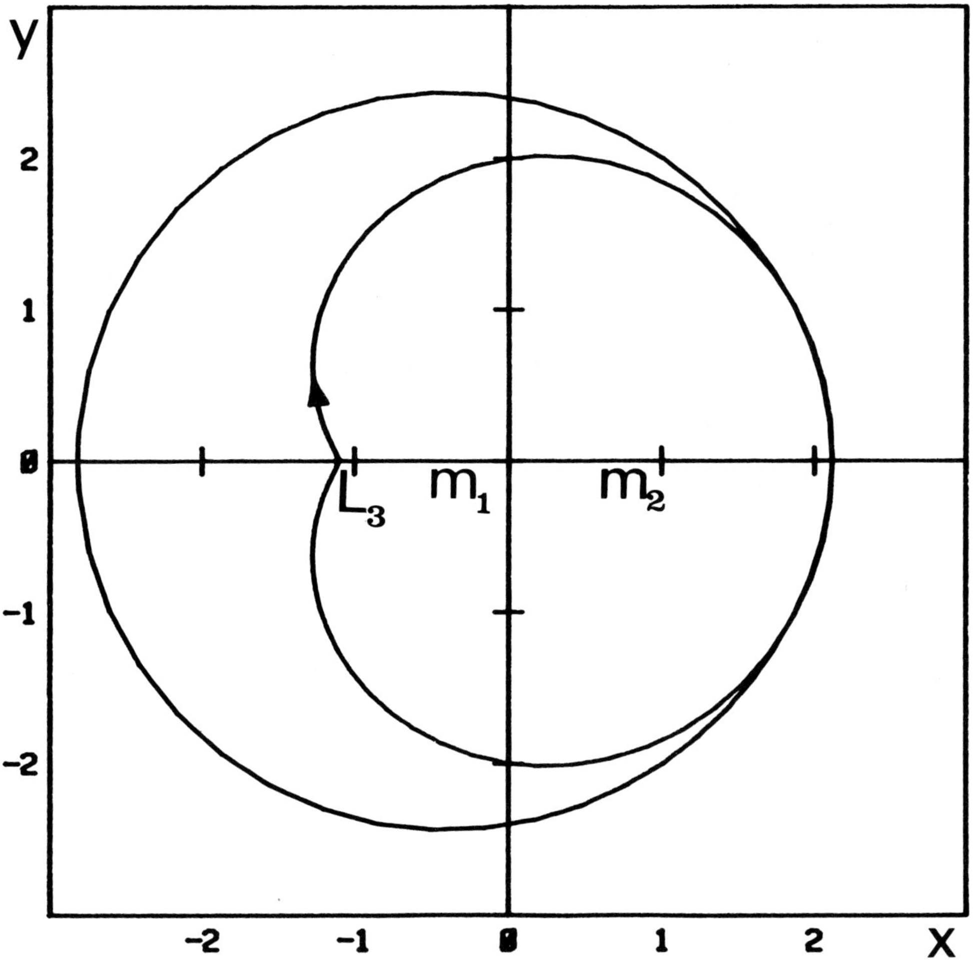

Figure 1. The fifth doubly asymptotic orbit of Table 1.

THE QUESTION OF ASYMPTOTIC TRAPPING

One of the motives behind the present work was to be able to experiment numerically with the idea that an asymptotic orbit in-coming toward the *unstable* equilibrium can perhaps be considered as representing a trapping mechanism, based on "duration of passage" rather than stability of motion.

With the use of some of the symmetric doubly asymptotic orbits, determined as described earlier, a number of "experimental" numerical integrations "toward the equilibrium" have been performed and the results seem to be encouraging in that, despite the high instability of the integrated orbits, the anavoidable deviations from the model orbits seem to result in "long duration of passage", i.e. slow motion near the unstable equilibrium.

ACKNOWLEDGEMENTS

I wish to thank V.V. Markellos for his help and advice on a number of topics during this work. The analysis of section 2 was done in collaboration with C. Georgiou.

REFERENCES

Bennett, A.: 1965, *Icarus* , 4 , 177

Bennett, A.: 1966, in R.L. Duncombe and V.G. Szebehely (eds.), *Methods in Astrodynamics and Celestial Mechanics* , Academic Press, New York, p. 101.

Deprit, A. and Henrard, J.: 1965, *Astron. J.*, 70, 271

PART VI

MISCELLANEOUS DYNAMICS

THE PLANAR INVERSE PROBLEM FOR AUTONOMOUS SYSTEMS

BASILIS C. XANTHOPOULOS
Astronomy Department, University of Thessaloniki, Thessaloniki and Department of Physics, University of Crete, Iraklion, Greece*
and
GEORGE BOZIS+
Department of Theoretical Mechanics, University of Thessaloniki, Thessaloniki, Greece.

Abstract: We study the general version of the inverse problem for planar trajectories and for autonomous dynamical systems possessing three integrals, i.e., for a given three-parametric family of curves $f(x,y,a,b)=c$ we find the potential $V(x,y)$ for which these curves are orbits of a unit mass. All possible cases, depending on the preassigned function f, are classified and in each case the necessary and sufficient conditions for the existence of a solution are established. Among the examples is the case of the Keplerian conic sections which is studied in detail.

1. INTRODUCTION

During the last few years there appeared a number of papers dealing with the following aspect of the inverse problem: A family of plane curves (depending on one or two parameters) is given in an inertial frame in Cartesian coordinates and required is the potential of a conservative dynamical system with two degrees of freedom (autonomous or not) for which all members of the given family are actual orbits.

For monoparametric families $f(x,y)=c$ and for autonomous systems the answer to the question whether such a potential does exist is, in general, affirmative; the potential $V=V(x,y)$ is given by Szebehely's linear partial differential equation of the first order in V (Shebehely, 1974). This equation is in fact associated with a certain dependence of the total energy E on the function $f(x,y)$ and in examples this dependence $E=E(f)$ has to be given in advance (Broucke, 1979, Broucke and Lass, 1977). Thus, in its general solution, there appear two arbitrary functions (Molnàr, 1981). One also finds in the literature applications, generalizations and modifications of Szebehely's equation. Thus, this equation was generalized by Bozis (1983) in order to include velocity dependent potentials and by Érdi (1982) to study three dimensional orbits. A modification of the same equation was presented by Szebehely and Broucke (1981) to account for non-inertial frames. A modification by Morrison (1976) uses the energy constant as the parameter of the family. Szebehely's equation was written in

*Present address
+Presented by G. Bozis

V. V. Markellos and Y. Kozai (eds.), Dynamical Trapping and Evolution in the Solar System, 353–367.

polar coordinates and an application was given recently to a problem of Galactic Dynamics by Szebehely, Lundberg and McGahee (1980).

For preassigned two-parametric families of curves $f(x,y,b)=c$ it is intuitevely expected that the picture regarding the existence of solutions changes. In fact potentials which give rise to such families do or do not exist depending on the function $f(x,y,b)$(Lass, 1972). Concrete criteria for this case were given recently by Bozis (1982) .

In the present paper we face the following problem: Given a three-parametric family of planar curves $f(x,y,a,b)=c$ in Cartesian coordinates x,y is there an autonomous dynamical system for which these curves are actual orbits of a unit mass? No assumption is made in advance for the dependence of the total energy E on the three parameters a,b and c. However, it is understood that when the problem admits a solution $V=V(x,y)$, the total energy $E=T+V$ is constant along each orbit, i.e., eventually $E=E(a,b,c)$.

The motivation for studying this problem is that the totality of the orbits of a two-dimensional autonomous dynamical system possessing three integrals of motion generally is a three-parametric family of planar curves. In this sense this appear to be the most general version of the inverse problem of this sort.

Since we now demand that a larger family of orbits results from a single potential, we expect more conditions to be necessary, and eventually necessary and sufficient, so that this problem admits a solution and, in fact, this is exactly what happens. At a first stage we find a set of necessary conditions for the case at hand to have a chance for an affirmative answer. This set also serves to classify each case at hand. Eventually we establish necessary and sufficient conditions for the problem to admit a solution.

2. ANALYSIS

We consider a three-parametric family of planar curves expressed in the form

$$f(x,y,a,b) = c \tag{1}$$

in the Cartesian x,y plane. We introduce the notation

$$\gamma=f_y/f_x \quad , \qquad \Gamma=\gamma\gamma_x-\gamma_y \quad , \tag{2}$$

$$\lambda=\Gamma^{-1}(-\Gamma_x+\gamma^{-1}\Gamma_y) \quad , \qquad \mu=\lambda\gamma+3\gamma^{-1}\Gamma \quad , \tag{3}$$

where the functions γ,Γ,λ and μ can be derived in a straightforward manner from any given $f(x,y,a,b)$ and the subscripts denote partial differentiations. The function γ vanishes or diverges identically when the family (1) represents the monoparametric families of straight lines which are parallel to the x or the y axis; and of course this is not the case here. Also $\Gamma\neq 0$, because $\Gamma=0$ represents in fact a two-parametric family of straight lines. Obviously, the functions γ,Γ,λ and μ depend on the parameters a and b as well. In fact $\gamma_a\neq 0$ and $\gamma_b\neq 0$

because otherwise the family (1) would essentially depend on at most two parameters. The above assumptions serve to guarantee that the family (1) is a genuine three-parametric family of planar curves. The force components $X=X(x,y)$ and $Y=Y(x,y)$, inasmuch as they exist, which are derived from a potential $V=V(x,y)$ and give rise to the family of orbits (1), satisfy the linear system of partial differential equations (Bozis 1982 b)

$$X_y = Y_x \tag{4}$$

and

$$-X_x+(1-\gamma^2)\gamma^{-1}X_y+Y_y = \lambda X+\mu Y \; . \tag{5}$$

We demand that the potential, therefore and the force components, are independent of the parameters a,b and c; this expresses the fact that the free parameters of the family (1) enumerate the totality of the orbits admitted by a single potential $V(x,y)$. We seek, therefore, solutions of the system of equations (4) and (5) which satisfy the conditions

$$X_a = X_b = Y_a = Y_b = 0 \; . \tag{6}$$

The analysis which follows is heavily based on and very much facilitated by this requirement.

First we look for necessary conditions on the function γ - which are, in effect, conditions on the function f - for the system of equations (4) and (5) to satisfy the conditions (6). By differentiating equation (5) with respect to a and b we obtain that

$$X_y = \ell X+mY \qquad \text{and} \qquad X_y = LX+MY \tag{7}$$

where

$$\ell = -\frac{\gamma^2}{(1+\gamma^2)\gamma_a}\lambda_a \; , \qquad L = -\frac{\gamma^2}{(1+\gamma^2)\gamma_b}\lambda_b \tag{8}$$

$$m = -\frac{\gamma^2}{(1+\gamma^2)\gamma_a}\mu_a \; , \qquad M = -\frac{\gamma^2}{(1+\gamma^2)\gamma_b}\mu_b \; . \tag{9}$$

Then by differentiating equations (7) with respect to a and b we further obtain that

$$\begin{aligned} \ell_a X+m_a Y = 0 \quad &, \quad \ell_b X+m_b Y = 0 \\ L_a X+M_a Y = 0 \quad &, \quad L_b X+M_b Y = 0 \; . \end{aligned} \tag{10}$$

Finally, by subtracting the two equations (7) we obtain

$$(\ell-L)X+(m-M)Y = 0 \; . \tag{11}$$

Since we disregard as trivial the solutions for which one of the force components vanishes, we demand that the linear homogeneous system of the algebraic equations (10) and (11) admits non-trivial solutions. This system has solutions different from the solution $X = Y = 0$ if and only if the determinant of any two of these five equations equals to zero. In view of the comments which follow, we shall express this requirements in the form

$$\rho = \frac{\ell - L}{m - M} = \frac{\ell_a}{m_a} = \frac{\ell_b}{m_b} = \frac{L_a}{M_a} = \frac{L_b}{M_b} \quad . \tag{12}$$

Comments: (i) The common ratio ρ must be different from zero and from infinity; otherwise one of the equations (10) and (11) would imply that either X or Y vanishes identically.

(ii) If some of the ratios (12) are indeterminate of the form 0/0, these are simply ignored; in fact the corresponding equations (10) and (11) are satisfied identically and give no additional information.

We next consider two cases, depending on whether the ratio ρ (of equation 12) is defined or it is indeterminate.

Case I: The ratio ρ is defined.

Since ρ must be equal, from equations (10) and (11), to $-Y/X$, it must also be independent of the parameters a and b. We obtain, therefore, the conditions

$$\rho_a = 0 \quad \text{and} \quad \rho_b = 0 \quad , \tag{13}$$

which, however, are immediate consequences of equations (12). In this case all equations (10) and (11) reduce to the single equation

$$\rho X + Y = 0 \tag{14}$$

which must be combined with one of the two equations (7), say

$$\ell X + mY = X_y \ . \tag{15}$$

Note that, in view of equations (11) and (15), the second of equations (7) is also satisfied. We have to satisfy, therefore, the system of equations (4),(5),(14) and (15). Depending on the form of the system of equations (14) and (15), we distinguish two subcases:

Subcase Ia: The determinant $\delta = \rho m - \ell$ is different from zero

We have, in this case, that

$$X = -\delta^{-1} X_y \quad , \quad Y = \rho\delta^{-1} X_y \ . \tag{16}$$

By combining now equations (4),(5),(14) and (15) we readily obtain that

$$X_x = \eta X \quad \text{and} \quad X_y = -\delta X, \tag{17}$$

where

$$\eta = (\gamma^2 - 1)\gamma^{-1}\delta + \rho\delta - \rho_y - \lambda + \mu\rho \ . \tag{18}$$

Provided that the integrability condition

$$\delta_x+\eta_y = 0 \tag{19}$$

of the system of equations (17) is satisfied, they can be solved and determine X uniquely, up to a multiplicative constant; the second of equations (16) then can be used to determine Y algebraically. The force components so obtained satisfy equation (4) and therefore they arise from a potential. The solutions are acceptable provided that they are independent of the parameters a and b of the family of orbits. This is satisfied provided that

$$\delta_a = \delta_b = \eta_a = \eta_b = 0\,. \tag{20}$$

However, it is straightforward to show that the conditions (20) are immediate consequences of the conditions (12) and (13) and therefore they are not additional necessary conditions. Note that in this case we also have that $\rho\neq\ell/m\neq L/M\neq\rho$. The final conclusion is that in the subcase Ia the necessary and sufficient conditions are the conditions (12) and (19). Whenever these conditions are satisfied, the force components are determined uniquely up to a constant factor.

Subcase Ib: The determinant $\delta=\rho m-\ell$ equals to zero

We have, in this case, that

$$\rho=\ell/m = L/M \tag{21}$$

as well. It is seen from equations (4),(14) and (15) that $X_y=Y_x= 0$ and therefore the force components are of the form

$$X=X(x) \quad \text{and} \quad Y=Y(y)\ ; \tag{22}$$

the corresponding potential is separable. The equations that have to be satisfied reduce in this case to the equations

$$-X_x+Y_y = \lambda X+\mu Y \quad \text{and} \quad \rho X+Y = 0\ . \tag{23}$$

By differentiating the first of these equations with respect to a and b we obtain that

$$\lambda_a X+\mu_a Y = 0 = \lambda_b X+\mu_b Y \tag{24}$$

from which we obtain the necessary conditions

$$\rho = {}^{\lambda_a}/_{\mu_a} = {}^{\lambda_b}/_{\mu_b} \tag{25}$$

which however are satisfied by virtue of the equations (8),(9),(12) and (21). The two equations (23) now give that

$$X_x+\Theta X = 0 \tag{26}$$

where

$$\Theta = \lambda - \mu\rho + \rho_y \ . \tag{27}$$

Equation (26) determines uniquely, up to a multiplicative constant, an acceptable solution of our problem provided that

$$\Theta_y = 0 \tag{28}$$

and

$$\Theta_a = \Theta_b = 0 \ . \tag{29}$$

The conditions (29) are immediate consequences of the conditions (25) while equation (28) gives an additional independent necessary condition. Then Y is determined algebraically from the second equation (23) and the resulting solution is acceptable, provided that $Y_x = 0$ which, since $Y = -\rho X$, leads to the additional and independent necessary condition

$$\rho_x = \rho\,\Theta \ . \tag{30}$$

The final conclusion then is in the subcase Ib that the necessary and sufficient conditions are given by equations (12),(21),(28) and (30) and that, whenever these conditions are satisfied, the force components are determined uniquely, up to a constant factor.

Case II: The ratio ρ is indeterminate

Since all the ratios of equations (12) are of the form 0/0, the equations (10) and (11) are satisfied and therefore they can be omitted. We are left with equations (4),(5) and one, say the first, of equations (7). These are written as follows

$$\begin{aligned} -X_x+Y_y &= \{\lambda+(\gamma^2-1)\gamma^{-1}\ell\}X+\{\mu+(\gamma^2-1)\gamma^{-1}m\}Y \ , \\ X_y &= \ell X+mY \ , \\ X_y &= Y_x \ . \end{aligned} \tag{31}$$

The coefficients of X and Y in the right hand sides of the first two of equations (31) are independent of the parameters a and b. Therefore, all the solutions of the above system are independent of a and b and, as such, they are acceptable solutions of our problem. The necessary and sufficient conditions, therefore, in this case are

$$\ell - L = m - M = \ell_a = \ell_b = m_a = m_b = 0 \ . \tag{32}$$

The example (35) of the next section shows that generally the force components are not uniquely determined in this case.

3. EXAMPLES

(3A): As an example for the subcase Ia of the analysis of the previous section we consider the (planar) Kepler problem: We assume that a system admits as orbits all the members of the three-parametric family of ellipses in the same plane and with common one of their focal points, but with arbitrary eccentricity, magnitude, and orientation of their major axes.

To describe this family we choose a Cartesian coordinate system whose origin coincides with the common focal point. The family then is described by the equation

$$f=(x^2+y^2)\{1+e\cos(\vartheta-\vartheta_0)\}^2 = \text{const.} \quad , \tag{33}$$

where e and ϑ_0 are the two parameters and $\tan\vartheta=y/x$. From the expression (33) we obtain that

$$\gamma=(\sin\vartheta+a)/(\cos\vartheta+b) \quad , \tag{34}$$

where, here and henceforth, we shall consider, instead of e and ϑ_0, the $a=e\sin\vartheta_0$ and $b=e\cos\vartheta_0$ as the two free parameters of the family. To simplify the computations, we introduce the notation

$$\tan\omega=(b\sin\vartheta-a\cos\vartheta)/(1+a\sin\vartheta+b\cos\vartheta) \tag{35}$$

in which γ simplifies to

$$\gamma=\tan(\vartheta-\omega) \; . \tag{36}$$

Note that the parameters a and b appear in γ only through the combination ω. Then we obtain that

$$\gamma+{}^x/_y = \cos\omega/\sin\vartheta\cos(\vartheta-\omega), \tag{37}$$

$$\Gamma=\mp(x^2+y^2)^{-1/2}\cos^2\omega\,\cos^{-2}(\vartheta-\omega)(\cos\vartheta+b)^{-1} \; . \tag{38}$$

In order to obtain the last expression we have used equation (A.4) of the appendix and that

$$\vartheta_x-\omega_x=-\sin\vartheta\cos\omega\cos(\vartheta-\omega)(x^2+y^2)^{-1/2}(\cos\vartheta+b)^{-1} \; . \tag{39}$$

From the first of equations (3) we obtain, after a lengthy calculation, that

$$\lambda=3\{\sin(\vartheta-\omega)-b\sin\omega\}/(x^2+y^2)^{1/2}(\cos\vartheta+b)\sin(\vartheta-\omega) \; , \tag{40}$$

and a similar expression for μ.

Because it is very complicated to evaluate all the quantities which appear in the ratios (12), we follow an indirect approach to establish the validity of the necessary and sufficient conditions (12), (13) and (19).

It turns out that the combination

$$x\lambda+y\mu=y\lambda(\gamma+xy^{-1})+3y\Gamma\gamma^{-1}=-3\sin2\omega/\sin(2\vartheta-2\omega) \tag{41}$$

is very simple and it depends on the parameters a and b only through

ω. Hence from equations (36) and (41) we immediately obtain that

$$(x\lambda+y\mu)_a = -6\sin2\vartheta\ \sin^{-2}(2\vartheta-2\omega)\omega_a$$
$$\gamma^2(1+\gamma^2)^{-1}=\sin^2(\vartheta-\omega)\ ,\ -\gamma^2(1+\gamma^2)^{-1}\gamma_a^{-1}=\sin^2(2\vartheta-2\omega)(4\omega_a)^{-1} \qquad (42)$$

and therefore that

$$x\ell+ym = -\frac{\gamma^2(x\lambda+y\mu)_a}{(1+\gamma^2)\gamma_a} = -\frac{3}{2}\sin2\vartheta. \qquad (43)$$

Obviously, we similarly obtain that

$$xL+yM = -\frac{3}{2}\sin2\vartheta\ . \qquad (44)$$

Hence $x\ell+ym=xL+yM$ which implies that

$$\rho = \frac{\ell-L}{m-M} = -\frac{y}{x}\ . \qquad (45)$$

Since the right hand sides of equations (43) and (44) are independent of the parameters a and b we also have that

$$x\ell_a+ym_a=x\ell_b+ym_b=xL_a+yM_a=xL_b+yM_b= 0 \qquad (46)$$

which, combined with equation (45), implies the validity of the conditions (12) and (13).

Finally for $\rho=-y/x$ we readily obtain that

$$\delta=-(x\ell+ym)/x = 3y/(x^2+y^2) \qquad (47)$$

and

$$\eta\dot{=}(y^2-2x^2)/x(x^2+y^2) \qquad (48)$$

from which the last condition (19) is also verified. For these expressions for δ and η, equations (17) and (14) are readily integrated and give

$$X=kx(x^2+y^2)^{-3/2}\ ,\ Y=ky(x^2+y^2)^{-3/2}\ ,\ k \text{ a constant}, \qquad (49)$$

i.e., the Newtonian force. It should be noted that we have here derived Newton's force law by using only the first of the three laws of Kepler's and the weaker assumption that the motion is conservative, not central as stated by Kepler's second law.

(3B). As an example of a three parametric family of curves which arise from a separable potential we consider the family

$$f=b\sqrt{ax^2-1} - a\sqrt{by^2-1} = c\ . \qquad (50)$$

For this family we readily obtain that

$$\gamma=-yx^{-1}(ax^2-1)^{1/2}(by^2-1)^{-1/2},\quad \Gamma=Ax^{-1}(by^2-1)^{-1} \qquad (51)$$

where

$$A=y^2x^{-2}-(ax^2-1)^{1/2}(by^2-1)^{-1/2} \tag{52}$$

and, after a long calculation, that

$$\lambda = \frac{1}{x} + \frac{2y^2}{x^3A} - \frac{2x}{A(y^2-x^2A)} + \frac{3bx}{\sqrt{(ax^2-1)(by^2-1)}} + \frac{1}{xA}\{\frac{y^2-x^2A}{x^2} - \frac{x^2}{y^2-x^2A}\} . \tag{53}$$

What turns out to be very simple is the combination

$$y^3\lambda+x^3\mu=(y^3+x^3\gamma)\lambda+3x^3\Gamma\gamma^{-1} = 3(y^4-x^4)(xy)^{-1} . \tag{54}$$

Therefore this family satisfies the relationships

$$\lambda_a/\mu_a= \lambda_b/\mu_b= - x^3/y^3 . \tag{55}$$

In addition we readily obtain that

$$y^3\ell+x^3m=0=y^3L+x^3M \tag{56}$$

and therefore all the ratios (12) are equal to $\rho = -x^3/y^3$ and the conditions (12),(13) and (21) are satisfied. Moreover, we obtain that $\theta=3/x$ which checks the validity of the final conditions (28) and (30). The family (50) represents the totality of orbits of the autonomous conservative system with potential

$$U = k(x^{-2}+y^{-2}),$$

where k is a constant.

(3Γ). For the case II we present a two-parametric worth of examples, characterized by the two arbitrary constants p and q which are distinct from the parameters a,b and c of the family of orbits.

The three-parametric family of curves is

$$f(x,y,a,b) = y+\int\{pg^3-3qg^2-3pg+q\}^{-1}dg = c , \tag{58}$$

where $g=g(x,y,a,b)$ is any two-parametric family of solutions of the equation

$$gg_x-g_y= pg^3-3qg^2-3pg+q . \tag{59}$$

We shall give the presentation in three steps.

(i) First we show that the family (58) satisfies the conditions (32). By using equations (58) and (59) we readily obtain that the function $\gamma=f_y/f_x$ of the present family equals to the solution g of equation (59) i.e., that $\gamma=g$. Therefore, $\Gamma=pg^3-3qg^2-3pg+q$ from which we can readily obtain that

$$\lambda=-\gamma^{-1}\, d\Gamma/d\gamma = -3g^{-1}(pg^2-2qg-p) \;,$$
$$\mu=3\Gamma\gamma^{-1}- d\Gamma/d\gamma = -3g^{-1}(qg^2+2pg-q) \;. \tag{60}$$

It is the fact that λ and μ depend on the parameters a and b only implicitly, through the function g, which makes the evaluation of ℓ,L,m and M rather simple. In fact we obtain that

$$\ell=L=3p \quad , \qquad m=M=3q \tag{61}$$

and therefore the necessary and sufficient conditions (32) are satisfied.

(ii) Second we determine the corresponding potential for a given choice of p and q. The force components satisfy the linear system of equations

$$X_x-Y_y+6qX-6pY = 0 \quad , \tag{62}$$

$$X_y = 3pX+3qY \quad , \tag{63}$$

$$X_y = Y_x \;. \tag{64}$$

By taking the x derivative of eq. (62) and using equation (64) to eliminate Y we obtain a second order linear partial differential equation in X with constant coefficients which generally is of the irreducible type. Therefore its solutions are of the form

$$X = e^{3Ax+3By} \tag{65}$$

for suitably chosen constants A and B. Equation (63) then gives that

$$Y = (B-p)q^{-1}e^{3Ax+3By} \tag{66}$$

while equations (62) and (64) give two algebraic equations in A and B which, after some manipulations, become

$$A^3-3(p^2+q^2)A+2q(p^2+q^2) = 0, \quad B=pA(A-q)^{-1} \;. \tag{67}$$

It turns out that the first of equations (67) has three different real roots when $pq \neq 0$, and obviously, to any of these roots there corresponds an acceptable solution of our problem. We conclude, therefore, that in this case the force components are not determined uniquely from the three-parametric family of curves. In fact, since the equations (62)-(64) are linear, we can also consider arbitrary superpositions of solutions; thus for a given choice of p and q (with $pq \neq 0$) we can construct a two-parametric family of force components which accept the three parametric family of orbits (58), where we have not counted the arbitrary overall multiplication factor as a free parameter. The corresponding potential is

$$V = -(3A)^{-1}e^{3Ax+3By} \quad . \tag{68}$$

(iii) Third, we describe how the family (58) was obtained. The condition $\ell=L$ demands that $\lambda_a/\gamma_a = \lambda_b/\gamma_b$ which is satisfied when $\lambda=\lambda(\gamma)$. This last condition is satisfied when $\Gamma=\Gamma(\gamma)$, which also guarantees that $m=M$. In this case we obtain that

$$\ell=L=\frac{\gamma\ddot{\Gamma}-\dot{\Gamma}}{1+\gamma^2}\ , \qquad m=M=\frac{\gamma^2\ddot{\Gamma}-3\gamma\dot{\Gamma}+3\Gamma}{1+\gamma^2}\ , \tag{69}$$

where the dots denote differentiations with respect to γ. The only way for ℓ and m to be independent of a and b is that the two expressions in (69) are constants, say $3p$ and $3q$ respectively. By solving the resulting equations we obtain that $\Gamma=p\gamma^3-3q\gamma^2-3p\gamma+q$. The family described by equations (58) and (59) is obtained by reconstructing f from a given Γ.

(3Δ). Finally as an example of a three parametric family of curves which does not arise from any autonomous conservative system we consider the family of all possible circles (with arbitrary center and radius) in the plane

$$f=f(x,y,a,b)=(x-a)^2+(y-b)^2 = c\ . \tag{70}$$

For this family we easily obtain that

$$\gamma=\frac{y-b}{x-a}\ , \quad \Gamma=-\frac{f}{(x-a)^3}\ , \quad \lambda=\frac{3}{x-a}\ , \quad \mu=-\frac{3}{y-b} \tag{71}$$

and therefore

$$\ell=-\frac{3(y-b)}{f}\ , \qquad m=0, \qquad L=0, \qquad M=-\frac{3(x-a)}{f}\ . \tag{72}$$

Obviously some of the ratios (12) become infinite, so no solution to our problem exists

4. INVARIANCE PROPERTIES OF THE THEORY

The analysis of the present paper and the necessary and sufficient conditions (12), (19), (21), (28), (30) and (32) at which we arrive on section 2, depend explicitly on the parameters a and b of the family of orbits (1). However, it is intuitively expected that one should have the freedom to reparametrize the original family in an arbitrary manner, say,

$$\tilde{a}=\tilde{a}(a,b) \quad \text{and} \quad \tilde{b}=\tilde{b}(a,b) \tag{73}$$

and that this reparametrization will not alter the classification and the conclusions of the analysis of section 2. The freedom in the choice of the parameters of the family (1) represents the gauge freedom of the problem considered in this paper. We now show that the above mentioned conditions are indeed gauge invariant.

It is straightforward to see that under the change of gauge (73) the functions f, γ, λ and μ remain invariant while ℓ, L, m and M change according to

$$\tilde{\ell} = \ell-(\ell-L)\gamma_b{}^{b}{}_{\tilde{a}}/\gamma_{\tilde{a}} \ , \quad \tilde{L}=L+(\ell-L)\gamma_a{}^{a}{}_{\tilde{b}}/\gamma_{\tilde{b}}$$
$$\tilde{m}=m-(m-M)\gamma_b{}^{b}{}_{\tilde{a}}/\gamma_{\tilde{a}} \ , \quad \tilde{M}=M+(m-M)\gamma_a{}^{a}{}_{\tilde{b}}/\gamma_{\tilde{b}} \qquad (74)$$

from which we readily obtain that

$$\tilde{\rho} = \frac{\tilde{\ell}-\tilde{L}}{\tilde{m}-\tilde{M}} = \frac{\ell-L}{m-M} = \rho \ . \qquad (75)$$

Hence the condition (13) implies that $\tilde{\rho}_{\tilde{a}} = \tilde{\rho}_{\tilde{b}} = 0$ which means that it is a gauge invariant condition. Then by using equations (12) and (74) we can easily show that

$$\tilde{\ell}_{\tilde{a}}/\tilde{m}_{\tilde{a}} = \tilde{\ell}_{\tilde{b}}/\tilde{m}_{\tilde{b}} = \tilde{L}_{\tilde{a}}/\tilde{M}_{\tilde{a}} = \tilde{L}_{\tilde{b}}/\tilde{M}_{\tilde{b}} = \rho = \tilde{\rho} \ , \qquad (76)$$

which establishes the gauge invariance of the condition (12). Next we easily see that

$$\tilde{\delta}=\tilde{\rho}\tilde{m}-\tilde{\ell}=\rho m-\ell=\delta \qquad (77)$$

which, among others, shows the gauge invariance of the classification of section 2 and of condition (21). Finally, equations (75) and (77) imply that $\tilde{\eta}=\eta$ and $\tilde{\theta}=\theta$ which establish the gauge invariance of the conditions (19),(28),(30) and (32), Q.E.D.

The original definitions, given by equation (3), of the two basic quantities λ and μ seem unrelated. However, the subsequent analysis is completely symmetrical in λ and μ. Here we establish the existence of a simple relationship between λ and μ which explains the symmetrical form of the theory in λ and μ. Precisely we shall show that under the change of coordinates

$$\tilde{x} = y \ , \quad \tilde{y} = \varepsilon x, \qquad (78)$$

where $\varepsilon^2=1$, the quantities λ and μ transform according to

$$\tilde{\lambda} = -\mu \ , \quad \tilde{\mu} = -\varepsilon\lambda \ . \qquad (79)$$

(For $\varepsilon=+1$ the transformation (78) represents the interchange of the x and y axis, while for $\varepsilon=-1$ it represents a rotation in the x-y plane by 90° degrees).

The proof is straightforward. For the same family of curves $\tilde{f}=f=$ $=$constant we obtain that

$$\tilde{\gamma} = f_{\tilde{y}}/f_{\tilde{x}} = \varepsilon\gamma^{-1} \qquad (80)$$

and therefore that

$$\tilde{\Gamma}=\tilde{\gamma}\tilde{\gamma}_{\tilde{x}}-\tilde{\gamma}_{\tilde{y}} = \Gamma\gamma^{-3} \quad . \tag{81}$$

By using equations (80) and (81) and performing the required differentiations we obtain that

$$\tilde{\lambda}=\tilde{\Gamma}^{-1}(-\tilde{\Gamma}_{\tilde{x}}+\tilde{\gamma}^{-1}\tilde{\Gamma}_{\tilde{y}}) = -\mu \tag{82}$$

and

$$\tilde{\mu}=\tilde{\lambda}\tilde{\gamma}+3\tilde{\gamma}^{-1}\tilde{\Gamma}=-\varepsilon\lambda, \tag{83}$$

Q.E.D

5. DISCUSSION

Newton's law of gravitation is derived in the literature on the assumptions that (i) the orbits are ellipses with common focal point (Kepler's first law) and (ii) the areal velocity is constant (Kepler's second law). As a byproduct of the present analysis we have derived Newton's law by using only the first law and the assumption that the forces are conservative, which is weaker than being central .

A possible generalization of the analysis of this paper, on which we are presently working, refers to non-conservative dynamical systems. In the corresponding analysis we no longer have equation (4), while equation (5) is slightly modified. Since the majority of the necessary and sufficient conditions derived in section 2 arises from the successive differentiations of equation (5) with respect to the parameters a and b, we expect that the lack of information which results from the omission of equation (4) will be easily substituted from the information arising from the remaining equations.

Currently there is a lot of interest in the precise determination of the gravitational field of the earth from the observed motions of artificial satellites (Szebehely 1980). The theory developed in the present paper might be modified to account for such trajectories. Some preparatory numerical work would of course be necessary to fit into the theory.

APPENDIX

The evaluation of the basic quantities λ and μ of the theory of the present paper for a typical three-parametric family of planar curves

$$f(x,y,a,b) = c \tag{A.1}$$

is rather lengthy. We here obtain some useful general expressions for them with the additional assumption that the function f of equation (A.1) is homogeneous in x and y of degree n. When $n=0$ equation (A.1) represents straight lines passing through the origin and this is rather uninteresting. When $n=0$ the degree of homogeneity is irrelevant since it can be altered by raising eq. (A.1) to a suitable power. It is expected therefore that n will not appear explicitly in the expressions for λ and μ.

In fact it can be argued that any three-parametric family of curves can be put in the form (A.1) with f homogeneous. Indeed, by expressing the equation of the family of curves in polar coordinates and solving it for "r" one can always write it in the form

$$r = cg(\vartheta, a, b) \tag{A.2}$$

where one of the constants (c) is made to determine the scaling of r. Since $r = \sqrt{x^2+y^2}$ and $\vartheta = \tan^{-1}(y/x)$ are homogeneous of degree one and zero respectively, the family (A.2) is of the form (A.1) with f homogeneous of degree one. It should be pointed out, however, that the use of the homogeneous form of a given family of curves is not always the most convenient computationally. For instance, in the Example (3B) it was found more convenient to consider the non-homogeneous presentation (50) of the family of curves.

For a homogeneous f one can use Euler's theorem (stating that $xf_x+yf_y = nf$) to simplify some of the computations. Moreover in this case the function $\gamma = f_y/f_x$ is homogeneous of zero degree and therefore it can be viewed as a function of the single independent variable $z=y/x$. Hence

$$\gamma_x = -z\dot{\gamma}x^{-1} \quad , \qquad \gamma_y = \dot{\gamma}x^{-1} \quad , \tag{A.3}$$

where the dot denotes differentiation with respect to z. By expressing all the partial derivatives in terms of ordinary derivatives of γ it is straightforward to obtain that

$$\Gamma = \gamma\gamma_x - \gamma_y = (\gamma + xy^{-1})\gamma_x = -(\gamma z+1)\dot{\gamma}x^{-1} \tag{A.4}$$

and then that

$$\lambda = (x\gamma)^{-1}\{(\gamma z+1)\ddot{\gamma}\dot{\gamma}^{-1} + z\dot{\gamma} + 2\gamma\} \tag{A.5}$$

and

$$\mu = x^{-1}\{(\gamma z+1)(\ddot{\gamma}\dot{\gamma}^{-1} - 3\dot{\gamma}\gamma^{-1}) + z\dot{\gamma} + 2\gamma\} \; . \tag{A.6}$$

Equations (A.5) and (A.6) imply the useful relation

$$x\lambda + y\mu = (\gamma z+1)\gamma^{-1}\{(\gamma z+1)\ddot{\gamma}\dot{\gamma}^{-1} - 2z\dot{\gamma} + 2\gamma\} \; . \tag{A.7}$$

REFERENCES

Bozis, G.: 1982 "Two parametric families of plane orbits of a dynamical system", Int. J. Engin. Sci. (submitted).

Bozis, G.: 1983, "Generalization of the Szebehely's equation" Celes. Mech. 29, 329.

Broucke, R.: 1979, Int. J. Engin. Sci. 17, 1151.

Broucke, R. and Lass, H.: 1977, Celes. Mech. 16, 215.

Érdi, B.: 1982, "A generalization of Szebehely's equation for three dimensions", Celes. Mech., 28, 209.

Lass, H.: 1972, "The field of force for a prescribed family of curves and for a prescribed first integral", JPL-TM391-370.

Molnàr, S.: 1981, Celes. Mech. 25, 81.

Morrison, F.: 1976, Celes. Mech. 16, 227.

Szebehely, V.: 1974, "On the Determination of the Potential by Satellite Observations" in E. Proverbio (ed), Proceedings of the International Meeting on Earth's Rotations by Satellite Observations. University of Cagliari, Bologna, Italy.

Szebehely, V.: 1980, "Analysis of Lageos' altitude decrease" Paper presented at the XXIIIrd COSPAR Meeting, Budapest, 1980.

Szebehely, V. and Broucke R.: 1981, Celes. Mech. 24, 23.

Szebehely, V., Lundberg, J. and McGahee, W.: 1980, Astroph. J. 239, 880.

ANALYTICAL THEORY OF A TRAPPING IN A TWO-BODY PROBLEM OF VARIABLE MASS.

T.B.Omarov, M.J.Minglibaev.

The Astrophysical Institute of the Academy of Sciences of Kazakh SSR, Alma-Ata, USSR.

SUMMARY.

The new nonstationary model problem is considered. Its solution generalizes by form the known particular Mestschersky-Vinti solution in a two-body problem of variable mass. The equations of the corresponding perturbed motion are deduced. In the case of a two-body problem of variable mass μ the perturbing force is proportional to second temporal derivative from the value μ^{-1}. It is possible to describe with a good approximation such qualitative effects in this problem as a trapping and disintegration on a basis of properties of the model problem. Let us consider the example of a trapping.

INTRODUCTION.

Let us consider the two-body problem of variable mass $\mu(t)$

$$\frac{d^2\vec{r}}{dt^2} = -\mu(t)\frac{\vec{r}}{r^3} \tag{1}$$

Well known is the Mestschersky-Vinti solution (Mestschersky 1893, Vinti 1974) for the following particular form of a mass function

$$\mu^* = \frac{1}{\alpha + \beta t} \tag{2}$$

where α, β - constants. In polar coordinates r, ϑ in this case we have:

$$r = \frac{p}{\mu^*(1 + e\cos\varphi)}, \quad r^2\frac{d\vartheta}{dt} = \sqrt{p}, \quad \varphi = \vartheta - \omega,$$

V. V. Markellos and Y. Kozai (eds.), Dynamical Trapping and Evolution in the Solar System, 369–375.

where ρ , e , ω - constants, accordingly

$$\vec{r} = \frac{\rho}{\mu^* (1 + e \cos\varphi)} \vec{e_r}$$

$$\frac{d\vec{r}}{dt} = \frac{-1}{\mu^*} \frac{d\mu^*}{dt} r \vec{e_r} + \frac{\mu^*}{\sqrt{\rho}} e \sin\varphi \, \vec{e_r} + \frac{\mu^*}{\sqrt{\rho}} (1 + e \cos\varphi) \vec{e_n} \tag{3}$$

where $\vec{e_r}$ - single radius-vector, $\vec{e_n}$ - transversal single vector.

If we expand the inverse value of arbitrary mass $\mu(t)$ into Tailor series by the time and restrict it by the linear part of this expansion,

$$\frac{1}{\mu} = \left(\frac{1}{\mu}\right)_{t=t_0} + \left[\frac{d}{dt}\left(\frac{1}{\mu}\right)\right]_{t=t_0} (t - t_0) , \tag{4}$$

then we'll have the law (2) for μ . Let us generalize the solution (3) by form for the case of arbitrary mass $\mu(t)$:

$$\vec{r} = \frac{\rho}{\mu \ (1 + e \cos\varphi)} \vec{e_r}$$

$$\frac{d\vec{r}}{dt} = - \frac{1}{\mu} \frac{d\mu}{dt} r \vec{e_r} + \frac{\mu}{\sqrt{\rho}} e \sin\varphi \, \vec{e_r} + \frac{\mu}{\sqrt{\rho}} (1 + e \cos\varphi) \vec{e_n} \tag{5}$$

It is possible to expect that there will be a solution of the equation of the following appearance:

$$\frac{d^2\vec{r}}{dt^2} = - \mu \frac{\vec{r}}{r^3} + \vec{f} \tag{6}$$

where the vector-function $\vec{f}$ is proportional to second derivative of the value μ^{-1} .

MODEL PROBLEM.

Let us consider the equation:

$$\frac{d^2\vec{r}}{dt^2} = - \mu \frac{\vec{r}}{r^3} + \mu \vec{r} \frac{d^2}{dt^2}\left(\frac{1}{\mu}\right) \tag{7}$$

It is easy to find its integrals:

$$\left[\vec{r} \times \frac{d\vec{r}}{dt}\right] = \overrightarrow{const} = \vec{K} \tag{8}$$

$$\frac{1}{\mu^4}\left[\left(\frac{d}{dt}\mu\vec{r}\right)^2 - 2\frac{\mu^3}{r}\right] = const = h \ , \tag{9}$$

$$\frac{1}{\mu^4}\left[\frac{d}{dt}\mu\vec{r} \times \left[\mu\vec{r} \times \frac{d}{dt}\mu\vec{r}\right]\right] - \frac{\vec{r}}{r} = \overrightarrow{const} = \vec{q} \tag{10}$$

Let us rewrite the equation (7) in the following way:

$$\frac{d^2}{dt^2}\mu\vec{r} = -\mu^4\frac{\mu\vec{r}}{(\mu r)^3} + \frac{1}{2\mu^4}\cdot\frac{d\mu^4}{dt}\cdot\frac{d}{dt}\mu\vec{r} \tag{11}$$

In polar coordinates r , ϑ equations (8) and (11) have the form:

$$\mu^2 r^2 \frac{d\vartheta}{dt} = \mu^2 K \ , \tag{12}$$

$$\frac{d^2}{dt^2}\mu r - \mu^2 r^2\left(\frac{d\vartheta}{dt}\right)^2 = -\frac{\mu^4}{(\mu r)^2} + \frac{1}{2\mu^4}\cdot\frac{d\mu^4}{dt}\cdot\frac{d}{dt}\mu r \tag{13}$$

From here the equation follows:

$$\frac{d^2}{d\vartheta^2}\left(\frac{1}{\mu r}\right) + \frac{1}{\mu r} = \frac{1}{K^2} \tag{14}$$

We can write the general solution of this equation:

$$\mu r = \frac{p}{1 + e\,\mathrm{Cos}(\vartheta - \omega)} \ , \qquad p = K^2 \tag{15}$$

It is easy to determine the connection between constants of integration e and ω and values (9) and (10). In particular, for the constant e we have:

$$e = q \ , \quad e^2 - 1 = hq \tag{16}$$

Let us introduce the designations :

$$V_r = \frac{dr}{d\varphi}\cdot\frac{d\varphi}{dt} \ , \quad V_n = r\frac{d\varphi}{dt} \ , \quad \varphi = \vartheta - \omega \tag{17}$$

From the equation of orbit (15) and integral (12) we find:

$$V_r = -\frac{1}{\mu}\frac{d\mu}{dt}r + \frac{\mu}{\sqrt{p}}e\,\mathrm{Sin}\,\varphi \ , \quad V_n = \frac{\mu}{\sqrt{p}}(1 + e\,\mathrm{Cos}\,\varphi) \tag{18}$$

Hence, the equation (7) possesses the general solution of

the form (5).

PERTURBED MOTION.

Let us consider now the perturbed motion

$$\frac{d^2\vec{r}}{dt^2} = -\mu \frac{\vec{r}}{r^3} + \mu \vec{r} \frac{d^2}{dt^2}\left(\frac{1}{\mu}\right) + \vec{F} \tag{19}$$

where vector-function $\vec{F}$ in a general case depends on $\vec{r}$, $d\vec{r}/dt$ and t.

Changing constants in a solution of the form (5), we have:

$$\frac{dp}{dt} = \frac{2p^{3/2}}{\mu(1+e\cos\varphi)} F_n , \tag{20}$$

$$\frac{de}{dt} = \frac{\sqrt{p}}{\mu} \sin\varphi\, F_r + \frac{\sqrt{p}}{\mu}\left(\cos\varphi + \frac{e+\cos\varphi}{1+e\cos\varphi}\right) F_n , \tag{21}$$

$$\frac{d\omega}{dt} = -\frac{\sqrt{p}}{\mu e}\cos\varphi\, F_r + \frac{\sqrt{p}}{\mu e}\left(\sin\varphi + \frac{\sin\varphi}{1+e\cos\varphi}\right) F_n - \frac{\sqrt{p}}{\mu}\, \frac{\sin\vartheta\, \mathrm{Ctg}\, i}{1+e\cos\varphi} F_3 , \tag{22}$$

$$\frac{d\varphi}{dt} = \frac{\mu^2}{p^{3/2}}(1+e\cos\varphi)^2 + \frac{\sqrt{p}}{\mu e}\cos\varphi\, F_r - \frac{\sqrt{p}}{\mu e}\left(\sin\varphi + \frac{\sin\varphi}{1+e\cos\varphi}\right) F_n \tag{23}$$

where F_r, F_n, F_3 - projections of $\vec{F}$ in a mobile orthogonal trihedron $\vec{e}_r$, $\vec{e}_n$, $\vec{e}_3 = \vec{e}_r \times \vec{e}_n$. Let us add here equations describing the variation of an ascending node (Ω) and inclination (i) of orbit:

$$\frac{d\Omega}{dt} = \frac{\sqrt{p}}{\mu}\, \frac{\sin\vartheta}{1+e\cos\varphi} F_3 , \tag{24}$$

$$\frac{di}{dt} = \frac{\sqrt{p}}{\mu}\, \frac{\cos\vartheta}{1+e\cos\varphi} F_3 \tag{25}$$

For the equation (1) of the two-body problem of variable mass the perturbing force has the appearance:

$$\vec{F} = -\mu \vec{r} \frac{d^2}{dt^2}\left(\frac{1}{\mu}\right) \tag{26}$$

Accordingly, we have:

$$\frac{dp}{dt} = 0 , \tag{27}$$

$$\frac{de}{dt} = -\frac{\sin\varphi}{1+e\cos\varphi}\cdot\frac{p^{3/2}}{\mu}\frac{d^2}{dt^2}\left(\frac{1}{\mu}\right), \tag{28}$$

$$\frac{d\omega}{dt} = \frac{\cos\varphi}{e(1+e\cos\varphi)}\cdot\frac{p^{3/2}}{\mu}\cdot\frac{d^2}{dt^2}\left(\frac{1}{\mu}\right), \tag{29}$$

$$\frac{d\varphi}{dt} = \frac{\mu^2}{p^{3/2}}\left(1+e\cos\varphi\right)^2 - \frac{d\omega}{dt}\,. \tag{30}$$

ON THE TRAPPING IN A TWO-BODY PROBLEM OF VARIABLE MASS.

There is known the following system of elements of osculating conic section in a two-body problem of variable mass (Hadjidemetriou, 1967):

$$\frac{d\tilde{p}}{dt} = -\frac{\tilde{p}}{\mu}\cdot\frac{d\mu}{dt}, \tag{31}$$

$$\frac{d\tilde{e}}{dt} = -\left(\tilde{e}+\cos\tilde{\varphi}\right)\cdot\frac{1}{\mu}\cdot\frac{d\mu}{dt}, \tag{32}$$

$$\frac{d\tilde{\omega}}{dt} = -\frac{\sin\tilde{\varphi}}{\tilde{e}}\,\frac{1}{\mu}\cdot\frac{d\mu}{dt}, \tag{33}$$

$$\frac{d\tilde{\varphi}}{dt} = \sqrt{\frac{\mu}{\tilde{p}}}\cdot\frac{(1+\tilde{e}\cos\tilde{\varphi})^2}{\tilde{p}} - \frac{d\tilde{\omega}}{dt} \tag{34}$$

Model problem here is aperiodical motion along a conic section

$$\vec{r} = \frac{\tilde{p}}{1+\tilde{e}\cos\tilde{\varphi}}\,\vec{e}_r \tag{35}$$

$$\vec{V} = \sqrt{\frac{\mu}{\tilde{p}}}\left[\tilde{e}\sin\tilde{\varphi}\,\vec{e}_r + \left(1+\tilde{e}\cos\tilde{\varphi}\right)\vec{e}_n\right] \tag{36}$$

which is described by the equation of the form:

$$\frac{d^2\vec{r}}{dt^2} = -\mu\,\frac{\vec{r}}{r^3} + \frac{1}{2\mu}\cdot\frac{d\mu}{dt}\cdot\frac{d\vec{r}}{dt} \tag{37}$$

The corresponding perturbed force has the following structure:

$$\vec{\tilde{F}} = -\frac{1}{2\mu}\cdot\frac{d\mu}{dt}\,\frac{d\vec{r}}{dt} \tag{38}$$

This interpretation of the two-body problem of variable mass has been made simultaneously and independently by Hadjidemetriou (1963) and Omarov (1963).

The parameter $\tilde{p}$ of osculating conic section changes inverse proportionally to mass μ. As it is clear from the formula (27), the value p in the two-body problem of variable mass is constant. Besides that, values de/dt and $d\omega/dt$ in this problem have a higher order of small magnitude comparatively with velocities of variation of the osculating eccentricity $\tilde{e}$ of the element $\tilde{\omega}$. In consequence of that fact unperturbed motion (5) and its integrals (8)-(10) can be considered as good approximate correlations of the two-body problem of variable mass when describing such qualitative effects, as trapping and disintegration.

Let us consider examples of trapping for the case when in the solution (5) for the increasing mass the constant value $e = 0$, and, accordingly, we have:

$$h = -\frac{1}{p}\,, \quad r = \frac{p}{\mu}\,, \quad p = \mu_0 r_0 \tag{39}$$

Let us rewrite the integral (9) in the following form:

$$V^2 - 2\frac{\mu}{r} = h\mu^2 - 2\frac{\dot{\mu}}{\mu}\dot{r}r - \frac{\dot{\mu}^2}{\mu^2}r^2 \tag{40}$$

where the point designates a differentiation by time. For our case we have:

$$V^2 - 2\frac{\mu}{r} = -\frac{\mu^2}{p}\left(1 - p^3\frac{\dot{\mu}^2}{\mu^6}\right) \tag{41}$$

Let, in a general initial moment of time t_0, the energy of the corresponding binary system of increasing mass be a positive one

$$V_0^2 - 2\frac{\mu_0}{r_0} > 0\,. \tag{42}$$

With due regard for $e = 0$, simultaneously, the inequality must be carried out that

$$\frac{\dot{\mu}_0^2}{\mu_0^3} > \frac{1}{r_0^3} \tag{43}$$

As it is clear from the formula (41), for the next trapping

$$V^2 - 2\frac{\mu}{r} < 0 \qquad (44)$$

accomplishment of the following condition is necessary

$$\frac{d}{dt}\left(\frac{\dot{\mu}^2}{\mu^6}\right) < 0 . \qquad (45)$$

If the mass μ increases according to Eddington-Jean´s law

$$\dot{\mu} = \alpha \mu^n , \qquad \alpha > 0 \qquad (46)$$

then the necessary condition of a trapping (45) is fulfilled for meaning $n < 3$. The following expression results from the law (46):

$$\mu(t) = \left[\alpha(1-n)(t-t_0) + \mu_0^{1-n}\right]^{\frac{1}{1-n}} , \qquad n \neq 1 \qquad (47)$$

and, hence, for the case $n < 1$, a trapping really takes place in the time interval

$$\Delta t = \frac{\left(\alpha r_0^{3/2} \mu_0^{3/2}\right)^{\frac{n-1}{n-3}} - \mu_0^{1-n}}{\alpha(1-n)} \qquad (48)$$

When $n = 1$, for the time of trapping we have:

$$\Delta t = \frac{1}{2\alpha} \ln\left(\frac{\mu_0^{1/2}}{\alpha r_0^{3/2}}\right) \qquad (49)$$

Finally, the restriction (43), connected with the condition $e = 0$, excluded a trapping for cases, when $1 < n < 3$.

REFERENCES.

1. Mestschersky I., 1893, Astron.Nachr., 132, N3159, p.9.
2. Vinti J., 1974, Mont.Not., 169, N3, p.417.
3. Hadjidemetriou J., 1963, Icarus, 2, p.404.
4. Hadjidemetriou J., 1967, Advances in Astron. and Astrophys., 5, p.131.
5. Omarov T.B., 1963, Astron. Journal of Academy of Sciences of USSR, 40, N4, p.921.

LOW VELOCITY ENCOUNTERS OF MINOR BODIES WITH THE OUTER PLANETS

A. Carusi, E. Perozzi and G.B. Valsecchi
I.A.S-C.N.R, Reparto di Planetologia, Roma (Italy)

Previous studies of close encounters of minor bodies with Jupiter have shown that the perturbations are stronger either if the encounter is very deep or if the velocity of the minor body relative to the planet is low. In the present research we investigate the effects of low velocity encounters between fictitious minor bodies and the four outer planets. Two possible outcomes of this type of encounter are the temporary satellite capture of the minor body by the planet, and the exchange of perihelion with aphelion of the minor body orbit. Different occurrence rates of these processes are found for different planets, and the implications for the orbital evolution of minor bodies in the outer Solar System are discussed.

INTRODUCTION

Close encounters with the outer planets were shown to be one of the most important factors determining the orbital evolution of those comets which move entirely within the planetary region by the classical studies of Kazimirchak-Polonskaya (1967) and Belyaev (1967). They integrated the motion of many short-period comets for a time span of 400 years, and found that encounters especially with Jupiter and Saturn caused drastic changes of the cometary orbits. Using fictitious objects, Kazimirchak-Polonskaya (1972) showed that also Uranus and Neptune could have an important effect on those orbital evolutions.

In fact, cometary orbits of this type are called "chaotic" (Oikawa and Everhart, 1979; Everhart, 1979, 1982), since they can pass through various orbital forms, such as unstable

V. V. Markellos and Y. Kozai (eds.), Dynamical Trapping and Evolution in the Solar System, 377–395.

Trojans, horseshoes, generalized Trojans and horseshoes of some planets, Chiron type orbits, short-period cometary orbits, temporary satellite captures and others (Everhart, 1973, 1979). The entrance into, and the exit from, any of these types of motion are caused by an encounter with a major planet. Chiron is presently in a typical chaotic orbit (Everhart, 1979): its motion has been studied by Kowal et al. (1979), by Oikawa and Everhart (1979) and by Scholl (1979). The two latter studies, in particular, showed that encounters with Saturn and, to a lesser extent, with Uranus play the major role in the evolution of Chiron's orbit, which will likely evolve into an orbit typical of short-period comets in about 10^5 years.

Even if the motion of many different objects in chaotic orbits has been integrated over such time spans, a better knowledge of these orbits is needed in order to obtain a reliable quantitative description of the evolution of comets into short-period orbits. A key point appears to be the knowledge of all the possible outcomes of planetary close encounters, with the relative a priori probabilities, given certain ranges of initial conditions.

Here we will concentrate on two types of events which can occur when the relative velocity at the encounter is low: temporary satellite capture of the minor body, and transformation of its orbit from outside to inside that of the planet or vice-versa. The data that we will use for the discussion come from a numerical research on close encounters with the outer planets described in Carusi and Valsecchi (1982c). In essence, 1000 fictitious minor bodies were followed during a single close encounter with each outer major planet; thus, a total of 4000 close encounters have been computed and stored on magnetic tape. The initial orbits are chosen so as to span a wide range of initial conditions. The form of the distributions is identical for the four populations having encounters with the four outer planets, except for a scaling factor multiplying the 1/a distributions. The initial distribution functions of e and i were:

$$P(e)=\sin(\pi e) \quad \text{for} \quad 0<e<1$$

$$P(i)=\sin(6i) \quad \text{for} \quad 0^\circ<i<30^\circ$$

The whole intervals of variation of e and i were divided in 40 classes, and for each class of eccentricity a flat distribution in 1/a between the limits:

$$\frac{1+e}{q-R} > \frac{1}{a} > \frac{1-e}{Q+R}$$

was generated, where q and Q are the planet's perihelion and aphelion distances, and a and e refer to the minor body; R is 2/3 AU for Jupiter (as in the previous research), Saturn, Uranus, and 4/3 AU for Neptune. In order to take into account the lack of low eccentricity short-period comets of high inclination, an additional constraint was added, namely:

$$i < 2.4^\circ + 80.8^\circ e$$

This condition gives a line that roughly divides, in the e-i plane, the region populated by known short-period comets from the empty one.

The values of ω and Ω were chosen at random between 0° and 360° and it was then checked if the resulting orbits had a minimum distance from the planet's orbit of less than R. If not, they were discarded and replaced by new orbits also conforming to the given constraints. The initial values of the true anomalies were chosen with the same procedure used in Carusi and Pozzi (1978a), moving the planet and the minor body backwards, along their unperturbed orbits, from the point of minimum distance to a relative distance of 4R.

Each sample was named using the first three letters of the name of the corresponding planet; each object of a sample was identified with a number. So, JUP 393 is a fictitious object having an encounter with Jupiter, and URA 802 one having an encounter with Uranus.

TEMPORARY SATELLITE CAPTURES

The subject of temporary satellite captures (TSC) of minor bodies by the planets has received some attention, especially in connection with the origin of some natural satellites (e.g. the retrograde satellites of Jupiter, Saturn and Neptune) and with the orbital evolution of objects in chaotic orbits, like comets and meteoroids (Everhart, 1982). It is generally agreed upon the statement that definitive satellite captures, either in the restricted or the general 3-body problem, are impossible without the help of a dissipative mechanism (see, e.g., Pollack et al., 1979). Therefore, in the framework of purely gravitationally interacting mass-points, we are concerned with satellite captures only of the temporary type; in fact, they have been found in many investigations on the orbital evolution of real and fictitious short-period comets (Chebotarev, 1967; Kazimirchak-Polonskaya, 1972; Everhart, 1973; Dvorak, 1976; Carusi and Pozzi, 1978b; Rickman, 1979).

In most of the cases cited above, Jupiter has been the planet involved; in all of them the gravitational model was either a n-body or an elliptical 3-body problem. The use of a planar, circular restricted 3-body model, as done for example by Horedt (1976), Hayashi et al. (1977) and Heppenheimer and Porco (1977), often leads to results different from those obtained with the more realistic models cited above (Everhart, 1973; Carusi et al., 1979).

Systematic work on TSC's of minor bodies by Jupiter has been presented in Carusi et al. (1979) and Carusi and Valsecchi (1979, 1980, 1981). It was shown that TSC's are more likely to occur when the pre-encounter orbit of the minor body is nearly tangent to that of the planet; in this case, because of the high value of the Tisserand invariant of such orbits, a low velocity of the minor body relative to the planet may be expected, leading more easily to the satellite capture. It was also shown that in a sample composed by the majority of the short-period comets with Tisserand invariant (relative to Jupiter, and in jovian semiaxis units) greater than 2.9, 7 out of 22 underwent a TSC in the last 120 years. Most of these captures, when the trajectories of the comets were plotted in a frame centred on Jupiter and rotating with its instantaneous orbital angular velocity, were found to be simple fly-bies, although one of the comets (P/Gehrels 3) made a complete revolution about the planet during its 7 years long TSC occurred between 1967 and 1974 (Rickman, 1979; Rickman and Malmort, 1981, 1982; Carusi and Valsecchi, 1979, 1981, 1982b).

Depending on the duration of the TSC, more or less complicated planetocentric orbital patterns of the small body may be expected. Carusi et al. (1981a,b, 1982) have given a first look into this problem, starting from the orbit on which P/Oterma moved before its 1937 encounter with Jupiter, and using also, for comparison, some fictitious model objects. Other studies of this type seem necessary in order to understand more deeply the phenomena occurring at close encounters, including TSC's.

So far we have spoken of TSC's using implicitly the definition of them given in Carusi and Valsecchi (1979, 1981, 1982a,b,c): the minor body is a temporary satellite if its planetocentric orbital elements are elliptical for some time during an encounter. This is essentially the definition used by Kazimirchak-Polonskaya (1972), Everhart (1973) and Dvorak (1976); Rickman and Malmort (1982), however, require in addition that a planetocentric orbit, from pericentre to pericentre, is to be completed by the minor body, and that the motion in a planetocentric elliptical orbit has to last for at least 1000 days.

In the present investigation we have used for comparison, together with the data referred to in the Introduction, also some of the cometary evolutions described in Carusi and Valsecchi (1981, 1982a). The number of TSC's found in our samples of fictitious objects are (Carusi and Valsecchi, 1982c): 46 in JUP, 16 in SAT, 4 in URA and 4 in NEP.

Some qualitative information can be obtained plotting the values of the heliocentric and planetocentric energies of the minor bodies, computed from the osculating orbits, for a time span including the encounter (Carusi and Valsecchi, 1982 b). This is especially true since in such a plot the relative influences of the Sun and of the planet on the motion of the small body can be easily recognized. Therefore we analyzed the motion of all the objects undergoing a TSC in any of our four samples, drawing for each of them a figure, like Figs. 1 and 2, composed of the trajectory in the rotating planetocentric frame plus the energy plot just described.

About 2/3 of the TSC's found in JUP and SAT, and all of those found in URA and NEP, turned out to last considerably less than the total duration of the interaction. The period spent with elliptical planetocentric parameters appeared in those cases as a minor, and almost incidental, part of the planet-dominated orbital evolution. A typical example of this type of satellite capture is the object JUP 63, shown in Fig. 1.

On the other hand, the remaining 1/3 of the TSC's found in the samples JUP and SAT appear much more interesting. Although a clear-cut distinction from those that we have just

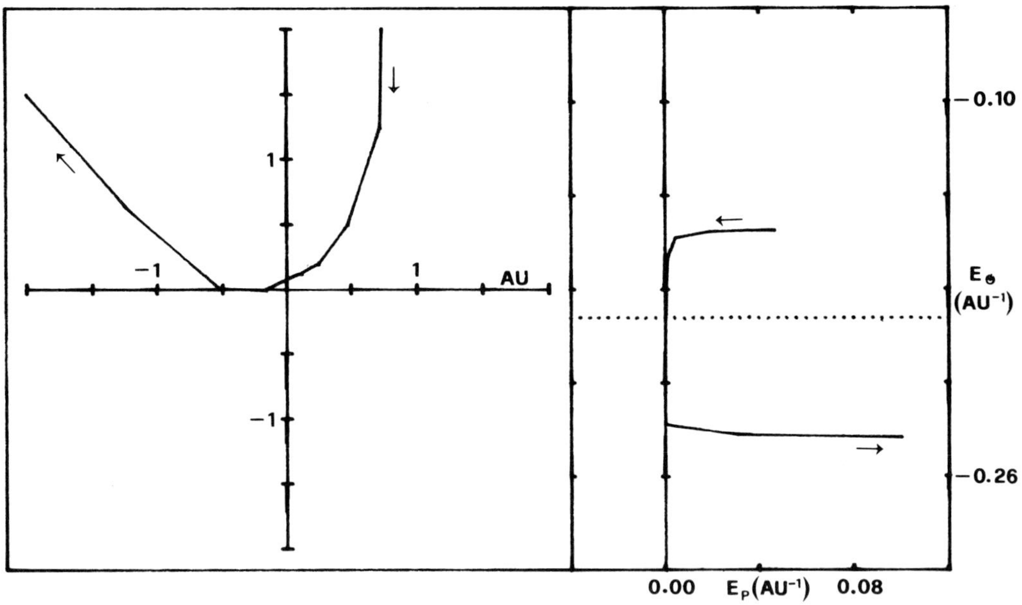

Fig. 1 - Left: ecliptic projection of the planetocentric path followed by JUP 63 in a rotating frame with the Sun on the negative x-axis.
- Right: energy plot of the evolution of JUP 63; $E_\odot = -1/a_\odot$, $E_p = -m_p/(m_\odot a_p)$ ($m_\odot, m_p$: masses of the Sun and the planet; $a_\odot$, a_p: heliocentric and planetocentric semiaxes of the minor body). Arrows show the direction of motion.

discussed cannot be made, this second type of TSC's is characterized by the fact that all, or almost all, the planetocentric motion is bound. Figure 2 shows an example, the object JUP 105; its trajectory in the energy plot resembles that of comet P/Oterma (see Carusi and Valsecchi, 1982b, Fig. 2), and in fact this comet, together with comet P/Gehrels 3, represents the best example of TSC of a short-period comet known so far. Carusi et al. (1981b) and Rickman and Malmort (1981, 1982) have examined the effects of varying one or more orbital parameters of the pre-encounter orbit (for P/Oterma) or the post-encounter one (for P/Gehrels 3): in both cases, among the varied orbits, some have been found that lead to long lasting (more than 50 years) TSC's. It is conceivable that also in the vicinity of the orbits of those objects in JUP and SAT undergoing these deeper and longer captures of the second type such long lasting satellite evolutions can be found. An example of an object of the sample SAT undergoing a TSC of the second type is shown in Fig. 3.

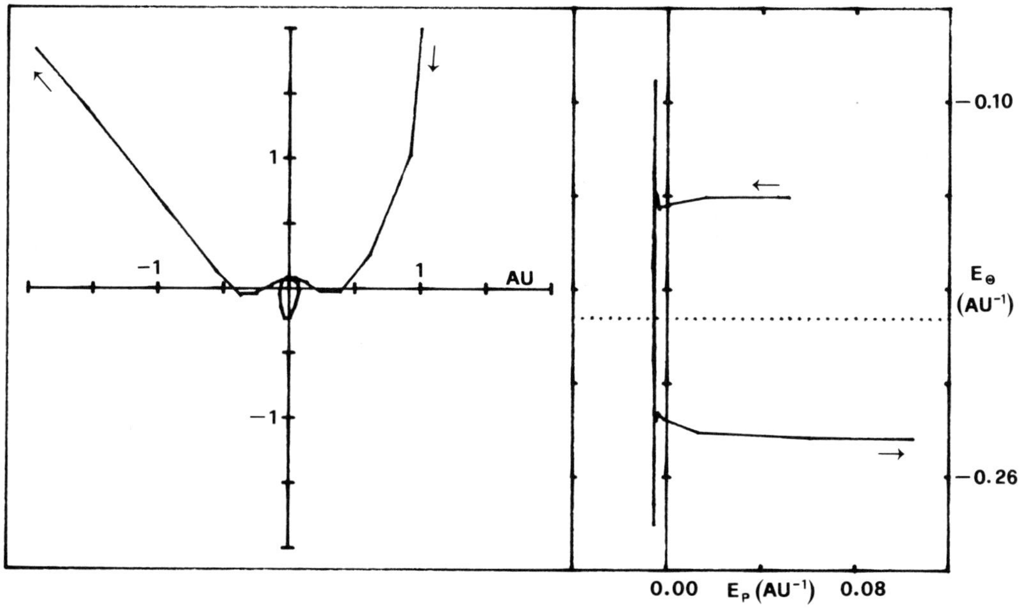

Fig. 2 - Same as Fig. 1 for the object JUP 105.

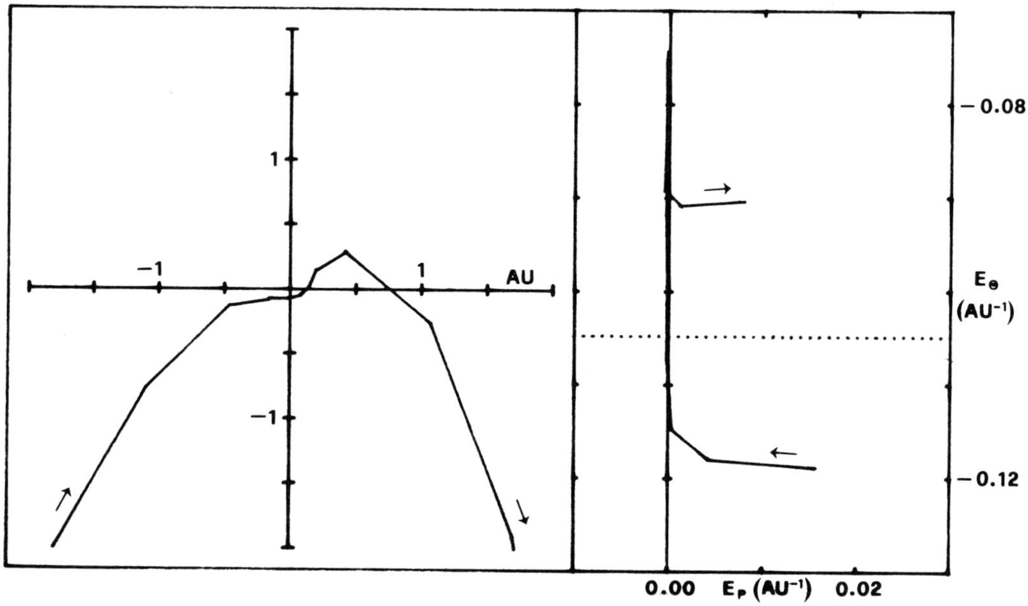

Fig. 3 - Same as Fig. 1 for the object SAT 33.

We can compare these results, relative to fictitious objects, with the findings regarding short-period comets. Carusi and Valsecchi (1981) reported 11 cases of TSC by Jupiter shared among 7 comets, within the last 120 years. Of them, 3 are of the second type (the one of P/Gehrels 3 between 1967 and 1974, and the two consecutive ones of P/Oterma between about 1935 and 1965; see Figs. 1 and 2 in Carusi and Valsecchi, 1982b), in rough agreement with the proportion found for the fictitious objects. Two examples of captures of the first type of real short-period comets, showing also rather interesting patterns in the energy plot, are given in Figs. 4 and 5.

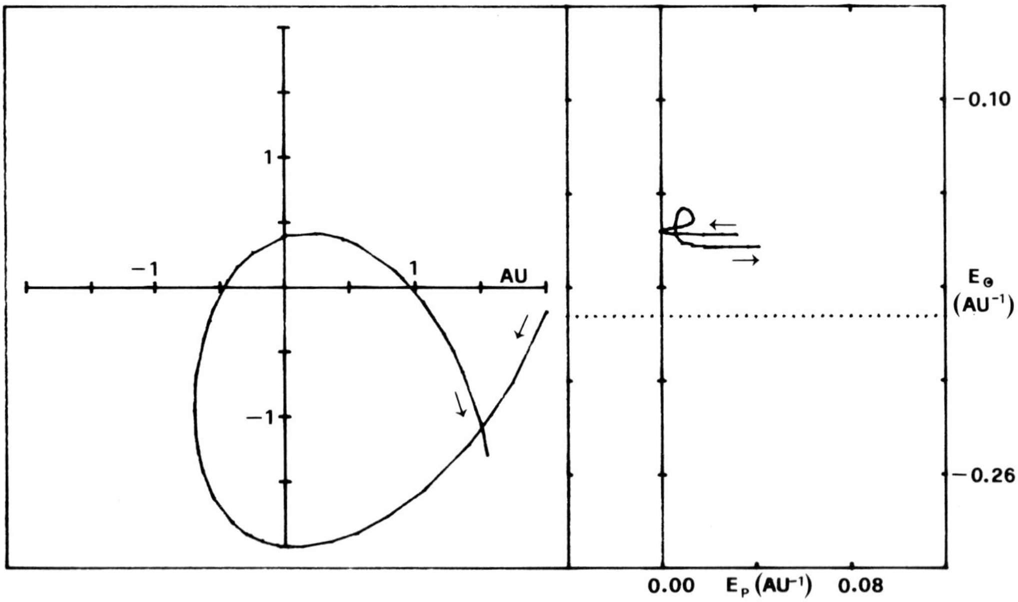

Fig. 4 - Same as Fig. 1 for the 1946-1954 encounter of P/Kowal with Jupiter.

Let us return to the question of the definition of TSC given by Rickman and Malmort (1982). Of their two additional requirements, one is rather general, and is that the small body has to perform a complete revolution about the planet; the second, the minimum duration of the bound motion of at least 1000 days, is related to captures by Jupiter, since that time lenght is roughly the period of a jovian satellite having semimajor axis of about 0.2 AU, and longer minimum du-

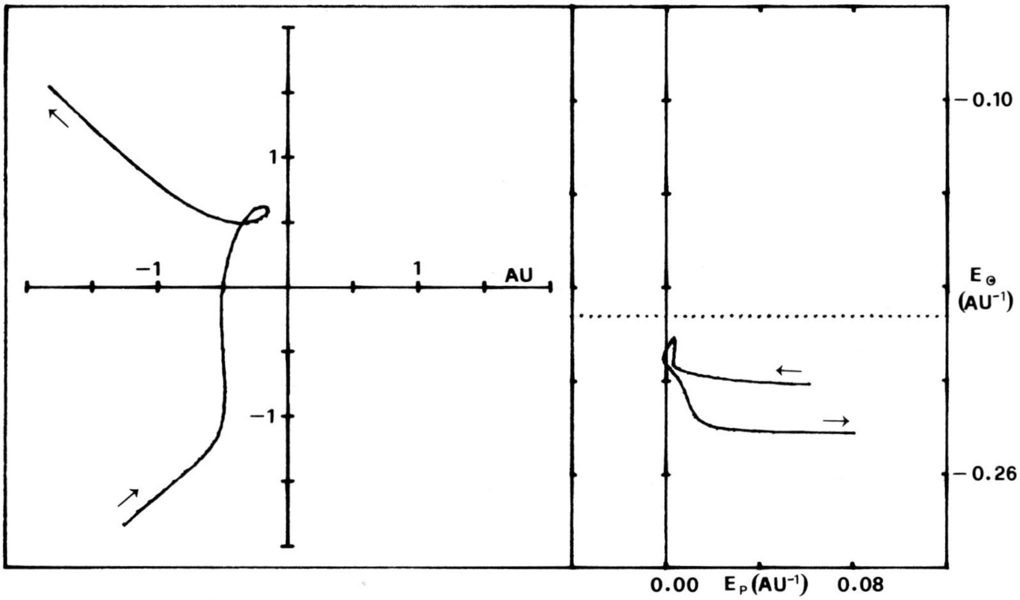

Fig. 5 - Same as Fig. 1 for the 1875-1885 encounter of P/Gunn with Jupiter.

rations should be applicable to planets less massive and more distant from the Sun. It appears that only some of the TSC's of the second type can meet these conditions, the duration of the "incidental" ones like that of Fig. 1 being too short. Satellite captures like those of P/Oterma and P/Gehrels 3 are not equivalent, according to Rickman and Malmort; however these two comets exhibited a similar behaviour in the energy plot, and the finding by Carusi et al. (1982) of a jovicentric orbital pattern of a "varied" P/Oterma very similar to the "true" P/Gehrels 3 shows that the TSC's of these two comets are to be considered as equivalent; it was only by chance that the "true" P/Oterma did not perform a complete revolution, and a definition of TSC according to which these two comets are not both temporary satellites seems a bit artificial. We suggest that, if any further refinement of the definition has to be done, starting from the simple one that has been used by the majority of the researchers, it should rather take into account the behaviour of the minor body in the energy plot.

The initial conditions leading to TSC's of the second type discussed are rather restrictive both in the case of encounters with Jupiter and with Saturn. Table I gives the approximate widths of the "capture zones" of the two types of TSC for JUP and SAT, and of the first type, the only one found, for URA and NEP. The pre-encounter orbits, as well as the post-encounter ones, are of low inclination, of high Tisserand invariant and of small to moderate eccentricity; the

Table I: Ranges of initial and final orbital parameters for the TSC's found in the present research.

		T	a_P/a	e	i max
JUP	first type	2.82-3.06	0.46-1.41	0.03-0.52	17°
	second type	2.98-3.04	0.66-1.38	0.03-0.33	7°
SAT	first type	2.97-3.02	0.76-1.29	0.03-0.26	8°
	second type	2.98-3.02	0.81-1.27	0.01-0.23	4°
URA	first type	2.99-3.00	0.89-1.11	0.01-0.07	6°
NEP	first type	2.99-3.00	0.89-1.11	0.01-0.08	5°

"capture zones" for the TSC's of the first type shrink considerably passing from Jupiter to Saturn, and decrease even much more for Uranus and Neptune.

We do not find TSC's of the second type in URA and NEP because the corresponding "capture zones" are very small (likely, they are smaller than those for TSC's of first type, as for Jupiter and Saturn) and due to the limitations of our samples, they are presumably not populated enough. Therefore a more specific study of TSC's of minor bodies by Uranus and Neptune requires samples of, say, 100 at least initial orbits confined in the region of the phase space given in Table I; also satellite captures of the second type should then be found in such samples.

In a study on the limits of stability of the outer jovian satellites Hunter (1967a,b) published the final heliocentric orbital elements of those fictitious satellites that had escaped from the jovian system. Table II has been compiled from Tables I and II of Hunter (1967b), and is to be compared to the second row of Table I, which contains the TSC's of the second type found in JUP.

Table II: Ranges of final heliocentric elements of escaped jovian satellites found by Hunter (1967b).

T	a_p/a	e	i max
3.00-3.07	0.56-1.41	0.03-0.40	5°

An indirect confirmation that the phenomena connected with TSC's are essentially the same for the various outer planets, the difference being in the suitable initial conditions, is given by the fact that in the examination of the 70 captured objects of this research the main conclusions of the study of orbital patterns of minor bodies at close encounters with Jupiter by Carusi et al. (1982) were confirmed: the basic types of planetocentric trajectories were found to be essentially the same described by Carusi et al.; no hyperbolic ejection after a TSC was found because of the high value of the Tisserand invariant; reduction of the minimum distance of approach below the "unperturbed" value, that is the value obtained if the minor body orbit were not modified by the planet, was found in the majority of cases.

The implications of the previous considerations for the possibility of satellite captures of real short-period comets can be summarized in this way:

a) the initial conditions required for TSC's become more and more restrictive as we proceed outwards from Jupiter to Neptune; the probability of finding a comet temporarily bound to a planet at any given time of course depends on the still unknown density of comets in the appropriate regions of the phase space given in Table I;

b) Table I helps to understand why Carusi and Valsecchi (1981) found several TSC's among the short-period comets with T greater than 2.9; also the second type TSC's of P/Oterma and of P/Gehrels 3 are in agreement with Table I;

c) on the other hand, a similar research among real short-period comets looking for TSC's by Saturn is almost hopeless: no comets of those listed in Marsden's Catalogue (Marsden, 1979) meet the requirements, although P/van Houten would only need a somewhat less eccentric orbit, its T and i being in the acceptable range.

It must be added to point c) that some of the orbits of

comets in the "trans-jovian belt" (Kresák, 1972) meet the requirements of Table I for TSC's by Saturn; the problem is that the integrations backwards, before the transferring encounter with Jupiter, are rather unreliable over long time spans (Pittich, 1981), rendering the initial conditions of possible encounters with Saturn very uncertain. The present probable orbit of comet P/Oterma (Marsden and Roemer, 1982) is within the limits of Table I for TSC's both by Jupiter and by Saturn, a dynamical characteristic unique in the whole sample of known short-period comets.

EXCHANGES OF PERIHELION AND APHELION

To study this process, in the samples JUP, SAT, URA and NEP all the objects that, as a consequence of the encounter, changed their semimajor axis from one greater than that of the planet to one smaller, or vice-versa, were identified and individually examined. They amounted to 117 in JUP, 57 in SAT, 11 in URA and 17 in NEP. Fig. 6 shows all these objects in a $-1/a,e$ diagram in which each of them is represented by a straight line connecting its initial and final orbital elements; the initial orbit, moreover, is denoted by a black dot. The dotted lines enclose the regions of the diagram in which encounters with a specific planet, if not prevented by inclination, orientation of the orbit or libration mechanisms, can occur.

The figure discloses many interesting features of the processes that we are considering. First of all, notice the small displacements of objects on orbits with initial and final semimajor axes very close to that of the corresponding planet and moderate eccentricity. The interactions of these objects do not seem very interesting, since they do not imply substantial orbital transformations.

Much more interesting are all the other interactions. We can notice that their number increases very much passing from the two outermost planets to Saturn and then to Jupiter, and that the range of suitable initial conditions increases accordingly. From previous studies of close encounters of minor bodies with Jupiter (Carusi and Valsecchi, 1982b) we know that the interaction with the planet is stronger either if the minimum approach distance, along the unperturbed initial orbit, is very small, or if the relative velocity at the

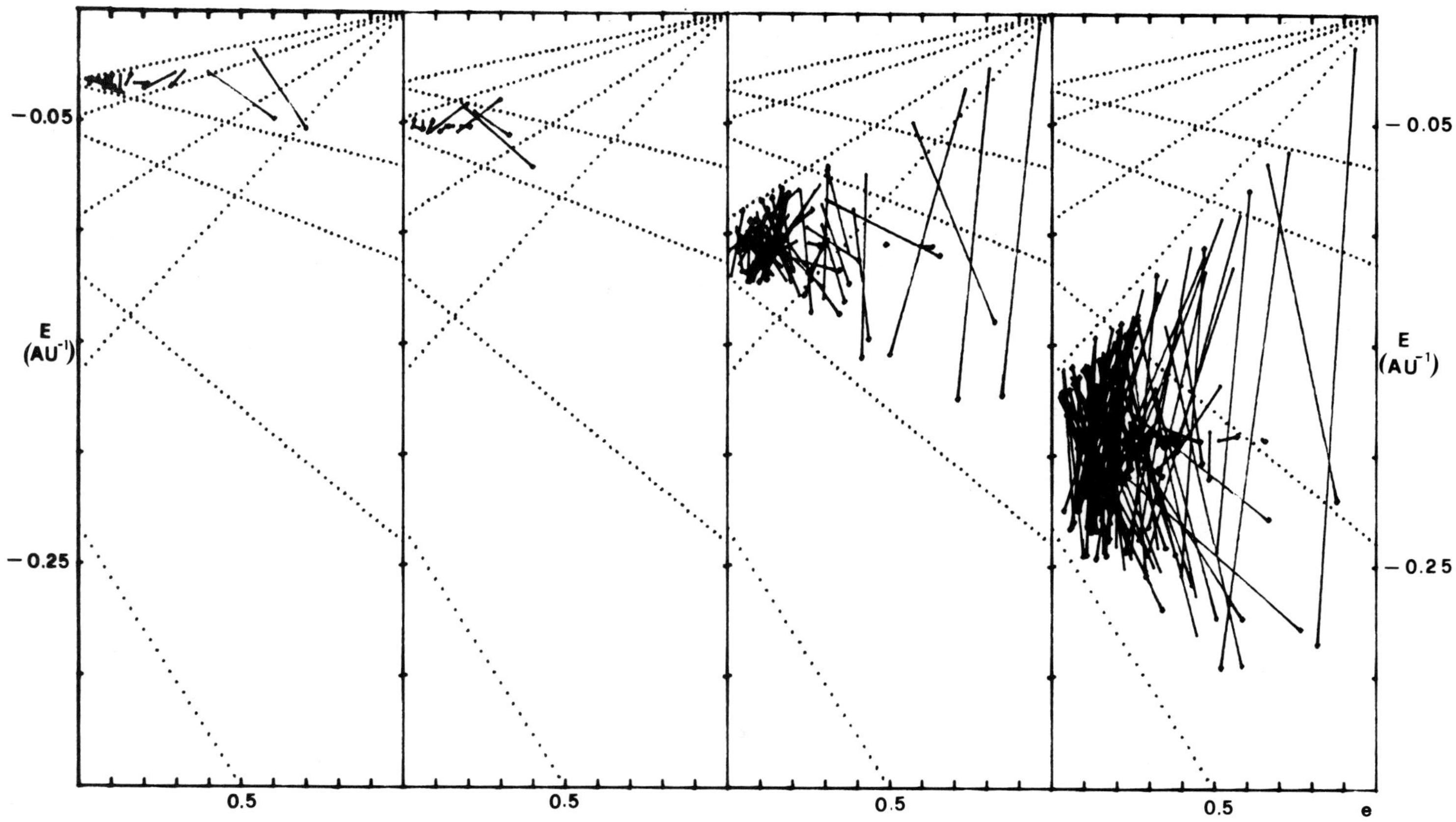

Fig. 6 - Each of the four boxes is a $-1/a, e$ diagram in which the pairs of dotted lines enclose the ranges of semimajor axis and eccentricity allowing close encounters with the corresponding planet; the uppermost pair refers to Neptune, the lowermost to Jupiter. Straight lines connect initial and final orbits of minor bodies encountering the planets (see text); dots denote the initial orbits. From left to right: encounters with Neptune; with Uranus; with Saturn; with Jupiter.

encounter is low. The first condition involves, in a complex way, all the orbital parameters of the planet and of the minor body; on the other hand, a good and simple way to check the second condition is to look at the value of the Tisserand invariant $T=1/a+2\sqrt{a(1-e^2)}\cos i$, where a is in units of the planet semimajor axis. A value of T close to 3 is typical of objects that can have encounters at low or very low relative velocity (Carusi and Valsecchi, 1982a).

Both the types of encounter have their representatives in Fig. 6; notice how many objects, as a consequence of a strong interaction, are transferred into regions in which one or more other planets can take the control of the object or, conversely, how many objects are subtracted to the control of other planets.

When the encounter velocity is very low it is possible that the pre-encounter orbit of the minor body does not cross the post-encounter one. This process requires an initial orbit nearly tangent to that of the planet; the ranges of orbital parameters of the cases that we have found in the present research are reported in Table III. These orbital transformations are a special case of the "transitions", defined by Carusi et al. (1982) as being exchanges of perihelion with aphelion or vice-versa, as a consequence of the close encoun-

Table III. Ranges of initial and final orbital parameters for the non-intersecting transitions found in this research.

	T	a_p/a	e	i max
JUP	2.93-3.03	0.48-1.40	0.03-0.52	15°
SAT	2.97-3.02	0.74-1.29	0.00-0.28	8°
URA	3.00-3.00	0.94-1.01	0.03-0.03	4°
NEP	2.99-3.00	0.95-1.15	0.03-0.14	4°

Note that the values for URA correspond to the only case found in that sample.

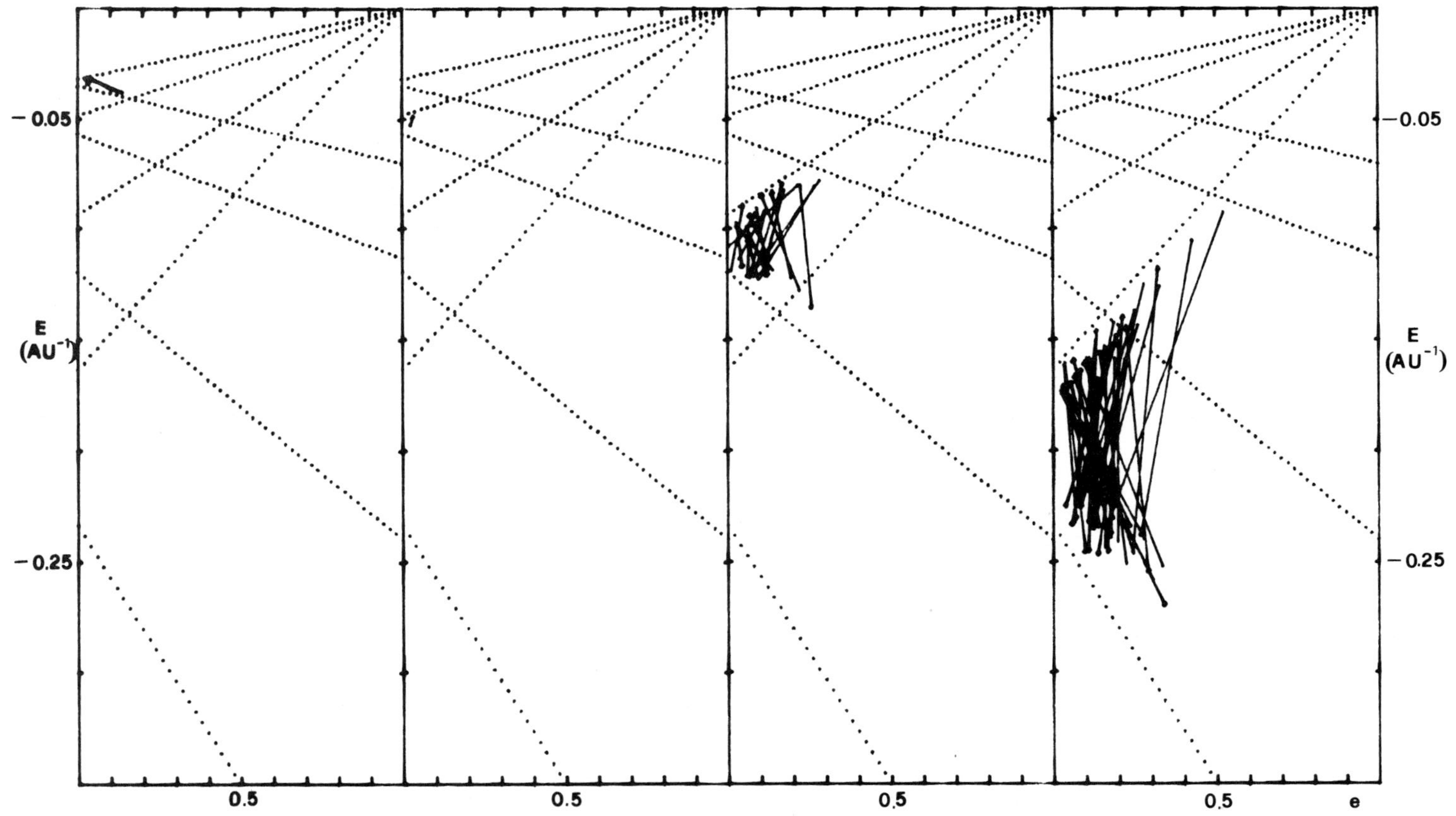

Fig. 7 - Same as Fig. 6 for objects of the four samples making "non-intersecting transitions (see text).

ter. Let us call them non-intersecting transitions. In only one case we have found an object whose pre-encounter and post-encounter orbits do not cross and the semimajor axes of both the orbits are less than that of the planet (Saturn, in this particular instance); this seems a limiting case, where the interaction has taken place with Saturn close to its perihelion - its distance from the Sun at the time of closest approach being 9.10 AU - and the initial and the final orbits of the minor body where nearly circular (e=0.06 for both).

Fig. 7 shows the cases of these "non-intersecting transitions" found in the four samples. Also in this figure the number of objects increases substantially passing from Neptune to Jupiter: the numbers (4 for NEP, 1 for URA, 19 for SAT and 43 for JUP) are comparable, except that of Uranus, to those of the temporary satellite captures found in the same samples. There are also other connections with the satellite captures:

a) the ranges of orbital parameters in which the two processes are found are very similar, as is shown by a comparison of Table III with Table I;

b) in the samples JUP and SAT, 22 out of 43 and respectively 8 out of 19 of the objects making a non-intersecting transition undergo also a temporary satellite capture; these captures are equally divided between the two types described above. For Uranus and Neptune, we have to remark that the size of the samples is not large enough to allow the finding of a sufficient number of events, thus rendering any conclusion only tentative.

The preceding considerations have some implications for the orbital evolution of small bodies in the outer Solar System. All the four outer major planets are able to transform effectively cometary orbits as a consequence of close encounters. All of them can change a cometary orbit in such a way that the pre-encounter and the post-encounter ellipses do not cross. However, Uranus and Neptune can do so only for initial orbits of the minor body very similar to that of the planet, a case which may be very rare in the orbital evolution of a comet. On the other hand, both Jupiter and Saturn display a great efficiency in such processes; we know of several comets that have undergone these orbital transformations: the first six entries of Table VI of Carusi et al. (1982) are known examples of this type, the sixth one, the 1886 encounter of P/Brooks 2 with Jupiter being an extreme case. The ranges of orbital parameters of Table III agree well with the parameters of the comets cited above. Moreover, as in the case of the fictitious objects, some of the six comets underwent a temporary satellite capture during the encounter; these are the two well known cases of P/Oterma and P/Gehrels 3. As regards Saturn, due to the limits on the orbital parameters given in Table III, the same considerations done in the case of the temporary satellite captures are applicable: some orbits of comets in the "trans-jovian belt" (Kresák, 1972) are suitable for the non-intersecting transitions caused by encounters with Saturn, although no example of this type has been found yet in numerical integrations of the past evolutions of short-period comets or in the numerical integrations of Chiron's orbit. In this last case, we can notice that the orbit is too eccentric, and the Tisserand invariant with respect to Saturn too low (see Oikawa and Everhart, 1979) in order to have very low velocity encounters with that planet.

A final remark concerns the planetocentric trajectories of the non-intersecting transitions. Direct inspection of these patterns has shown that most of them are of the very simple type reported in Figs. 1 and 3; this pattern has been encountered very frequently also by Carusi et al. (1982) in their study of orbital patterns at close encounters of minor bodies with Jupiter.

The authors thank the Accademia Nazionale dei Lincei for supporting the computing expenses, Dr. L. Kresák for useful discussions on the subject of this paper, and Drs. P. Farinella, H. Rickman and H. Scholl for critically reading an earlier version of the manuscript.

REFERENCES

Belyaev, N.A.: 1967, Sov. Astron. - A.J., 11, pp. 366-373.
Carusi, A., and Pozzi, F.: 1978a, Moon and Planets 19, pp. 65-70.
Carusi, A., and Pozzi, F.: 1978b, Moon and Planets 19, pp. 71-87.
Carusi, A., and Valsecchi, G.B.: 1979, in "Asteroids" (T. Gehrels ed.), Tucson, USA, pp. 391-416.

Carusi, A., and Valsecchi, G.B.: 1980, Moon and Planets 22, pp. 113-124.
Carusi, A., and Valsecchi, G.B.: 1981, Astron. Astrophys. 94, pp. 226-228.
Carusi, A., and Valsecchi, G.B.: 1982a, in "Comparative Study of the Planets" (A. Coradini and M. Fulchignoni eds.), D. Reidel, Dordrecht, Holland, pp.131-138.
Carusi, A., and Valsecchi, G.B.: 1982b, in "Sun and Planetary System" (W. Fricke and G. Teleki eds.), D. Reidel, Dordrecht, Holland, pp. 379-384.
Carusi, A., and Valsecchi, G.B.: 1982c, in "Sun and Planetary System" (W. Fricke and G. Teleki eds.), D. Reidel, Dordrecht, Holland, pp. 385-388.
Carusi, A., Kresák, L., and Valsecchi, G.B.: 1981a, I.A.S. Internal Report n. 2.
Carusi, A., Kresák, L., and Valsecchi, G.B.: 1981b, Astron. Astrophys. 99, pp. 262-269.
Carusi, A., Kresák, L., and Valsecchi, G.B.: 1982, Bull. Astron. Inst. Czechosl. 33, pp. 141-150.
Carusi, A., Pozzi, F., and Valsecchi, G.B.: 1979, in "Dynamics of the Solar System" (R.L. Duncombe ed.), IAU Symp. 81, D. Reidel, Dordrecht, Holland, pp. 185-189.
Chebotarev, G.A.: 1967, "Analytical and Numerical Methods of Celestial Mechanics", New York, USA, p. 239.
Dvorak, R.: 1976, Astron. Astrophys. 49, pp. 293-298.
Everhart, E.: 1973, Astron. J. 78, pp. 316-328.
Everhart, E.: 1979, in "Asteroids" (T. Gehrels ed.), Tucson, USA, pp. 283-288.
Everhart, E.: 1982, in "Comets" (L. Wilkening ed.), Tucson, USA, pp. 659-664.
Hayashi, C., Nakazawa, K., and Adachi, I.: 1977, Publ. Astron. Soc. Japan 29, pp. 163-196.
Heppenheimer, T.A., and Porco, C.: 1977, Icarus 30, pp. 385-401.
Horedt, Gp.: 1976, Astron. J.81, pp. 675-678.
Hunter, R.B.: 1967a, M.N.R.A.S. 136, pp. 245-265.
Hunter, R.B.: 1967b, M.N.R.A.S. 136, pp. 267-277.
Kazimirchak-Polonskaya, E.I.: 1967, Sov. Astron. - A.J. 11, pp. 349-365.
Kazimirchak-Polonskaya, E.I.: 1972, in "The Motion, Evolution of Orbits and Origin of Comets" (G.A. Chebotarev, E.I. Kazimirchak-Polonskaya and B.G. Marsden eds.), D. Reidel, Dordrecht, Holland, pp. 373-397.
Kresák, L.: 1972, in "The Motion, Evolution of Orbits and Or-

igin of Comets" (G.A. Chebotarev, E.I. Kazimirchak-Polonskaya and B.G. Marsden eds.), D. Reidel, Dordrecht, Holland, pp. 505-514.
Kowal, C.T., Liller, W., and Marsden, B.G.: 1979, in "Dynamics of the Solar System" (R.L. Duncombe ed.), D. Reidel, Dordrecht, Holland, pp. 245-250.
Marsden, B.G.: 1979, "Catalogue of Cometary Orbits", Smithson. Astrophys. Obs., Cambridge, USA.
Marsden, B.G., and Roemer, E.: 1982, in "Comets" (L. Wilkening ed.), Tucson, USA, pp. 707-733.
Oikawa, S., and Everhart, E.: 1979, Astron. J. 84, pp. 134-139.
Pittich, E.M.: 1981, Bull. Astron. Inst. Czechosl. 32, pp. 340-345.
Pollack, J.B., Burns, J.A., and Tauber, M.E.: 1979, Icarus 37, pp. 587-611.
Rickman, H.: 1979, in "Dynamics of the Solar System" (R.L. Duncombe ed.), D. Reidel, Dordrecht, Holland, pp. 293-298.
Rickman, H., and Malmort, A.M.: 1981, Astron. Astrophys. 102, pp. 165-170.
Rickman, H., and Malmort, A.M.: 1982, in "Sun and Planetary System" (W. Fricke and G. Teleki eds.), D. Reidel, Dordrecht, Holland, pp. 395-396.
Scholl, H.: 1979, Icarus 40, pp. 345-349.

TRAPPING TIME OF RESONANT ORBITS IN PRESENCE OF POYNTING-ROBERTSON DRAG.

R. Gonczi[*], Ch. Froeschlé[**] and C.Froeschlé[**]
[*] Laboratoire de Physique Théorique, Université de Nice and Observatoire de Nice, B.P. 252, 06007 Nice Cedex, France.
[**] Observatoire de Nice, B.P. 252, 06007 Nice Cedex, France.

ABSTRACT

We study numerically the competition between the Poynting-Robertson drag and the gravitational interaction of grains with Jupiter near orbital resonances. The computations are based on the plane elliptic restricted three body problem. Numerical investigations show that the grains always cross the resonance region without any oscillation, except in the special case where the grains were initially inside the resonance. Such grains are temporarily trapped, then due to the drag they are ejected out of the resonance. The trapping time of a particle turns out to be much more important in the 3/2 and 2/1 commensurabilities than in the others.
A numerical exploration of numerous orbits for different initial conditions and different sizes of grains has been performed. The trapping time appears to be closely connected to the size of the librator-type orbits regions; it increases with the initial eccentricity of the orbit, and is also proportional to the radius and the density of the particle.

1. INTRODUCTION

In a previous paper (R. Gonczi, Ch. Froeschlé and C. Froeschlé 1982) henceforward referred to as Paper I, we studied the effect of the Poynting-Robertson drag on grain orbits near the resonance 2/1 with Jupiter. Our calculations were based on the plane elliptic restricted three body problem. We found two kinds of orbital behaviour. If the grains were initially inside the resonance they were trapped temporarily, otherwise they crossed the resonance without any oscillation. We also explained the variation of the osculating elements of the orbits by Greenberg's and Schubart's theories. These results are summed up in Section 2 of the present paper. In Section 3, we calculate the size of the librator orbit region around the principal resonances, which we call the resonance width. We determine in Section 4, the trapping time of grains for the commensurabilities 3/2, 2/1, 3/1 and 5/2. We study

V. V. Markellos and Y. Kozai (eds.), Dynamical Trapping and Evolution in the Solar System, 397–410.

the variation of this time both according to the initial values of the eccentricity e_o and semimajor axis a_o, and according to the radius and density of the grains. It is shown that the trapping time is closely connected to the width of the resonance.

2. REVIEW OF THE PREVIOUS RESULTS .

Let us briefly recall the equations of the planar three body elliptic restricted problem in the presence of a non gravitational force $\vec{f}$. We consider a particle of negligible mass moving in the gravitational field of the sun (mass m_1) and of Jupiter (mass m_2), which are both assumed to be point masses orbiting around their common centre of mass G. Denoting by r_1, r_2 and r the distances of the particle respectively to the sun, Jupiter and G, the equation of motion for the particle reads as :

$$\ddot{\vec{r}} = - Km_1 \frac{\vec{r}_1}{r_1^3} - Km_2 \frac{\vec{r}_2}{r_2^3} + \vec{f} \qquad (1)$$

where K is the gravitational constant.

As we consider the effect of the Poynting-Robertson drag (1903,1937), $\vec{f}$ reads as :

$$\vec{f} = - \frac{\alpha \vec{V}}{r^2}$$

where $\vec{V}$ is the particle velocity, and α a parameter depending on the radius s and the density ρ of the particle :

$$\alpha = \frac{2.5 \times 10^{11}}{s\rho} \text{ cm}^2 \text{ s}^{-1}$$

The numerical integration of Eq (1) is performed by a Burlich-Stoer (1966) method. After each step the elements of the osculating orbits are computed with respect to the sun. In Paper I, numerical explorations were done for the resonance 2/1, considering a particle of radius $s = 10^{-3}$ cm and density $\rho = 2$g/cm3 (i.e. $\alpha = 2 \times 10^{-5}$ (a.u.)^2/year). The initial osculating elements of the orbit were : $e_o = 0.14$ and $a_o = 3.36$ a.u. and we took as variable the critical argument σ defined by :

$$\sigma = \ell_J \, (p + q)/q - \ell - \tilde{\omega}$$

where ℓ and ℓ_J are respectively the particle and the Jovian mean longitude, p and q are integers which determine the commensurability $\frac{p+q}{p}$ and ω the pericenter longitude of the particle.

It was found that the orbits always cross the resonance region without any oscillation (Fig. 1a) except in the special case where the grain

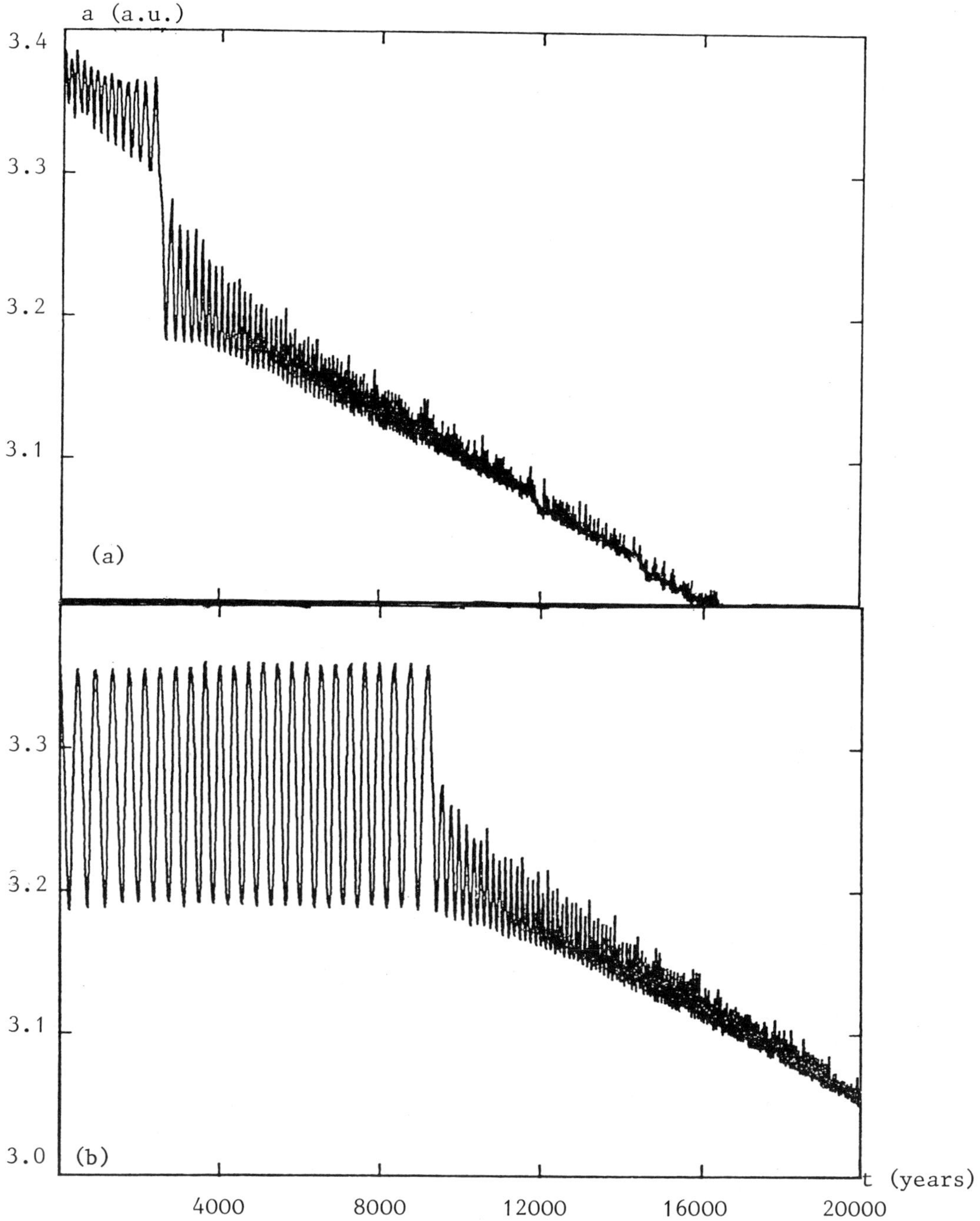

Fig. 1 : variation of the semimajor axis for an orbit near the 2/1 resonance with Jupiter; the initial conditions are :

a) $a_o = 3.36$ a.u.; $e_o = 0.14$; $\sigma_o = -2\pi/3$; $\alpha = 2\text{x}10^{-5}(\text{a.u})^2/\text{yr}$

b) $a_o = 3.36$ a.u.; $e_o = 0.14$; $\sigma_o = -2\pi/10$; $\alpha = 2\text{x}10^{-5}(\text{a.u.})^2/\text{yr}$

was initially inside the resonance. In that case (Fig. 1b) the particle remains inside the resonance for a long time, until the drag ejects it toward the sun. These two cases have been explained using Schubart's theory. The orbits of the second type (Fig. 1b) are librators, which become circulators after ejection of the particle from the resonance.

3. DETERMINATION OF THE WIDTH OF THE RESONANCES.

As we have seen before that only librators can be trapped into a resonance, in order to calculate the trapping time of such a particle, we have first to estimate the size of the libration region around the commensurabilities.

First a systematic exploration of the resonance region is performed : on the (a,e) diagram we construct a lattice defined by 8 values of "e" regularly spaced between 0.0 and 0.4 and 50 values of "a" regularly spaced between two values a_{min} and a_{max} surrounding the commensurability. Each point of this lattice represents the initial conditions a_o and e_o of one orbit; moreover, the initial critical argument σ_o takes the value σ^* corresponding to the most favourable geometrical configuration for getting a librator : $q\sigma^* = 0°$ in the 2/1, 3/2, 5/2 commensurabilities, and $q\sigma^* = 180°$ in the 3/1 case (Schubart, 1964). Each orbit is numerically integrated until $q\sigma$ crosses 6 times either 0° or 180°.

It is then considered as :

- a circulator if $q\sigma$ passes alternatively through 0° and 180°;
- a librator if $q\sigma$ only crosses the value $q\sigma^*$ and never its opposite $q\sigma^* + \Pi$;
- an alternator (C. Froeschlé and H. Scholl, 1977) in all the other cases.

The results for the resonances 3/2, 2/1, 3/1 and 5/2 are given on Figs.2. We see that the resonances 3/1 and 5/2 have very few librators, which is not surprising since they are known as gaps in the distribution of asteroids. On the other hand, the resonance 3/2 and 2/1 show a large librator region. This is again in agreement with the observations for the 3/2 resonance, but not for the 2/1 one.

The boundaries of the libration region turn out to be well defined, except in the 3/2 case where many alternators and hyperbolae appear, probably due to the influence of the closeness of Jupiter.

Figs. 2a, 2b, 2c also clearly show that for all the commensurabilities studied here, the resonance width increases with the orbit eccentricity; in particular, there are very few librators with $e_o < 0.1$ in the 3/2 resonance, which is confirmed by the observed asteroids in the Hilda group.

To complete this study and eventually eliminate the incidence of the

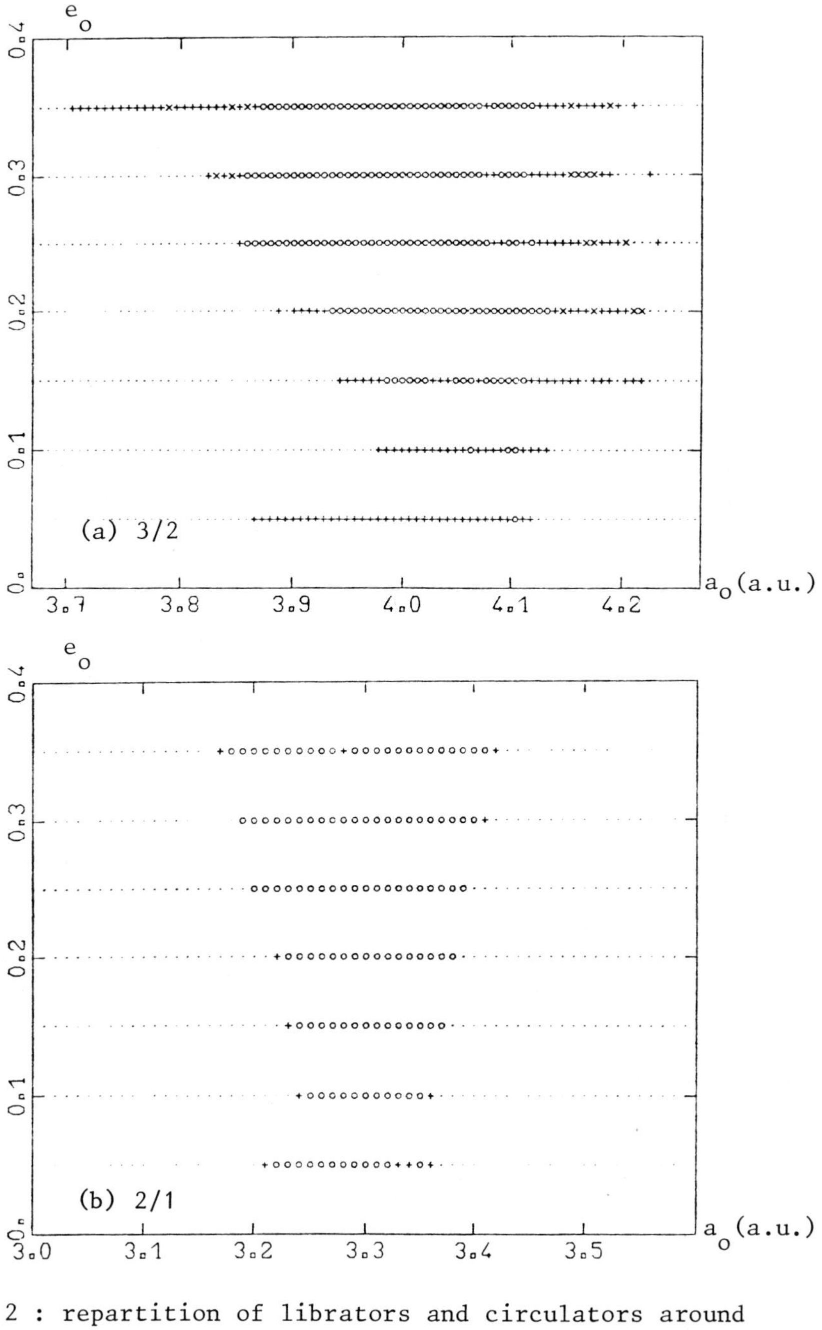

Fig. 2 : repartition of librators and circulators around resonance regions.
Initial conditions of each orbit : a_o, e_o and $\sigma_o = \sigma^*$ is the most favourable value for getting a librator.
$\circ$ librator; . circulator; + alternator; x orbit suffering a close approach.
a) orbits near the 3/2 commensurability; b) orbits near the 2/1 commensurability; c) orbits near the 3/1 and 5/2 commensurabilities

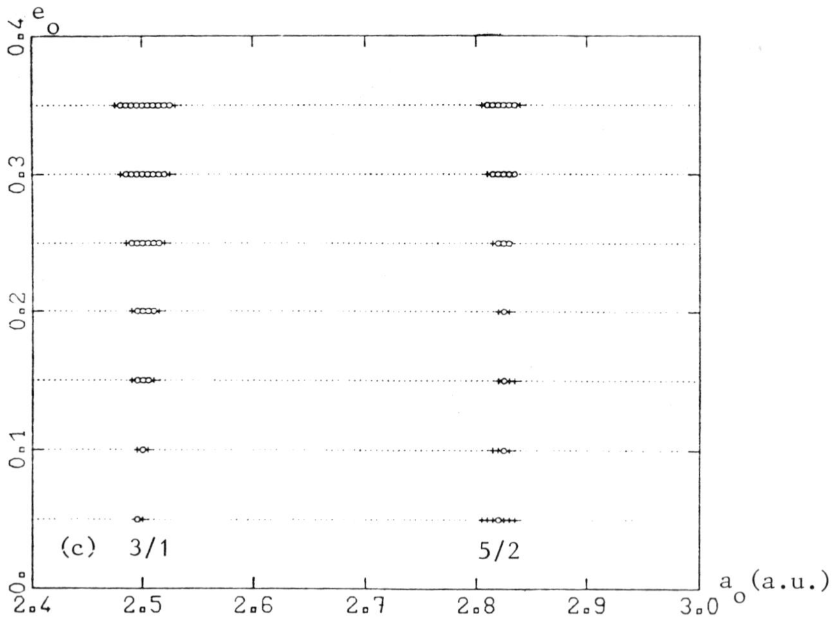

Fig. 2 - continued.

particular choice of initial condition $\sigma_o = \sigma^*$, we now investigate by a Monte-Carlo method, orbits whose initial conditions a_o, e_o and σ_o are randomly chosen within the extremes :

$$a_{min} < a_o < a_{max} \ ; \quad 0. < e_o < 1. \ ; \quad -\Pi < \sigma_o < \Pi$$

The integration and classification of each orbit into circulator, librator and alternator are performed as before and plotted on Figs. 3a, 3b, 3c.

For each resonance we have also calculated (Table 1) the percentage of each type of orbits obtained with both estimations. The qualitative comparison between Figs. 2 and 3, as well as the quantitative results of Table 1 do not show significant differences between the two numerical experiments. Indeed the peculiar case $\sigma_o = \sigma^*$ is a good representative of the global picture of the phase space.

4. DETERMINATION AND VARIATION OF THE TRAPPING TIME.

For a given resonance we consider orbits chosen among the librators determined in Section 3. We introduce now the Poynting-Robertson drag $\vec{f}$: we know that the particle will not librate indefinitely but

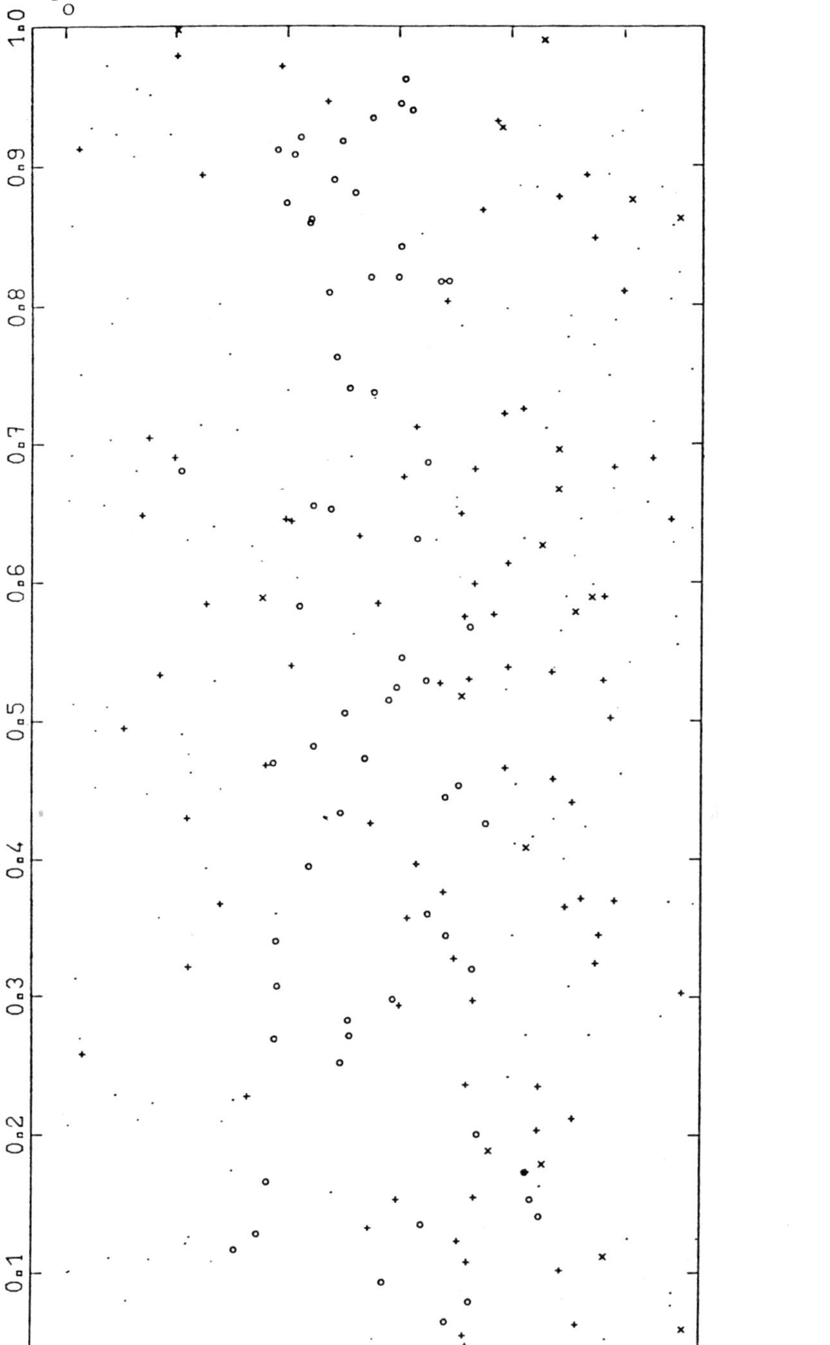

Fig. 3 : same as Fig. 2 with initial conditions a_o, e_o, σ_o chosen at random.

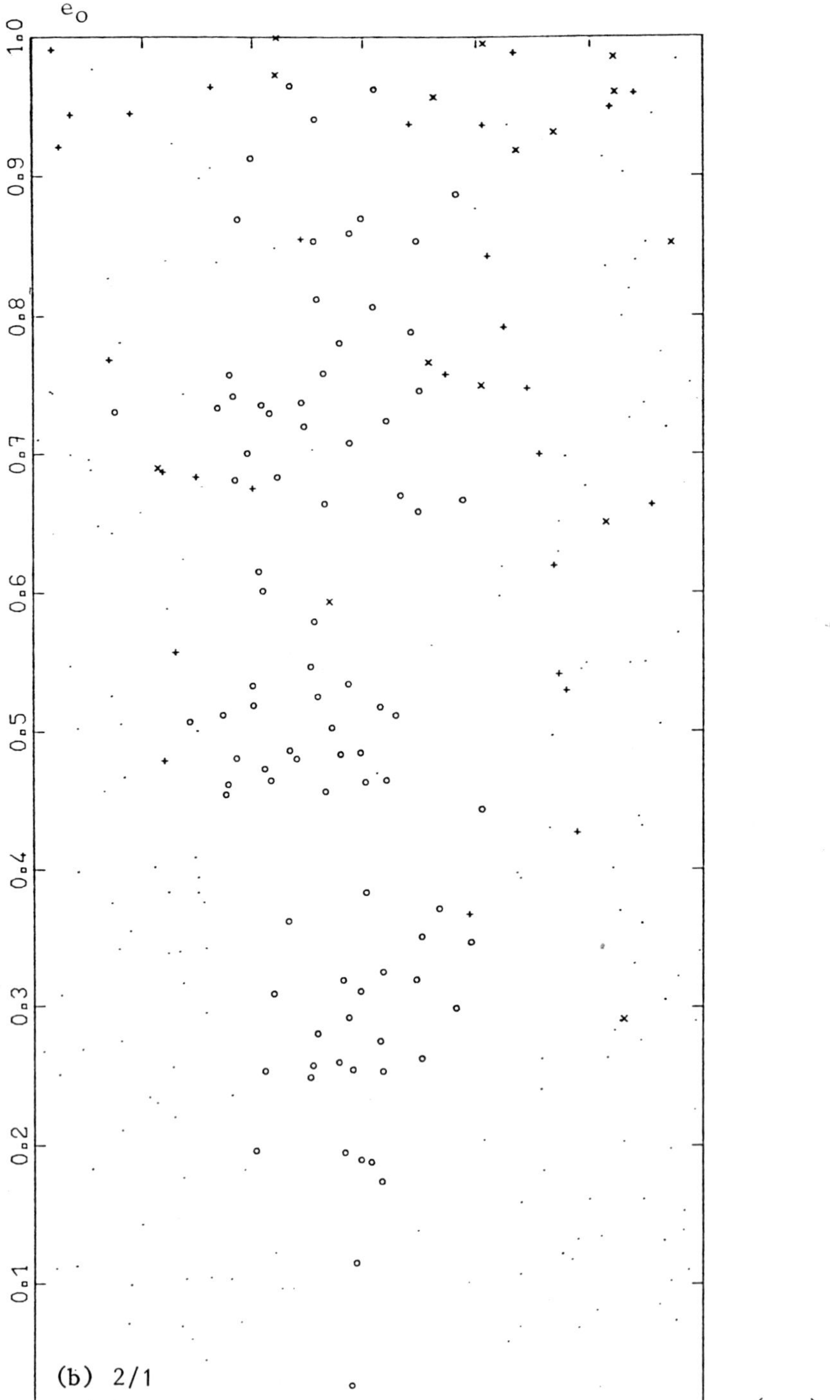

Fig. 3 - continued.

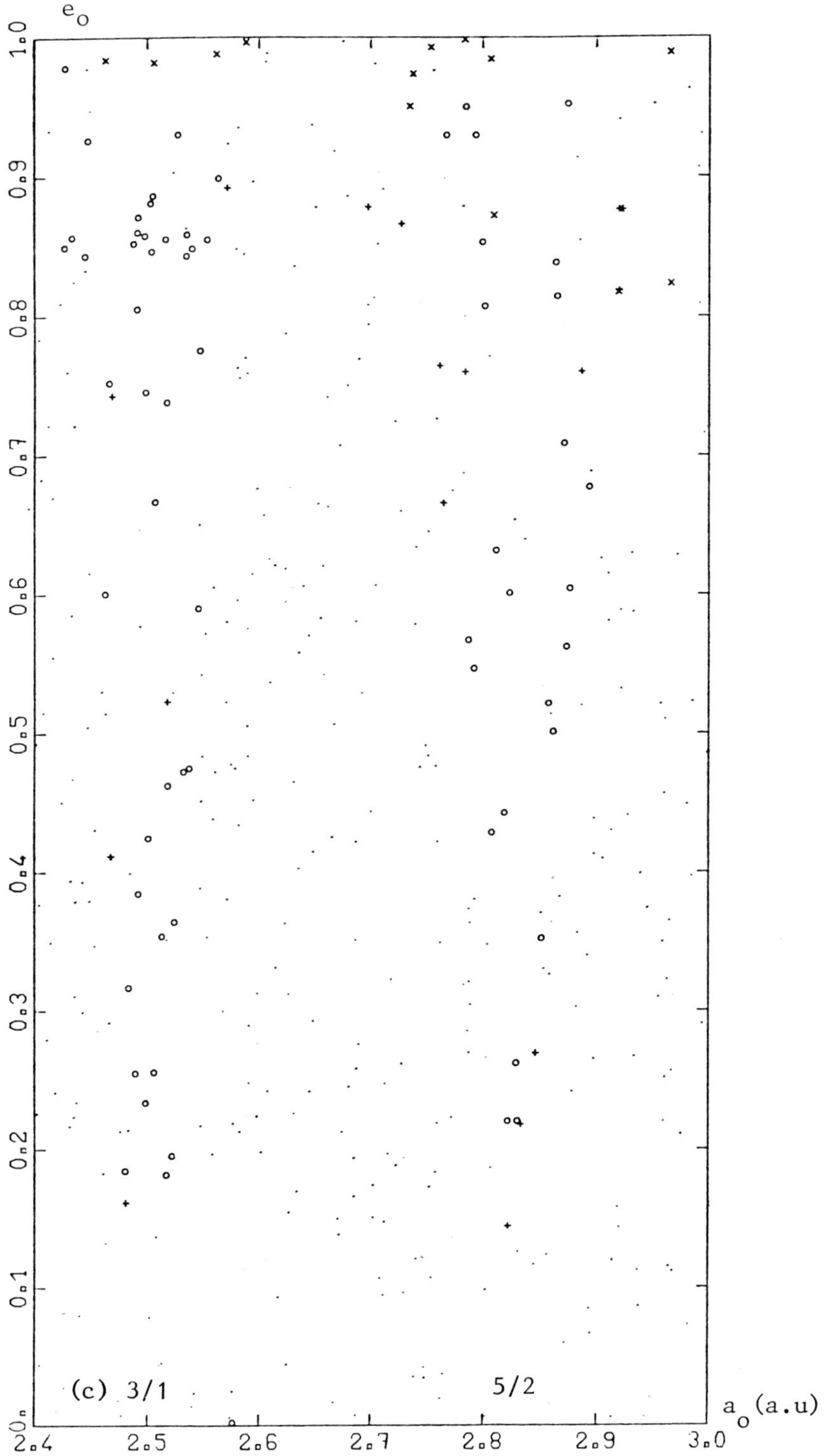

Fig. 3 - continued.

Com.	Initial conditions	Librators	Circulators	Alternators	Close approach	Total
3/2	case 1	25.8	45.0	26.0	3.2	100.
	case 2	21.3	44.4	30.6	3.7	100.
2/1	case 1	28.5	68.5	2.9	0.0	100.
	case 2	23.4	72.6	3.2	0.8	100.
3/1	case 1	8.2	88.9	2.9	0.	100.
	case 2	10.6	89.4	0.0	0.	100.
5/2	case 1	5.5	89.0	5.5	0.	100.
	case 2	5.1	91.0	3.8	0.	100.

Table 1 : Percentages of each type of orbit obtained around the commensurabilities 3/2, 2/1, 3/1 and 5/2.

case 1 : systematic exploration for the initial conditions e_o and a_o, σ_o is equal to σ^* (see Figs. 2)

case 2 : e_o, a_o and σ_o are chosen at random (see Figs. 3)

will be ejected out of the resonance and become a circulator. The time an orbit is trapped inside the resonance (trapping time) depends on several parameters. In this section we investigate the variation of the trapping time with p, q, a_o, e_o and α.

For each orbit, Eq. (1) is numerically integrated and we note the first time $t = t_1$ for which $q\sigma$ takes the value $q\sigma^*$. In order to be sure, that the orbit is a circulator we note also the time $t = t_n$ of the nth crossing of $q\sigma$ through $q\sigma^*$. We choose n = 6, and assume, that the trapping time t_T lies between t_1 and t_6, which are respectively the lower and the upper bounds of t_T.

4.1. Variation of t_T with a_o and e_o.

Let α and e_o be fixed, we calculate the trapping time for orbits close to the commensurabilities : 3/2, 2/1, 3/1 and 5/2, and for different values of the initial semimajor axis a_o.

On Fig. 4, the time t_T is plotted as a function of $d = a_o - a_{res.}$ (where a_{res} is the value corresponding to the exact commensurability), for the fixed values $e_o = 0.14$ and $\alpha = 2 \times 10^{-5}$ $(a.u.)^2$/year. We notice a great difference in magnitude of the time t_6 between the 3/2 and 2/1 resonance and the others, as already observed in the previous

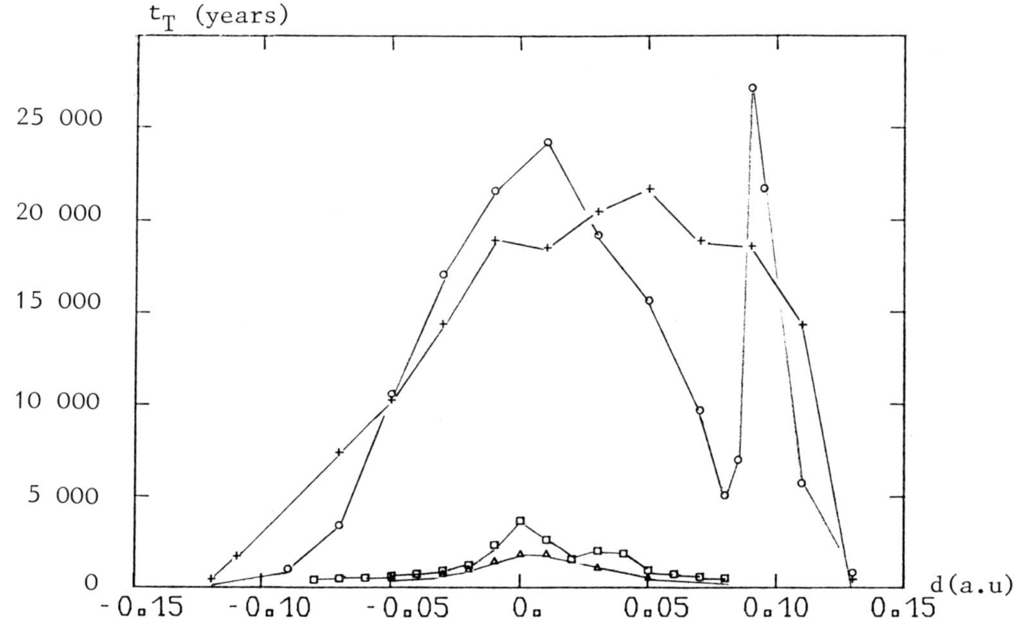

Fig. 4 : trapping time as a function of $d = a_o - a_{res}$ for the commensurabilities :
3/2 (+); 2/1 (o); 3/1 (□); 5/2 (△),
α, e_o and σ_o are fixed: $\alpha = 2\times10^{-5}$ (a.u.)^2/yr; $e_o = 0.25$; $\sigma_o = \sigma^*$

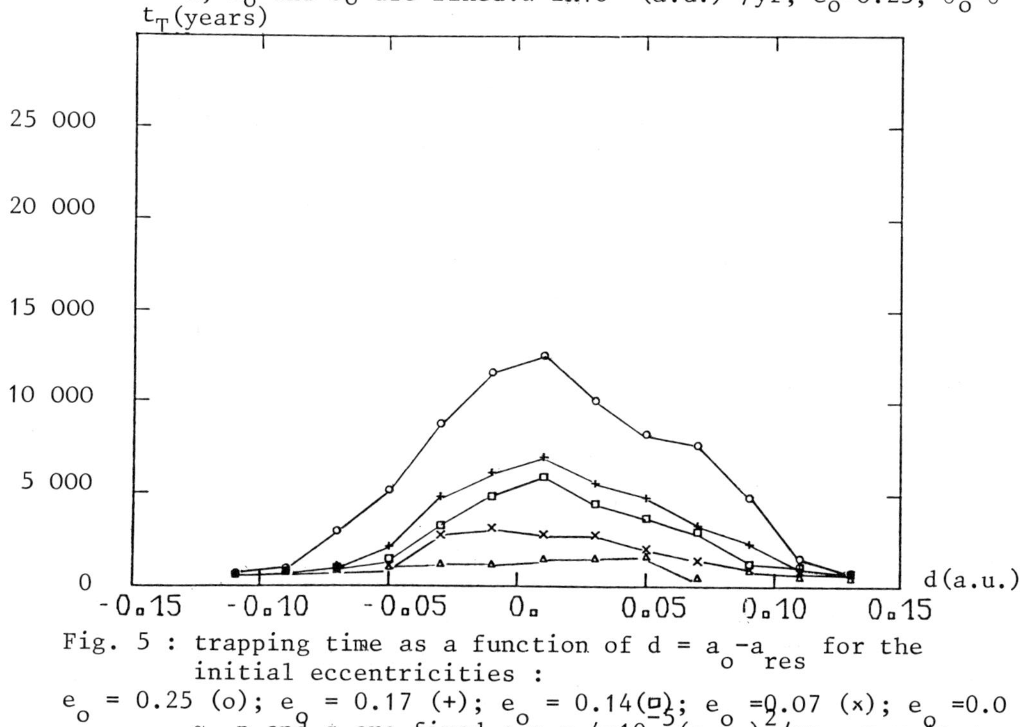

Fig. 5 : trapping time as a function of $d = a_o - a_{res}$ for the initial eccentricities :
$e_o = 0.25$ (o); $e_o = 0.17$ (+); $e_o = 0.14$ (□); $e_o = 0.07$ (x); $e_o = 0.0$ (△)
α, p and q are fixed : $\alpha = 4\times10^{-5}$ (a.u.)^2/yr, p = 1; q = 1.

section for the width of these resonances. Indeed in the 3/1 and 5/2 resonance which have a very small width, the trapping time is very short, while for the other two which have a comparably large width, the trapping time is longer and of the same order of magnitude.

The quasi symmetry of the curves is a consequence of the well known fact that the semimajor axis of librators oscillates from one edge of the resonance to the other. The centre of symmetry is slightly different from the exact resonance $d = 0$. It would rather correspond to the centre of the librator region in the Schubart's plot.

As $|d|$ increases, i.e. the initial value a_o moves away from the exact resonance, the time t_6 obviously decreases, to become almost zero near the outer edges of the resonances. We have no explanation for the secondary peak which appears in the resonance 2/1, we can only say that this peak disappears for other values of α (see Fig. 5).

We have done several similar numerical explorations for different values of e_o, and different resonances. On Fig. 5, we have plotted for the 2/1 commensurability the trapping time t_6 versus $d = a_o - a_{res}$ for five initial values of e_o. Again this is closely related to the width of the librator region (Figs. 2) which is larger for higher eccentricities. While for both the two body problem (Wyatt and Whipple (1950)) and the restricted three body problem (out of the resonance), the falling time of a particle into the sun decreases when e_o increases, we have here an opposite behaviour.

4.2. Variation of the trapping time with α.

It is known that for the two body problem, the Poynting-Robertson drag is more efficient for small particles than for larger ones, and that the time of fall into the sun is a linearly increasing function of the product $s\rho$. Here we study the trapping time t_T as a function of α between 10^{-6} and 10^{-4} $(a.u)^2/yr$ (i.e. between 10^{13} and 10^{15} cm^2/s).

As $\alpha = \dfrac{2.5\times10^{11}}{s\rho}\ cm^2/s$, these values correspond, for a particle of density $\rho \cong 2g/cm^3$, to a radius from 10^{-4} cm to 10^{-2} cm.

Several orbits have been computed. The results obtained for one orbit near the 3/2 resonance and two orbits near the 2/1 one are shown on Figs. 6a, 6b, 6c. The trapping time t_T (or more precisely the lower and upper bound t_1 and t_6) plotted versus α in a log-log scale shows a good linear behaviour, implying that the trapping time is proportional to the product $s\rho$.

$$\frac{\{J_n(2I_1)^{1/2}\}}{(2I_1)^{1/2}} \gtrsim \frac{1}{4n\varepsilon} \tag{5}$$

where n is the integer nearest to ν again.

Despite the fact that the chaos sets on in a different way when $\xi\simeq1$ or $\xi=0$, the similarity in the form of equs. (4) and (5) leads one to investigate whether these equations could be obtained as special cases of a more general relation. To do so at first we apply the usual arguments (e.g., Greene 1980) and we estimate the strength of the perturbation that causes the "overlap" of the primary islands with rotation number $\frac{1}{n^*}$ and $\frac{1}{n^*+1}$ in the case $\xi\simeq1$, where it is known that the "stochasticity" (the dissolution of the invariant curves) begins indeed with the interaction of the n^* and n^*+1 resonances. This "overlap" takes place when (Varvoglis and Papadopoulos 1982a)

$$\left\{\varepsilon|J_{n^*}\{(2I_1)^{1/2}\}|\{\frac{\xi^2}{n^{*2}} + \frac{\varepsilon}{2I_1}|J''_{n^*}\left[(2I_1)^{1/2}\right]|\}\right\}^{1/2} \gtrsim \frac{1}{4n^*} \tag{6}$$

Therefore a trajectory shows stochastic behavior if equ. (6) is satisfied at any of the trajectory points. Equ. (6) reduces to equ. (4) if the term $\frac{\varepsilon}{2I_1} J''_{n^*}\{(2I_1)^{1/2}\}$ is ignored compared to $\frac{\xi^2}{n^{*2}}$, and thus the Smith and Kaufman result is recovered.

If the chaotic behavior sets on in the degenerate case in the same way as in the non-degenerate one, the stochasticity criterion of equ. (3) for $\xi=0$ would be the limit for $\xi\to0$ of the equ. (6)

$$\frac{|J_n\{(2I_1)^{1/2}\}J''_n\{(2I_1)^{1/2}\}|^{1/2}}{(2I_1)^{1/2}} \gtrsim \frac{1}{4n\varepsilon} \tag{7a}$$

Notice that for $2I_1 \gg n^2$ equ. (7a) reduces to

$$\frac{|J_n\{(2I_1)^{1/2}\}|}{(2I_1)^{1/2}} \gtrsim \frac{1}{4n\varepsilon} \tag{7b}$$

which is identical to equ. (5). Now it is known (Fukuyama et al 1977, Karney 1978) that the dissolution of the invariant curves in the degenerate case begins with the interaction of secondary islands, so that the process of taking the limit of equ. (6) cannot be justified. In the special case $\nu=n+0.5$, however, it has been shown (Hsu 1982) that the cause of the invariant curve dissolution is indeed the interaction of the n and n+1 islands, and therefore, at least in this case, the limit of equ. (6) is justified.

It is worth to stress at this point that in the degenerate case the stochasticity criterion turns out to be unique, despite the fact that for various values of the quantity $\delta=|\nu-n|$ the stochasticity sets on following different paths. This fact has been recognized in the literature (Karney 1978) and advances the conjecture that perhaps equ. (6) is valid in the whole interval $0 \lesssim \xi \lesssim 1$, irrespectively of the value of δ (that is even for $\delta \neq \frac{1}{2}$); this conjecture has to be tested numerically to evaluate its usefulness.

In the case of small ξ and large ν it has been shown that the above conjecture gives results in agreement with numerical calculations (Varvoglis and Papadopoulos 1982a). In the small ν regime we have the interesting situation that equ. (7a) does not fully agree with equ. (5) (old result). Namely, both equs. (5) and (7a) imply that for $\nu<\nu_0$ the stochastic region in the action space has an upper bound and for $\nu>\nu_0$ it has an upper and a lower bound, but equ. (5) gives $\nu_0 = 1.5$ while equ. (7a) gives $\nu_0=2.5$. This qualitative difference in behavior can be attributed to the fact that for $\nu<\nu_0$ the origin (of the surface of section) is an unstable periodic orbit, while for $\nu>\nu_0$ it is a stable one, where ν_0 is equal to 2.5 (Abe et al. 1980, Hsu 1982, Varvoglis and Papadopoulos 1982b); this result is in agreement with equ. (7a).

IV. SUMMARY

The above discussion can be summarized as follows: If a certain degenerate dynamical system can be defined as the limit of a monoparametric family of non-degenerate systems, then it is possible under certain conditions to find its stochasticity criterion as the corresponding limit of the more easily found stochasticity criteria of the non-degenerate systems. In the case of the system of equ. (3) this process not only leads to the (already known) correct answer, but also gives some new information for the behavior of the system in question.

ACKNOWLEDGMENTS

Part of this work was done while the author was a Research Associate in the Astronomy Program of the University of Maryland. I would like to thank the Solar Terrestrial Theory Group of the Astronomy Program and especially Dr. K. Papadopoulos for the hospitality, the financial support and many useful discussions.

REFERENCES

Abe, H., Momota, H. and Itatani, R. 1980, Phys. Fluids 23, 2417.
Arnold, V. 1978 Mathematical Methods of Classical Mechanics, New York, Springer, p. 400.
Barbanis, B. 1966, Astron. J. 71, 415.
Berry, M.V. 1978 in Topics in Nonlinear Dynamics, AIP Conference Proceedings 46, New York, AIP.
Chirikov, B.V. 1969, Prepr. 267, Institute of Nuclear Physics, Novosibirsk (Engl. Transl. CERN Transl. 71-40 1971).
Chirikov, B.V. 1979, Phys. Rep. 52, 264.
Fukuyama, A., Momota, H., Itatani, R. and Takizuka, T. 1977, Phys. Rev. Lett. 38, 701.
Greene, J.M. 1980, Ann. N.Y. Acad. Sci. 357, 80.

Hadjidemetriou, J. and Ichtiaroglou, S. 1983, this volume, p. 141
Hénon, M. and Heiles C. 1964 Astron. J. 69, 73.
Hsu, J.Y. 1982 Phys. Fluids 25, 159.
Karney, C.F.F. 1978 Phys. Fluids 21, 1584.
Smith, G.R. and Kaufman, A.N. 1978 Phys. Fluids 21, 2230.
Varvoglis, H. and Papadopoulos, K., 1982a, University of Maryland preprint AP82-048.
Varvoglis, H. and Papadopoulos, K., 1982b, Ap.J. Letters (submitted).

INDEX OF NAMES

Abe, H. 414
Adachi, I. 394
A'Hearn, M.F. 100, 104
Allan, R.R. 19, 21, 23, 25
Arenstorf, R.F. 143, 152, 226, 233
Arnold, J.R. 117, 122
Arnold, V.I. 30, 35, 146, 152, 229, 233, 411, 414
Avez, A. 146, 152

Babadzhanov, P.B. 85, 87
Barbanis, B. 272, 274, 412, 414
Barrar, R.B. 226, 233
Belen'kii, I.M. 39, 40, 46
Belyaev, N.A. 160, 377, 393
Benest, D. 107-114
Bennett, A. 289, 290, 291, 299, 347, 350
Berry, M.V. 411, 414
Bien, R. 107-114, 153-161, 169, 175
Bigourdan, G. 53, 55, 59
Binney, J. 271, 274
Birkhoff, G.D. 143, 152, 302, 314
Bowell, E. 180, 187
Bozis, G. 302, 308, 314, 315, 353-367
Brandt, J.C. 89, 94
Bretagnon, P. 48, 50
Brjuno, A.D. 226, 233
Broucke, R.A. 142, 152, 353, 366, 367
Brouwer, D. 6, 17, 49, 50, 117, 119, 122, 138, 139, 205, 210
Brown, E.W. 153, 160
Buck, T. 236, 241, 247
Burlisch, R. 398, 409
Burns, J.A. 85, 87, 95, 395
Burns, T.J. 12, 17

Caballero, J.A. 39, 45, 46
Calame, O. 3, 17
Caranicolas, N. 271-274
Carusi, A. 105, 106, 179, 187, 377-395
Casasayas, J. 263-270
Cassen, P. 59
Cazenave, A. 3, 17
Chandrasekhar, S. 178, 187
Chapman, C.R. 187, 188
Chapman, R.D. 89, 94
Chapront 37
Chebotarev, G.A. 153, 160, 380, 394
Chen Zhen, 73-79
Chirikov, B.V. 411, 414 4
Cid, R. 39-46
Clemence, G.M. 49, 50
Cohen, C.J. 158, 160
Colombo, G. 142, 151, 152, 227, 233
Cowan, J.J. 100, 104
Contopoulos, G. 271, 272, 274

Daillet, S. 3, 17
Danby, J.M.A. 289-291, 299
Davis, D.R. 177, 187
Delva, M. 317-323
Deprit, A. 39, 46, 347, 350
Dermott, S.F. 141, 152
Devaney, R. 263-266, 270
Donnison, J.R. 84, 87
Drake, S. 48, 50, 51, 59
Dubois-Moons, M. 37
Dulinski, G. 27-35
Dvorak, R. 319, 323, 380, 381, 394

Easton, R. 277, 287, 301, 306, 314
Eckhardt 37
Edelman, C. 257-261
Elipe, A. 39-46
Emerson, B. 89, 95
Emslie, G. A. 315

Érdi, B. 153, 161, 165-176, 319, 323, 353, 367
Eremenko, R.P. 160
Evans, D.S. 304, 314
Everhart, E. 104, 377-381, 393-395

Farinella, P. 177-188, 308, 315
Faukner, J. 121, 122
Ferrari, A.J. 3, 17
Ferrer, S. 39-46
Fox, K. 86, 87, 89-95
Franklin, F.A. 142, 151, 152
Freeman, K.C. 271, 274
Froeschlé, C. 97, 104, 142, 152, 163, 397-410
Froeschlé, Ch. 397-410
Fujiwara, A. 178, 179, 185, 186, 188
Fukuyama, A. 411-414

Gehrels, T. 187
Giacaglia. G.E.O. 175, 317, 319, 323
Gillispie, C.C. 53, 55, 59
Goldreich, P. 19, 20, 25
Golubev, V.G. 302, 315
Gómez, G. 325-338
Gonczi, R. 397-410,
Goudas, C.L. 224
Gradie, J.C. 178, 186, 188
Greenberg, R. 19, 20, 22, 25, 52, 59, 187
Greene, J.M. 413, 414
Guillaume, P. 143, 152, 227, 233
Gutzwiller, M. 263, 270

Hadjidemetriou, J.D. 137-139, 141-152, 255, 256, 373-375, 411, 415
Hagihara, Y. 141, 152
Hale, J.K. 299
Harris, A.W. 187
Hartmann, P. 230, 233
Hayashi, C. 380, 394
Heiles, C. 412, 415
Hennawi, A. 260, 261
Hénon, M. 213, 214, 217, 218, 224, 412, 415
Henrard, J. 12, 17, 190, 191, 196, 201, 347, 350
Heppenheimer, T.A. 380, 394
Herrick, S. 86, 87
Hill, G.W. 39, 46, 136, 137, 139
Hirayama, K. 117, 119, 122
Horedt, Gp. 380, 394
Hori, G. 6, 17
Houlden, M.A. 48, 50
Hsu, J.Y. 412-415
Hubbard, E.C. 160
Hunter, R.B. 386, 387, 394
Hughes, D.W. 83, 86, 87, 89, 94, 95

Ichtiaroglou, S. 137-139, 141-152, 411, 415
Illingworth, G. 271,274
Itatani, R. 414
Ip, W.H. 178, 180, 181, 188

Johnson, G.A.L. 3, 17

Kaiser, T.R. 95
Kalogeropoulou, M. 213, 224
Kamimoto, G. 188
Kane, T.R. 27, 35
Karney, C.F.F. 413-415
Katsiaris, G.A. 224
Kaufman, A.N. 412, 413, 415
Kazimirchak-Polonskaya, E.I. 105, 106, 377, 380, 381, 394
Kevorkian, J. 167, 175
King, A. 83, 87
Kovalevsky, J. 3-17
Kowal, C.T. 48, 50, 51, 59, 378, 395
Kozai, Y. 22, 25, 117-122, 138, 139, 205, 210
Krein, M.G. 292, 299
Kresák, L. 84, 87, 97, 104, 179, 188, 388, 393, 394
Kresáková, M. 105, 106
Kustaanheimo, P. 39, 46

Lambeck, K. 3, 17
Lamy, P.L. 87, 95
Laplace, P.S. 139
Lass, H. 353, 354, 366, 367
Lecar, M. 142, 152
Lemaître, A. 189-201
Levin, B.J. 97, 104
Lieske, J.H. 51-59
Liller, W. 395
Lindblad, B.A. 117, 122

Liu, A.S. 73, 79
Llibre, J. 263-270, 325-338
Lukac, M.R. 59
Lundberg, J. 123-136, 139, 142, 152, 354, 367

MacDonald, G.J.F. 4, 17
McGahee, W. 354, 367
McGehee, R. 329, 337
Maciejewski, A.J. 27-35
McKenzie, R. 174, 176, 302, 315
Malmort, A.M. 380-382, 384, 395
Marchal, C. 277, 287, 302, 308, 309, 315
Markeev, A.P. 27, 30, 35
Markellos, V.V. 213-224, 235-247, 252, 256, 261
Markley, F.L. 27, 35
Marsden, B.G. 89, 95, 98, 99, 100, 101, 104, 108, 114, 387, 388, 395
Marsh, L.E. 27, 35
Matas, V. 299
Mayo, A.P. 73, 79
Meire, R. 289-299
Melchior, P. 3, 17
Mestschersky, I. 369, 375
Michaelidis, P. 271, 274
Mignard, F. 4, 17
Migus 37
Milani, A. 188, 225-233, 301-315
Minglibaev, M.J. 369-375
Molnàr, S. 353, 367
Momota, H. 414
Morrison, F. 353, 367
Morrison, L.V. 51, 53, 59
Moser, J.K. 229, 233, 326, 330, 337, 338
Mulholland, J.D. 3, 17
Muller, P.M. 48, 50
Munford, C.M. 142, 152
Murray, C.D. 86, 87, 95, 141, 152

Nakamura, T. 97-104
Nakazawa, K. 394
Newhall, X.X. 47, 49, 50
Nobili, A.M. 188, 301-315
Noerdlinger, P. 95
Nudds, J.R. 3, 17

Obrubov, Yu.V. 85, 87
Oesterwinter, C. 160
Oikawa, S. 377, 378, 393, 395
Omarov, T.B. 369-375
Öpik, E.J. 97, 104
Ovenden, M.W. 19, 25, 214, 215, 224, 226, 233, 279, 287

Paolicchi, P. 177-188
Papadopoulos, K. 413-415
Park, C. 95
Peale, S.J. 19, 20, 23, 25, 52, 59
Perdios, E. 339-350
Perozzi, E. 377-395
Pingre, A.G. 53, 59
Pittich, E.M. 388, 395
Plakhov, Yu.V. 73, 79
Plavec, M. 90, 95
Plummer, H.C. 71
Poincaré, H. 226, 233, 257, 260, 261
Pollack, J.B. 95, 379, 395
Poole, L.M.G. 92, 94, 95
Porco, C. 380, 394
Poynting, J.M. 398, 410
Pozzi, F. 379, 380, 393, 394
Presler, W.H. 169, 175

Rahe, J. 100, 104
Reynolds, R.T. 59
Rickman, H. 97, 104, 107-114, 380-382, 384, 385, 395
Robertson, H.P. 398, 410
Robin, I.A. 213-224
Rocher, P. 261
Roemer, E. 388, 395
Roseau, M. 257-259, 261
Rothschild, R.F. 48, 50
Roy, A.E. 19, 25, 214, 215, 224, 226, 233, 277-287, 302, 304, 315
Russel, J.A. 92, 95

Saari, D.G. 277, 287, 302, 315
Sampson, R.A. 52, 59
Scheifele, G. 39, 46, 326, 338
Schmidt, D. 143, 152
Scholl, H. 49, 50, 142, 152, 163, 378, 395, 400, 410
Schubart, J. 153-161, 197, 201, 400, 410
Sekanina, Z. 104

Serra, R. 336, 338
Shook, C.A. 153, 160
Siegel, C.L. 229, 233, 326, 338,
Simó, C. 327, 329, 330, 334, 337
Simonenko, A.N. 97, 104
Sinclair, A.T. 19-25, 142, 152
Sinclair, W.S. 17
Sjogren, W.L. 17
Smale, S. 277, 287, 302, 303, 315
Smart, W.M. 71
Smirnow, W.I. 319, 323
Smith, B.A. 22, 25
Smith, G.R. 412, 413, 415
Smoluchowski, R. 199, 201
Soter, S. 87, 95
Southworth, R.B. 117, 122
Standish, E.M. 47-50
Stellmacher, I. 257, 261
Stephenson, F.R. 48, 50
Stiefel, E.L. 39, 46, 326, 338
Stoer, J. 398, 409
Stumpff, P. 48, 50, 154, 161
Subbotin, M.F. 261
Szebehely, V. 39, 46, 123-136, 137, 139, 142, 143, 152, 174, 176, 259, 261, 299, 302, 315, 317, 319, 323, 353, 354, 365, 367

Takizuka, T. 414
Tauber, M.E. 395
Tedesco, E.F. 186-188
Tong Fu 73-79
Toon, O.B. 95
Torbett, M. 199, 201
Touzé 37
Tschauner, J. 289-291, 297, 299
Tsouroplis, A. 249-256
Tsukamoto, A. 178, 186, 188
Tung, C.C. 302, 315
Turco, R.P. 90, 95

Valsecchi, G.B. 105, 106, 179, 187, 377-395
van Woerkom, A.J.J. 139
Varvoglis, H. 411-415
Vicente, R.O. 123-139, 142, 152
Vinti, J. 369, 375
Voigt, H.H. 311, 315

Walker, I.W. 302, 304, 311, 315
Wartman, M. 83, 87
Weidenschilling, S.J. 177, 187, 188
Weissman, P.R. 100, 104
Wetherill. G.W. 97, 98, 102, 104
Whipple, F.L. 90, 95, 100, 104, 408, 410
Whitten, R.C. 95
Wickramasinghe, N.C. 84, 87
Wiesel, W. 178, 188, 260, 261
Williams, I.P. 83-87, 92, 95
Williams, J.G. 17, 50, 117, 121, 122, 179, 184, 188
Wintner, A. 219, 224
Wyatt, S.P. 408, 410

Xanthopoulos, B.C. 353-367

Yeomans, D. 89, 95, 104
Yoder, C.F. 17, 20, 24, 25, 52, 59, 160, 161
Yuasa, M. 203-210

Zagouras, C. 213, 224, 235-247, 249-256
Zappalà, V. 177-188
Zare, K. 277, 287, 302, 315
Zellner, B. 187
Zhang Jia-xiang 61-71
Zikides, M. 271, 274

INDEX OF SUBJECTS

Adiabatic invariant 12, 196
Ages of asteroid families 203-210
Area index 191
Asteroid belt
- perturbations due to 73-79
Asteroidal fragments
- reaccumulated 181
Asteroids
- families of 117-122, 177-188, 203-210
- masses of 180 *et seq*
- rotational properties of 179
- stability of 123-136 137-139, 141-152
- Trojan 153-161, 165-176
Asymptotic
- orbits 339-350
- trapping 349

Bifurcation
- parameter 308
- theorem 230, 232
- vertical 213-224

Capture
- escape boundary in the collinear restricted problem 325-338
- of comet P/Boethin 107-114
Central force field 39-46
Chaotic
- behaviour 150, 412, 413
- cometary orbits 377
Collision
- in the anisotropic Kepler problem 266, 267
- in the collinear restricted problem 326 *et seq*
- meteoroid 86
Collisional origin of asteroid families 177-188
Comet
- P/Boethin 107-114
- P/Lexell 105, 106
- Swift-Tuttle 89 *et seq*
Comets
- extinct in high inclination orbits 97-104
Dynamical systems
- degenerate 411-415
- hierarchical 277-287, 303 *et seq*
- stochasticity criterion of 411 *et seq*

Earth-Moon system
- evolution of 4
Eclipses of Galilean satellites
- collection of 51-59
- Delambre collection of 52 *et seq*
Ejection of particles
- from comet P/Lexell 105, 106
- from comet Swift-Tuttle 90, 91
Elliptical galaxies
- periodic orbits in 271-274
Ephemeris
- long by JPL 47-50
Equilibrium figures
- relaxation of asteroidal fragments toward 177
Equilibrium points
- asymptotic orbits at 339-350
- linear stability of 289 *et seq*
- orientation of satellite located at 27-35
- periodic orbits about 235-247, 289 *et seq*
Equipotential curves
- boundaries for 317-323
Extinct comets
- steady state number of 97-104

Final evolutions
- in the collinear restricted problem 334 *et seq*

Four-body problem 107, 158, 159, 309 *et seq*

Galilean satellites
- eclipses of 51-59

Gauge invariance
- of parametrization 364

Grain orbits
- trapping time of 397-410

Hierarchical systems 277-287, 303 *et seq*

Hierarchy breaking 309 *et seq*

High inclination orbits
- extinct comets in 97-104

Hill equation 241 *et seq*, 291 *et seq*

Hill stability
- criterion in general three-body problem 307 *et seq*
- of asteroids 123-139

Homothetic orbits
- in the anisotropic Kepler problem 266

Impact experiments
- for asteroidal collision 178

Integrals
- classical 302 *et seq*
- disappearance of 411 *et seq*

Invariant
- curves 146-150
- manifolds 334

Inverse problem
- for autonomous systems 353-367

Kepler problem
- anisotropic 263-270
- inverse 358 *et seq*
- symmetric periodic orbits in 265

Kirkwood gaps 141, 149, 163, 189-201

Level manifolds
- of classical integrals 302 *et seq*

Libration
- of Saturn´s satellites 21
- period of Trojans 154
- region of resonant asteroids 400, 408

Libration of perihelion
- of Trojan asteroids 165-176

Libration zone
- depletion of 199
- topological 191

Lunar orbit
- evolution of 14

Lunar theory
- resonant terms of 6

Measure of instability 145

Meteor streams
- computer model of 86, 92
- orbital evolution of 89-95
- physical processes affecting the motion of 83-87

Minor bodies
- encounters of/ with the outer planets 377-395

Mirror conditions
- asymptotic approach to 277 *et seq*

Mirror theorem 226,278

N-body systems 277-287, 301 *et seq*

Non-gravitational forces 84, 100
- combined with restricted problem 107 *et seq*, 397-410

Numerical integrations 15, 47, 68 92, 98, 100, 105, 107-114, 141-152, 153-161, 249-256, 260, 330 *et seq*, 347 *et seq*, 397-410

Observations
- computerized reduction of 48
- of Galilean satellite eclipses 51-59

Orbital evolution
- of comet P/Lexell 105
- of high inclination comets 101
- of meteor streams 89-95
- of small bodies in the outer solar system 392
- of Trojan asteroids 165-176

Orientation of a satellite
- located at equilibrium point 27-35

Outer planets
- encounters of minor bodies with 377-395

Perihelion-aphelion exchanges
- as a consequence of close encounters 388 *et seq*
Perihelion of Trojan asteroids
- libration of 165-176
- period of 154
Periodic orbits
- about equilibrium points 235-247, 296 *et seq*
- asymmetric 249-256
- bifurcations of 213-224, 225-233
- characteristics of 271-274
- construction of 257-261
- continuation of 143, 144, 225 *et seq*
- in elliptical galaxies 271-274
- resonant 143 *et seq*, 235-247
- stability of 144, 145, 225-233, 253 *et seq*, 257-261
- three-dimensional 213-224, 235-247, 257-261
- vertical self-resonant 213, 218 *et seq*, 242, 246
Perseid meteor stream 89 *et seq*
Perturbations
- due to asteroid belt 73-79
- planetary secular 61-71
- secular on the c^2h integral 311 *et seq*
Poincaré map 328, 330
Poynting Robertson effect 84, 397-410

Quadrantid meteor stream 92 *et seq*

Regularization
- Belen´kii 40 *et seq*
- in satellite theory 39-46
Resonance
- deep 229
- evolution through 191 *et seq*
- fundamental model of 189, 190
- Jupiter-Saturn 158
- of comet P/Boethin with Jupiter 107 *et seq*
- parametric 30
- secular of Trojan asteroids 160
- shallow 227
- with tidal terms in Lunar theory 9
Resonances
- among Saturn´s satellites 19-25
- in the evolution of the Lunar orbit 3-17
- width of 400
Resonant
- asteroid orbits 141-152, 163
- orbits in presence of Poynting-Robertson drag 397-410
- periodic orbits 143 *et seq*, 235 *et seq*

Satellite
- capture 378 *et seq*
- orientation of 27-35
- theory/ linearization in 39 *et seq*
- theory/ regularization in 39-46
Satellites
- of Jupiter/ eclipses of 51-59
- of Saturn/ resonances among 19-25
Secular period
- of Trojan asteroids 158
Stability
- linear of equilibrium points 289 *et seq*
- topological 301-315
- vertical 241 *et seq*
Stability regions 33, 123-136, 289 *et seq*
- in phase space 145 *et seq*
Surface of section 141-152, 226, 325 *et seq*
Szebehely´s equation 353

Three-body problem
- circular restricted 141-152, 197 *et seq*, 213-224, 235-247
- collinear restricted 325-338
- elliptic restricted 27-35, 163, 165-176, 257-261, 289-299, 317-323, 339-350, 397
- general 249-256, 278 *et seq*, 301-315
- restricted 107-114, 153-158, 225-233

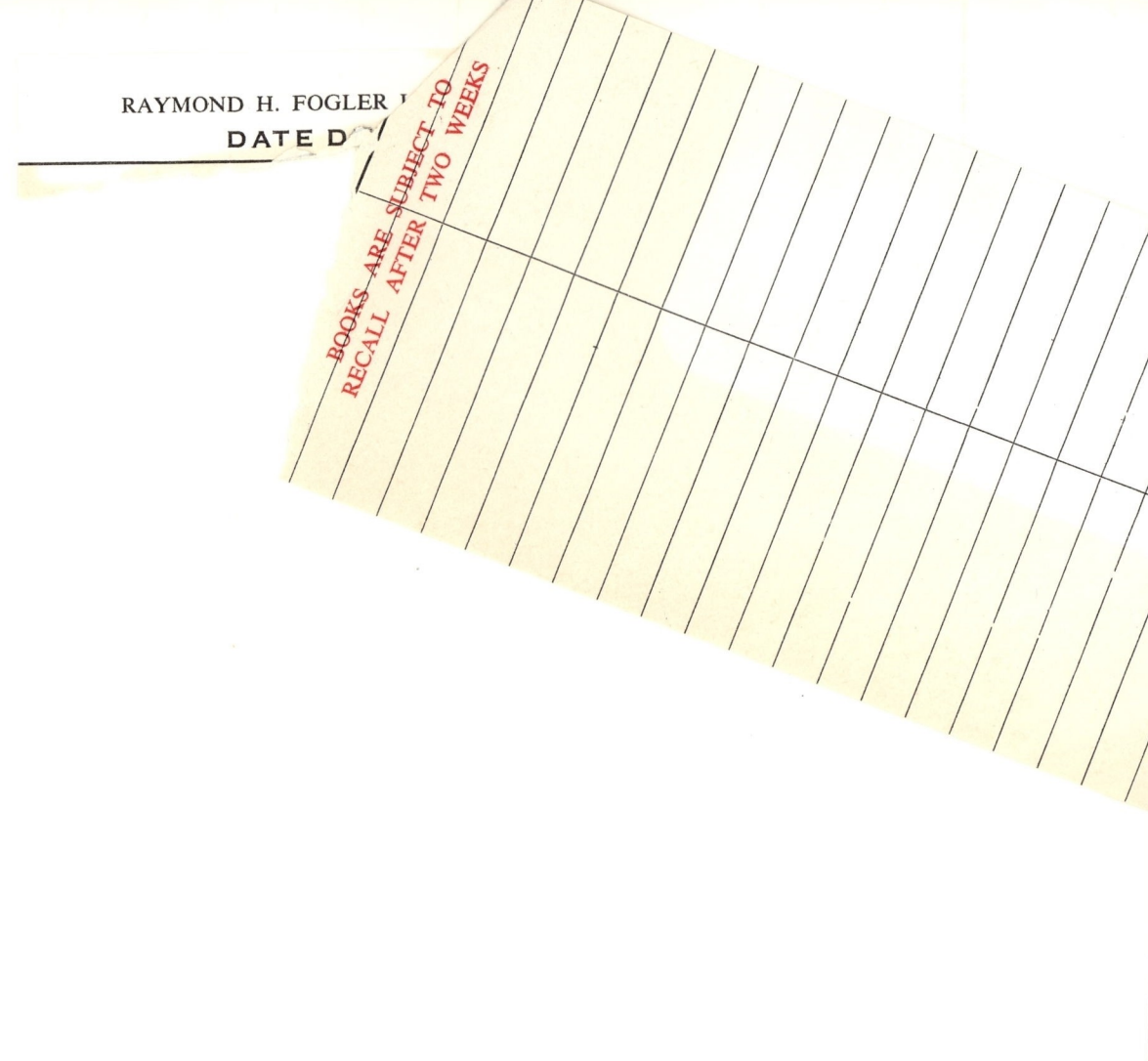